CALCULUS SUPPLEMENT

CALCULUS SUPPLEMENT

AN OUTLINE WITH SOLVED PROBLEMS

Robert Kurtz

Vassar College

W. A. BENJAMIN, INC.

New York

1970

CALCULUS SUPPLEMENT:

An Outline with Solved Problems

Standard Book Number: 8053-5663-0 (Paperback edition)
Library of Congress Catalog Card Number 79-99275
Manufactured in the United States of America
12345-KP-43210

W. A. BENJAMIN, INC.
New York, New York 10016

PREFACE

The primary, more overt, purpose of this book is to help the reader acquire a mastery of the basic techniques of analysis such as simple $\epsilon - \delta$ proofs together with the evalulation and manipulation of limits, integrals, derivatives, sums, products, and so on, as well as the determination of the convergence or divergence of sequences and series. I have tried to attain this end in two ways. First, the problems have been chosen and formulated so as to be nonroutine, for the most part, and simultaneously to display the various techniques across a broad spectrum of difficulty. Second, the outline material has been arranged so as to stress the computational approach. This does not mean that theory has been slighted. Indeed, all the standard theorems (as well as some that are nonstandard in a first course) are here, although, as befits a book of this nature, the great bulk of them are not proved. Furthermore, I have tried to indicate how the technical aspects of analysis stem naturally from, and are complemented by, the conceptual aspects, and how in turn technical considerations can illuminate and even suggest conceptual considerations. If the reader comes to realize that while it is important to distinguish the theoretical from the technical, these two aspects are also in many respects simply two sides of the same coin, then my secondary, more covert, objective will have been achieved.

I believe that this work will supplement most current calculus texts, either in the classroom or on an individual basis. In addition, experience with a preliminary edition has shown that it can serve as the sole text for students who have had almost one year of preparation, and that it is of value to students taking advanced calculus or introductory real analysis who want to brush up on their calculus. It could be used also as the sole text in a beginning course by an ambitious instructor willing to flesh out the somewhat skeletal explanatory sections.

In a field as old as calculus it is as difficult to devise a truly original problem as it is to trace a familiar problem to its source. Accordingly, let me say that the problems here are often well known. However, most of the solutions (and, needless to say, the responsibility for their accuracy) are my own, although I make no claim to either their originality or economy. The reader is cordially invited, and indeed encouraged, to furnish better solutions whenever possible.

Robert Kurtz

Poughkeepsie, New York
January 1970

ACKNOWLEDGMENTS

In the course of the various revisions and rewritings of this book I was aided by the efforts of others—both known and unknown to me. Among the latter are the readers of the several preliminary versions who suggested many worthwhile improvements. As for the former, it is a pleasure to thank the following: Donald Hight, professor of mathematics at Kansas State College, for testing a preliminary edition under classroom conditions and for his resulting comments, criticisms, and suggestions; Eileen Sprague, secretary of the Williams College Mathematics Department for lightening the burden of revision by her expert, rapid typing; and Carolyn Colburn, formerly of the Vassar College Mathematics Department, for helping with the dull job of proofreading the final manuscript.

A NOTE TO THE READER

If you are reading this book on your own, I would strongly advise, that you read it only in conjunction with a standard text.

Each chapter consists of an outline followed by a problem section. Interspersed in the outline are examples and exercises. The examples are generally typical of the problems while the exercises are somewhat simpler. I would advise you to first read the examples, then do the exercises, and then try the problems. Do not be discouraged if you cannot solve all the problems, as some of them are difficult. Hints for the more difficult ones are collected on page 151. An asterisk (*) before a problem indicates that it has a hint.

For best results you should resist your natural tendency to peek at the solutions. This is not to say you should never look at a solution if you are stumped, but that you should make a serious attempt at a solution. Solving one problem on your own will teach you much more than reading the solutions of several similar problems. Also, a solution will be more meaningful if you are familiar with the difficulties of the problem.

CONTENTS

Chapter 1

THE $\sum$–$\prod$ NOTATION, INDUCTION, AND ABSOLUTE VALUE

We shall begin by discussing four topics that can be collectively labeled as "precalculus." Two of these, *inequalities* and *absolute value*, are fundamental. The reader should make every effort to acquire skill and confidence in their manipulation as soon as possible since they are the natural language of mathematical analysis, of which calculus is the first phase. The other two, *mathematical induction* and the $\sum$ and $\prod$ notations, are less basic since they are not necessary for the study of calculus; nevertheless, experience has proved them useful.

Let us take up first the standard notations for the sum and product of several numbers. Let $a_1, a_2, \ldots, a_n$ be n numbers that are not necessarily different. We set

$$\sum_{k=1}^{n} a_k = a_1 + a_2 + \cdots + a_n \tag{1-1}$$

and

$$\prod_{k=1}^{n} a_k = a_1 a_2 \cdots a_n \tag{1-2}$$

Here $\sum$ is the upper case Greek letter sigma which is the counterpart of S, the first letter of *sum*; and $\prod$ is the upper case Greek letter pi whose counterpart is P, the first letter of *product*.

1

EXAMPLE 1-1   If $a_1 = 2$, $a_2 = 3$, $a_3 = 1$, and $a_4 = 3$, then $n = 4$ and

$$\sum_{k=1}^{4} a_k = 2 + 3 + 1 + 3 = 9, \qquad \prod_{k=1}^{4} a_k = 2 \cdot 3 \cdot 1 \cdot 3 = 18$$

EXAMPLE 1-2   Put $a_1 = 1/2$, $a_2 = -2/3$, $a_3 = -4/5$, $a_4 = 11/15$, $a_5 = 0$, $a_6 = -7/10$, and $a_7 = 1/2$.   Then $n = 7$,

$$\sum_{k=1}^{7} a_k = \frac{1}{2} - \frac{2}{3} - \frac{4}{5} + \frac{11}{15} + 0 - \frac{7}{10} + \frac{1}{2} = -\frac{13}{30}$$

and

$$\prod_{k=1}^{7} a_k = \frac{1}{2}\left(-\frac{2}{3}\right)\left(-\frac{4}{5}\right)\left(\frac{11}{15}\right)0\left(-\frac{7}{10}\right)\left(\frac{1}{2}\right) = 0$$

EXAMPLE 1-3   Set $a_k = 3.1204$ for $k = 1, 2, \ldots, 5607$; then $n = 5607$,

$$\sum_{k=1}^{5607} a_k = (5607)(3.1204)$$

and

$$\prod_{k=1}^{5607} a_k = (3.1204)^{5607}$$

It should be noted that the letter k plays no essential role in Eq. (1-1) or Eq. (1-2) and that we could just as well put

$$a_1 + a_2 + \cdots + a_n = \sum_{m=1}^{n} a_m$$

or

$$a_1 a_2 \cdots a_n = \prod_{m=1}^{n} a_m$$

Two consequences of this observation are the relations

$$\sum_{m=0}^{n} a_{m+k} = \sum_{m=k}^{n+k} a_m$$

and

$$\prod_{k=0}^{n} a_{m+k} = \prod_{m=k}^{n+k} a_m$$

Strictly speaking, nothing new is involved in either Eq. (1-1) or Eq. (1-2). They are simply compact ways of writing cumbersome expressions, and as such they often point up relationships that tend to be obscured by the details of long or intricate equations.   Thus, the familiar property that the insertion of parentheses in a sum or product does not alter its value, written symbolically as

$$a_1 + a_2 + \cdots + a_n = (a_1 + a_2 + \cdots + a_{m-1}) + (a_m + a_{m+1} + \cdots + a_n)$$

or

$$a_1 a_2 \cdots a_n = (a_1 a_2 \cdots a_{m-1})(a_m a_{m+1} \cdots a_n)$$

can be expressed in the $\sum$ and $\prod$ notation as follows:

$$\sum_{k=1}^{n} a_k = \sum_{k=1}^{m-1} a_k + \sum_{k=m}^{n} a_k \tag{1-3a}$$

and

$$\prod_{k=1}^{n} a_k = \prod_{k=1}^{m-1} a_k \prod_{k=m}^{n} a_k \tag{1-3b}$$

where m is any integer between 1 and n.   Similarly, the facts that

$$(a_1 + b_1) + (a_2 + b_2) + \cdots + (a_n + b_n) = (a_1 + a_2 + \cdots + a_n)$$
$$+ (b_1 + b_2 + \cdots + b_n)$$

and

$$(a_1 b_1)(a_2 b_2) \cdots (a_n b_n) = (a_1 a_2 \cdots a_n)(b_1 b_2 \cdots b_n)$$

are expressible as

$$\sum_{k=1}^{n} (a_k + b_k) = \sum_{k=1}^{n} a_k + \sum_{k=1}^{n} b_k \tag{1-4a}$$

and

$$\prod_{k=1}^{n} a_k b_k = \prod_{k=1}^{n} a_k \prod_{k=1}^{n} b_k \tag{1-4b}$$

while the relations

$$ca_1 + ca_2 + \cdots + ca_n = c(a_1 + a_2 + \cdots + a_n)$$

and

$$a_1{}^p a_2{}^p \cdots a_n{}^p = (a_1 a_2 \cdots a_n)^p$$

become

$$\sum_{k=1}^{n} ca_k = c \sum_{k=1}^{n} a_k \tag{1-5a}$$

and

$$\prod_{k=1}^{n} a_k{}^p = \left[\prod_{k=1}^{n} a_k \right]^p \tag{1-5b}$$

EXERCISE

1–1   Establish the identities

$$\sum_{k=1}^{n}(a_k + c) = \sum_{k=1}^{n} a_k + nc \qquad \text{(1-6a)}$$

and

$$\prod_{k=1}^{n} ca_k = c^n \prod_{k=1}^{n} a_k \qquad \text{(1-6b)}$$

by using Eqs. (1-4a) and (1-4b), or otherwise.

Recall that the numbers $a_0, a_1, \ldots, a_n$ form a *geometric progression* if

$$\frac{a_{k+1}}{a_k} = r \neq 1 \qquad \text{(1-7)}$$

for $k = 0, 1, \ldots, n-1$. Notice that Eq. (1-7) entails $a_1 = a_0 r$, $a_2 = a_1 r = a_0 r^2$, and $a_3 = a_2 r = a_0 r^3$. It is clear that this procedure can be repeated as many times as necessary; consequently, Eq. (1-7) is equivalent to

$$a_k = a_0 r^k \qquad \text{(1-8)}$$

for $k = 0, 1, 2, \ldots, n$.

EXAMPLE 1-4   As an illustration of the utility of the identities (1-3a) to (1-6a), we will derive the well-known formula for the sum of a geometric progression. By Eq. (1-8) we have

$$\sum_{k=0}^{n} a_k = \sum_{k=0}^{n} a_0 r^k$$

and by Eq. (1-5a) we have

$$\sum_{k=0}^{n} a_0 r^k = a_0 \sum_{k=0}^{n} r^k$$

To evaluate

$$\sum_{k=0}^{n} r^k$$

we use a trick:

$$(1 - r)\sum_{k=0}^{n} r^k = \sum_{k=0}^{n}(1 - r)r^k$$

again by Eq. (1-5a), and

$$\sum_{k=0}^{n}(1 - r)r^k = \sum_{k=0}^{n}(r^k - r^{k+1}) = \sum_{k=0}^{n} r^k - \sum_{k=0}^{n} r^{k+1}$$

by Eq. (1-4a) and Eq. (1-5a) with $c = -1$. Now by Eq. (1-3a)

$$\sum_{k=0}^{n} r^k = \sum_{k=0}^{0} r^k + \sum_{k=1}^{n} r^k = 1 + \sum_{m=1}^{n} r^m$$

and

$$\sum_{k=0}^{n} r^{k+1} = \sum_{k=0}^{n-1} r^{k+1} + \sum_{k=n}^{n} r^{k+1} = \sum_{m=1}^{n} r^m + r^{n+1}$$

(see also the comments following Example 1-3).

Therefore,

$$(1 - r) \sum_{k=0}^{n} r^k = 1 + \sum_{m=1}^{n} r^m - \sum_{m=1}^{n} r^m - r^{n+1} = 1 - r^{n+1}$$

and so,

$$\sum_{k=0}^{n} a_0 r^k = a_0 \frac{1 - r^{n+1}}{1 - r} \tag{1-9}$$

It is natural to ask if there is a formula corresponding to Eq. (1-9) for the sum of an *arithmetic progression*. The numbers $a_0, a_1, \ldots, a_n$ form an arithmetic progression if

$$a_{k+1} - a_k = d \neq 0 \tag{1-10}$$

for $k = 0, 1, \ldots, n - 1$. Equation (1-10) means that $a_1 = a_0 + d$, $a_2 = a_1 + d = a_0 + 2d$, and $a_3 = a_2 + d = a_0 + 3d$. In the same way that Eq. (1-8) followed from Eq. (1-7), we see that Eq. (1-10) is equivalent to

$$a_k = a_0 + kd \tag{1-11}$$

for $k = 0, 1, 2, \ldots, n$. For the sum of the arithmetic progression $a_0, a_1, \ldots, a_n$, we thus have

$$\sum_{k=0}^{n} a_k = \sum_{k=0}^{n} (a_0 + kd) = (n + 1)a_0 + d \sum_{k=1}^{n} k$$

and so, the problem is reduced to finding

$$\sum_{k=1}^{n} k = S_n$$

It is not difficult to discover, by examining S_n for several values of n, the formula

$$S_n = \frac{n(n + 1)}{2} \tag{1-12}$$

and to verify it for any value of n. But verification of Eq. (1-12) for a particular n cannot hold for *all* n, although such checking does make it more plausible. The question then is how a formula such as Eq. (1-12) is proved for all n since infinitely many equations are involved.   One method of proof is to give a derivation valid for all n as we did in establishing Eq. (1-9), which also involves infinitely many equations.   Another method is *mathematical induction*, called "induction" for short.

The method of induction is a logical scheme for proving an infinity of propositions $P_1, P_2, \ldots, P_n, \ldots$ to be true; Eqs. (1-9) and (1-12) would be examples of such sequences of proportions or statements.   To emphasize that we are dealing with a collection of propositions, we introduce the notation $\{P_n\}$ to represent the set of all the statements $P_1, P_2, \ldots, P_n, \ldots$.   With this notation we can say that induction is a method by means of which every member of the set $\{P_n\}$ can be proved true.   The method consists of two steps:

Step 1.   Prove that P_1 is true.

Step 2.   For any integer n, prove that if P_n is true, then P_{n+1} is true.

To see that this is sufficient to ensure the truth of every member of $\{P_n\}$, suppose that they are not all true.   Then as we look successively at $P_1, P_2, \ldots$, we must encounter at least one false statement.   Let P_k be the *first* false statement so encountered.   Because of step 1, k must be larger than 1, and because P_k is the first false statement, all of $P_1, P_2, \ldots, P_{k-1}$ are true.   But then step 2 implies that P_k is true as well.   This contradiction means that our supposition is false, or that every member of $\{P_n\}$ is true.

EXAMPLE 1-5   Let Eq. (1-12) be denoted by P_n.   We will prove by induction that all members of $\{P_n\}$ are true.   To prove P_1 is true, simply notice that $1 = 1 \cdot \frac{2}{2}$.   Assume that P_n is true, and consider P_{n+1}.   Then $S_{n+1} = (n + 1) + S_n$, and the hypothesis "P_n is true" means that

$$S_{n+1} = (n + 1) + \frac{n(n + 1)}{2} = \frac{(n + 1)(n + 2)}{2}$$

In other words, we have shown that P_{n+1} is true if P_n is true.   The proof by induction is completed.

Notice that we have established the formula

$$\sum_{k=0}^{n} (a_0 + kd) = (n + 1)a_0 + d\,\frac{n(n + 1)}{2} \tag{1-13}$$

for the sum of the arithmetic progression (1-11).

Remark   The phrase in this proof "assume that P_n is true..." is called the induction hypothesis, and it often brings forth from the beginner the cry, "But you're assuming what you want to prove!"   This is, of course, not the case since what we are trying to do is to prove that all members of the set $\{P_n\}$ are true.   In doing this we have to carry out step 2; that is, prove a statement of the form: If A, then B.   Since all mathematical statements are of this form, and the hypothesis A is assumed true as a matter of course, there is nothing illegitimate in making use of the induction hypothesis when attempting to complete step 2.

EXAMPLE 1-6   In most of the inductions encountered in calculus the statements involved are equations or identities.   Indeed, this was the case in the previous example.   To show that induction is more flexible, we apply it to a collection of statements $\{P_n\}$ each of which corresponds more to the everyday usage of the word "statement" than did Eq. (1-12).

Let P_n be the statement: Given a line segment of unit length in a plane, it is possible to construct by ruler and compass a line segment of length $(n + 1)^{1/2}$.   As a preliminary observe that if line segments of length a and b are given, then it is possible to construct by ruler and compass a right triangle having legs of length a and b.

Step 1.   We have to show that a line segment of length $2^{1/2}$ can be constructed.   By our preliminary observation we know that we can construct a right triangle having both legs of length 1.   But then the Pythagorean relation means that the hypotenuse has length $(1^2 + 1^2)^{1/2} = 2^{1/2}$.   Thus P_1 is true and step 1 is completed.

Step 2.   Assume that P_n is true.   Since this means we have a line segment of length $(n + 1)^{1/2}$, our preliminary observation permits us to construct a right triangle having legs of length $(n + 1)^{1/2}$ and 1; by the Pythagorean relation again, its hypotenuse has length $\{[(n + 1)^{1/2}]^2 + 1^2\}^{1/2} = (n + 2)^{1/2}$. This completes step 2.

EXAMPLE 1-7   We must emphasize that induction is a two-step procedure and that both steps must be carried out.   If one step holds without the other, there is no proof.   To show that care is needed, consider the statement

$$S_n = \sum_{k=1}^{n} k = (\tfrac{1}{8})(2n + 1)^2 \tag{1-14}$$

Since Eq. (1-14) contradicts Eq. (1-12), which we know to be true from Example 1-5, it follows that Eq. (1-14) must be *false for all n*.   Nevertheless, if we denote Eq. (1-14) by P_n and assume that P_n is true, we can prove that P_{n+1} is also true.   In other words, step 2 can be carried out for these false propositions.   The proof is as follows:

If P_n is true, then $S_n = (\tfrac{1}{8})(2n + 1)^2$, and hence

$$S_{n+1} = S_n + (n + 1) = (\tfrac{1}{8})(2n + 1)^2 + (n + 1) = \frac{(4n^2 + 4n + 1) + 8n + 8}{8}$$

$$= (\tfrac{1}{8})[4n^2 + 12n + 9] = (\tfrac{1}{8})(2n + 3)^2 = (\tfrac{1}{8})[2(n + 1) + 1]^2$$

EXERCISES

1-2 Prove by induction that

$$\sum_{k=1}^{n} (2k - 1) = n^2$$

1-3 Prove by induction that

$$\prod_{k=2}^{n} \left(1 - \frac{1}{k}\right) = \frac{1}{n}$$

Remark Some of the problems in this chapter, as well as others throughout the book, involve induction. We leave it to the reader to discover which ones do. However, be warned that some beginners tend to overuse induction. Many propositions that can be proved by induction can also be proved directly, and the direct proof is often shorter or more illuminating. Equation (1-9) is a case in point, and so is Eq. (1-12), which was proved by induction in Example 1-5. It can be derived elegantly as follows:

$$\sum_{k=1}^{n} k = S_n = \sum_{k=1}^{n} (n + 1 - k)$$

and so,

$$2S_n = \sum_{k=1}^{n} k + \sum_{k=1}^{n} (n + 1 - k) = \sum_{k=1}^{n} [k + (n + 1 - k)]$$

$$= \sum_{k=1}^{n} (n + 1) = n(n + 1)$$

The number analogous to S_n in that we also put $a_k = k$ for $k = 1, 2, \ldots, n$ but use Eq. (1-2) in place of Eq. (1-1) is called *n factorial* and is denoted by $n!$ Thus

$$n! = \prod_{k=1}^{n} k = 1 \cdot 2 \cdot 3 \cdots (n - 1) \cdot n \tag{1-15}$$

We put $0! = 1$. Notice that

$$(n + 1)! = (n + 1)n! \tag{1-16}$$

Closely associated with $n!$ are the *binominal coefficients*

$$\binom{n}{m} = \frac{n!}{m!(n - m)!} \tag{1-17}$$

where m is not larger than n. It will be convenient to put

$$\binom{n}{0} = 1$$

The usefulness of the binominal coefficients will be apparent later; for now, let us establish the identity

$$\binom{n+1}{m} = \binom{n}{m-1} + \binom{n}{m} \tag{1-18}$$

We have

$$\binom{n}{m-1} + \binom{n}{m} = \frac{n!}{(m-1)![n-(m-1)]!} + \frac{n!}{m!(n-m)!}$$

$$= \frac{m(n!) + n!(n+1-m)}{m!(n+1-m)!} = \frac{(n+1)!}{m!(n+1-m)!} = \binom{n+1}{m}$$

Equation (1-18) is the rule of formation of *Pascal's triangle*, the first few rows of which follow

$$
\begin{array}{ccccccccc}
 & & & & 1 & & & & \\
 & & & 1 & & 1 & & & \\
 & & 1 & & 2 & & 1 & & \\
 & 1 & & 3 & & 3 & & 1 & \\
1 & & 4 & & 6 & & 4 & & 1
\end{array}
\tag{1-19}
$$

If the first row is regarded as the zeroth row, then the $(m+1)$st member, counting from left to right, of the nth row is

$$\binom{n}{m}$$

Thus

$$\binom{3}{2} = 3, \quad \binom{4}{2} = 6, \text{ and } \binom{4}{3} = 4$$

The triangle is extended one row by writing a 1 at each end in such a way as to preserve the trianglar array and then, using Eq. (1-18), writing the sum of two adjacent members of the last completed row in the row to be completed so that it appears between them as in Eq. (1-19). The next row in Eq. (1-19) would be 1 5 10 10 5 1; this would be followed by 1 6 15 20 15 6 1, and so forth. It is not difficult to extend Pascal's triangle to 20 or so rows; it is a practical computational tool.

As we indicated at the outset of this chapter, the topics of inequalities and absolute value are of great technical importance in calculus; for this reason they are generally treated adequately. We shall accordingly only give their definitions and point out those properties that are most useful in computing and problem solving.

The notion of inequality between real numbers is simply a restatement of the fact that any number is positive, negative, or zero. We say that *a is less than b*, symbolically $a < b$, if $b - a$ is positive, and that *a is less than or equal to b*, symbolically $a \leq b$, if $b - a$ is not negative. Then *a is greater than b*, symbolically $a > b$, if $b < a$, and *a is greater than or equal to b*, symbolically $a \geq b$, if $b \leq a$.

EXAMPLE 1-8   Because $5 - 3 = 2$ is positive, $3 < 5$; because $-2 - 1 = -3$ is negative, $1 > -2$; because $4 - 4 = 0$, $4 \geq 4$ (or $4 \leq 4$); because $1.577 - 1.576 = 0.001$ is positive, $1.576 \leq 1.577$.

Inequalities may be algebraically manipulated by employing the following rules.
(1) if $a < b$, then $a \pm c < b \pm c$ for any number c.
(2) If $a < b$ and $0 < c$, then $ac < bc$.
(3) If $a < b$ and $c < 0$, then $bc < ac$.
To see that rule (1) is true, notice that if $b - a$ is positive, so is $(b \pm c) - (a \pm c)$. The truth of rule (2) follows from the fact that if $b - a$ is positive and c is positive, then so is $(b - a)c = bc - ac$. Rule (3) is established by observing that if $b - a$ is positive and c is negative, then $(b - a)c$ is negative. These rules continue to hold if $<$ is replaced by $\leq$, $>$, or $\geq$ provided that c remains positive in rule (2), negative in rule (3).

EXAMPLE 1-9   We will solve the inequality $(-x + 1)/2 \geq -3$ for x. Because $2 > 0$, rule (2) means that $-x + 1 \geq -6$. Then rule (1) allows us to subtract 1 from each side and obtain $-x \geq 7$. Finally, we may apply rule (3) and conclude that $x \leq 7$.

Remark   It can be seen that the steps of this solution are similar to the steps involved in the solution of the corresponding equality $(-x + 1)/2 = -3$. That this is so should not be surprising because of the similarity of rules (1)–(3) to the familiar rules of algebra that equals added to or multiplied by equals are still equal. This similarity can make the solution of inequalities seem more natural to the beginner, but it can also lead him to commit errors by forgetting that inequalities, as opposed to equalities, become reversed when they are multiplied by negative numbers.

EXAMPLE 1-10   Consider the inequality $(x - 2)(x + 3) < 0$. The only way it can be true is if $x - 2$ and $x + 3$ have opposite signs. In other words, if either $x - 2 < 0$ and $x + 3 > 0$, or $x - 2 > 0$ and $x + 3 < 0$. In the first case we have $x < 2$ and $x > -3$; in the second case we have $x > 2$ and $x < -3$. Since it is impossible for x to satisfy both $x > 2$ and $x < -3$, the solution of this inequality is $-3 < x < 2$.

EXAMPLE 1-11   The method of induction can, of course, be applied to establish inequalities.   An example is

$$\sum_{k=0}^{n-1} k^2 < \frac{n^3}{3} < \sum_{k=0}^{n} k^2$$

for $n \geq 2$, which we will denote by P_{n-1}.   Since P_1 is the obvious inequality $1^2 < (2^2)/3 < 1^2 + 2^2$, we may proceed to step 2.   Suppose then that P_{n-1} is true.   To prove that P_n is then true as well, we consider the right-hand and left-hand inequalities of P_n separately.   The left-hand inequality is given by

$$\sum_{k=0}^{n} k^2 = \sum_{k=0}^{n-1} k^2 + n^2 < \frac{n^3}{3} + n^2 = \frac{n^3 + 3n^2}{3}$$

$$< \frac{n^3 + 3n^2 + 3n + 1}{3} = \frac{(n+1)^3}{3}$$

The right-hand inequality is similarly established

$$\sum_{k=0}^{n+1} k^2 = \sum_{k=0}^{n} k^2 + (n+1)^2 > \frac{n^3}{3} + (n+1)^2$$

$$= \frac{n^3 + 3n^2 + 6n + 3}{3} > \frac{n^3 + 3n^2 + 3n + 1}{3}$$

EXERCISES

1-4   If $a > 0$, prove that $a/(1+a) < 3a/(1+2a)$.
1-5   Prove that $x^2 \geq 0$ for all x.
1-6   If $a < b$, and $b < c$, show that $a < c$.

The *absolute value* of a number x, denoted by $|x|$, can be defined by either of the following equations

$$|x| = \begin{cases} x \text{ if } x \geq 0 \\ -x \text{ if } x < 0 \end{cases} \tag{1-20}$$

or

$$|x| = (x^2)^{1/2} \tag{1-21}$$

In Eq. (1-21), $(x^2)^{1/2}$ is the positive square root of x^2.

EXAMPLE 1-12   We have $|4| = 4$, $|-3/4| = 3/4$, $|0| = 0$, and $|-2.13| = 2.13 = |2.13|$.

EXAMPLE 1-13   We will show that Eqs. (1-20) and (1-21) are indeed the same.   Suppose first that $x \geq 0$.   Then Eq. (1-21) gives us $|x| = (x^2)^{1/2} = x$, but this is the same as Eq. (1-20).   On the other hand, if $x < 0$, then Eq. (1-21) becomes $|x| = (x^2)^{1/2} = -x$ since $-x > 0$; this too is the same as Eq. (1-20).   Thus in either case Eq. (1-20) and Eq. (1-21) are identical.

In dealing with absolute values the following theorem is basic because it enables us to replace inequalities involving absolute values by ordinary inequalities.

Theorem 1-1   Let $b > 0$.   *Then $|a| < b$ if and only if $-b < a < b$.   The same is true if $<$ is replaced by $\leq$.*

EXAMPLE 1-14   We solve the inequality $|x - 2| < 3$.   By Theorem 1-1 it is equivalent to $-3 < x - 2 < 3$, or $-1 < x < 5$.

EXAMPLE 1-15   The same methods can be used to solve the inequality $|2x - 1| > 5$.   Indeed, this inequality is the logical negation of $|2x - 1| \leq 5$, which is equivalent to $-5 \leq 2x - 1 \leq 5$, or $-2 \leq x \leq 3$.   By negating the last inequality, we see that the solution of $|2x - 1| > 5$ is either $x < -2$ or $x > 3$.

EXAMPLE 1-16   The technique of Example 1-15 can be used to prove the following companion to Theorem 1-1: *If $b > 0$, then $|a| > b$ if and only if either $a < -b$ or $a > b$.*   As above we negate the given inequality and obtain $|a| \leq b$, which is equivalent to $-b \leq a \leq b$ by Theorem 1-1.   The negation of the last inequality must then be equivalent to $|a| > b$, and so we have that $|a| > b$ if and only if either $a < -b$ or $a > b$.

Remark   It is clear that our assertion continues to hold if $>$ is replaced by $\geq$.

EXERCISES

1-7   Solve for x: $|5 - 1/x| \leq 1$.
1-8   If $b > 0$, prove that $|x - a| \leq b$ if and only if $a - b \leq x \leq a + b$.
1-9   Show that if $a \neq 0$, then $|1/a| = 1/|a|$.

SOLUTIONS TO EXERCISES

1-1   To establish these identities, put $b_k = c$ for $k = 1, 2, \ldots, n$ in both Eq. (1-4a) and Eq. (1-4b).   Then

$$\sum_{k=1}^{n} (a_k + c) = \sum_{k=1}^{n} a_k + \sum_{k=1}^{n} c = \sum_{k=1}^{n} a_k + nc$$

and

$$\prod_{k=1}^{n} a_k c = \prod_{k=1}^{n} a_k \prod_{k=1}^{n} c = c^n \prod_{k=1}^{n} a_k$$

1-2　As usual, denote this statement by P_n.　Since P_1 reads $(2-1) = 1^2$, we see that P_1 is true.　If P_n is true, then

$$\sum_{k=1}^{n+1}(2k-1) = \sum_{k=1}^{n}(2k-1) + [2(n+1)-1]$$
$$= n^2 + (2n+1) = (n+1)^2$$

Thus P_{n+1} is also true, and the proof by induction is complete.

1-3　If the statement is denoted by P_n, then P_1 is true since it is the identity $(1 - \frac{1}{2}) = \frac{1}{2}$.　If P_n is true, then

$$\prod_{k=2}^{n+1}\left(1 - \frac{1}{k}\right) = \left(1 - \frac{1}{n+1}\right)\prod_{k=2}^{n}\left(1 - \frac{1}{k}\right)$$
$$= \left(1 - \frac{1}{n+1}\right)\frac{1}{n} = \frac{(n+1)-1}{n(n+1)} = \frac{1}{n+1}$$

and we see that P_{n+1} is true as well.

1-4　We prove this inequality by reducing it to an equivalent inequality that is obviously true.　Observe first that since $a > 0$, both $1 + a > 0$ and $1 + 2a > 0$.　This fact means that we can multiply the given inequality by $(1+a)(1+2a)$ and obtain $a(1+2a) < 3a(1+a)$.　But this last inequality is the same as $a + 2a^2 < 3a + 3a^2$, $2a + a^2 > 0$, or $2 + a > 0$.　Since all the above steps can be reversed, the original inequality must be true.

1-5　If $x = 0$, then $x^2 = 0^2 = 0$.　If $x > 0$, rule (2) gives us $x \cdot x > x \cdot 0$, or $x^2 > 0$.　Finally, if $x < 0$, rule (3) gives us $x \cdot x > x \cdot 0$, or $x^2 > 0$.　Thus for all cases $x^2 \geq 0$.

1-6　To show that $a < c$, it is enough to show that $c - a$ is positive.　But $c - a = (c - b) + (b - a)$, and both $c - b$ and $b - a$ are positive because $b < c$ and $a < b$, respectively.　Hence $c - a$ is expressed as the sum of two positive numbers, and this means that $c - a$ is itself positive.

1-7　By Theorem 1-1 this inequality is equivalent to $-1 \leq 5 - 1/x \leq 1$, or $6 \geq 1/x \geq 4$.　Now, the last inequality means that $1/x$ must be positive; but that can be true only if x is positive.　Hence we can multiply the inequality $6 \geq 1/x \geq 4$ by x without changing its direction and obtain $6x \geq 1$ and $1 \geq 4x$.　These reduce to $x \geq \frac{1}{6}$ and $\frac{1}{4} \geq x$ or $\frac{1}{6} \leq x \leq \frac{1}{4}$.

1-8　By Theorem 1-1 we have $|x - a| \leq b$ equivalent to $-b \leq x - a \leq b$ or $a - b \leq x \leq a + b$.

1-9　From Eq. (1-21) we have

$$\left|\frac{1}{a}\right| = \left[\left(\frac{1}{a}\right)^2\right]^{1/2} = \left[\frac{1}{a^2}\right]^{1/2} = \frac{1}{[a^2]^{1/2}} = \frac{1}{|a|}$$

PROBLEMS

1-1   Establish each of the given identities.

(a) $\sum_{k=1}^{n} (a_k - a_{k-1}) = a_n - a_0$

(b) $\prod_{k=1}^{n} \frac{a_k}{a_{k-1}} = \frac{a_n}{a_0}$

where $a_k \neq 0$ for all k

*1-2   Show that

(a) $\sum_{k=1}^{n} \frac{1}{k(k+1)} = 1 - \frac{1}{n+1}$

(b) $\sum_{k=1}^{n} (-1)^{k+1} \frac{2k+1}{k(k+1)}$
$= 1 + \frac{(-1)^{n+1}}{n+1}$

(c) $\prod_{k=2}^{n} \left(1 - \frac{1}{k^2}\right) = \frac{n+1}{2n}$

(d) $\sum_{k=1}^{n} (1 + a^{2k-1})$
$= \frac{1 - a^{2^n}}{1 - a}, \quad a \neq 1$

*1-3   Use induction to establish the *bionomial theorem*

$$(a + b)^n = \sum_{k=0}^{n} \binom{n}{k} a^k b^{n-k}$$

*1-4   Show, by induction, that

$$\sin \theta = 2^n \sin \frac{\theta}{2^n} \prod_{k=1}^{n} \cos \frac{\theta}{2^k}$$

*1-5   Use induction to show that

$$\left(\sum_{k=1}^{n} k\right)^2 = \sum_{k=1}^{n} k^3$$

*1-6   Use induction to prove that

$$\sum_{k=1}^{n} (-1)^{k-1} \frac{1}{k} \binom{n}{k} = \sum_{k=1}^{n} \frac{1}{k}$$

1-7   In each of the following show that the first inequality implies the second.
Give an example to show that the second inequality does not imply the first.
(a) (i) $x < y$ (ii) $x - 2 < y + 3$
(b) (i) $x \geq y$ (ii) $-x + 4 < -y + 5$
(c) (i) $x > y$ (ii) $2x + 5 > 2y$

1-8   Solve the given inequalities for x.
(a) $3x + 1 < x + 5$ (b) $-2x + 1 \geq x + 4$

(c) $\frac{3x - 5}{2} \leq 0$ (d) $\frac{1}{x + 4} > 0$

(e) $\dfrac{x-1}{x+3}<0$ (f) $4x+\dfrac{25}{x}>20$

(g) $\dfrac{1}{x+1}<\dfrac{2}{3x-1}$ (h) $(x-6)(x-2)(x+3)(x+8)<0$

1-9 Solve the given inequalities for x.
(a) $|x-1|<4$ (b) $|1-2x|\le1$

(c) $|x+2|>3$ (d) $\left|\dfrac{3x-2}{4}\right|\ge5$

1-10 (a) If $a>0$, show that $1/a>0$.
(b) If $a<0$, show that $1/a<0$.
(c) if $0<a<b$, show that $0<1/b<1/a$.

1-11 Establish the following properties of absolute value.
(a) $|-x|=|x|$ (b) $|xy|=|x||y|$
(c) $|x^2|=x^2$ (d) $-|x|\le x\le|x|$
*(e) $|x+y|\le|x|+|y|$ *(f) $|x-y|\ge||x|-|y||$

1-12 If $0<x^2<a$, show that $0<|x|<a^{1/2}$.

1-13 (a) If $a<b$ and $c\le d$, prove that $a+c<b+d$.
(b) If at least three of a,b,c, and d are positive, $a<b$ and $c<d$, show that $ac<bd$.

*1-14 If $a>b>0$ and n is a positive integer, show that $a^n>b^n$. Use this to prove that $a^r>b^r$ for any positive rational number r.

1-15 (a) If $a>0, b>0$, and $b<a^2$, show that $(a^2+b)^{1/2}-a<a-(a^2-b)^{1/2}$.
(b) If $a>0, b>0$, and $ab<1$, show that $a-a/(1+ab)<a/(1-ab)-a$.
(c) If $0<a<b$ and $c>0$, show that $a^2/(1+ac)<b^2/(1+bc)$.
(d) $\dfrac{1}{(b^2+c)^{1/2}+b}<\dfrac{1}{(a^2+c)^{1/2}+a}$

1-16 (a) If $a\ne b$, prove that $(a+b)/2$ is between a and b.
(b) If $a>0, b>0$, and $a\ne b$, prove that $(ab)^{1/2}$ and $2/(1/a+1/b)$ are between a and b.
(c) If $a>0$ and $b>0$, prove that $(ab)^{1/2}\le(a+b)/2$.
(d) if $a>0$ and $b>0$, prove that $2/(1/a+1/b)\le(ab)^{1/2}$.

*1-17 Let a,b, and c be positive. Show that
(a) $(a+b)(b+c)(a+c)\ge8\,abc$
(b) $a^2+b^2+c^2\ge ab+bc+ac$
(c) $a^2b^2+b^2c^2+a^2c^2\ge abc(a+b+c)$

1-18 Show that $|x+1/x|\ge2$ for all $x\ne0$.

1-19 If $c>0$, show that $|a+b|^2\le(1+c)|a|^2+\left(1+\dfrac{1}{c}\right)|b|^2$.

*1-20 If a,b, and c are positive, show that $3abc\le a^3+b^3+c^3$.

1-21   Show that

$$\sum_{k=0}^{2n} x^{2n-k} y^k \geq 0$$

for all positive integers n.

*1-22   If $a_1 \geq a_2 \geq \cdots \geq a_n$ and $b_1 \geq b_2 \geq \cdots \geq b_n$, prove that

$$n \sum_{k=1}^{n} a_k b_k \geq \left(\sum_{k=1}^{n} a_k \right) \left(\sum_{k=1}^{n} b_k \right)$$

1-23   Show that $(a_1 b_1 + a_2 b_2)^2 \leq (a_1^2 + a_2^2)(b_1^2 + b_2^2)$.

*1-24   Prove *Minkowski's inequality*

$$\left[\sum_{k=1}^{n} (a_k + b_k)^2 \right]^{1/2} \leq \left[\sum_{k=1}^{n} a_k^2 \right]^{1/2} + \left[\sum_{k=1}^{n} b_k^2 \right]^{1/2}$$

*1-25   If $a_1, a_2, \ldots, a_n$ are positive, prove that

(a) $\displaystyle\sum_{k=1}^{n} a_k \sum_{k=1}^{n} \frac{1}{a_k} \geq n^2$ (b) $\displaystyle\left(\sum_{k=1}^{n} a_k b_k \right)^2 \leq \sum_{k=1}^{n} a_k \sum_{k=1}^{n} a_k b_k^2$

*1-26   Establish the inequality

$$\left(\sum_{k=1}^{n} a_k b_k c_k \right)^4 \leq \left(\sum_{k=1}^{n} a_k^4 \right) \left(\sum_{k=1}^{n} b_k^2 \right)^2 \left(\sum_{k=1}^{n} c_k^4 \right)$$

*1-27   If $a > 0$, $b > 0$, and $a + b = 1$, prove that

$$\left(a + \frac{1}{a} \right)^2 + \left(b + \frac{1}{b} \right)^2 \geq \frac{25}{2}$$

*1-28   Prove that

$$\prod_{k=1}^{n} \frac{2k-1}{2k} < \frac{1}{(3n+1)^{1/2}}$$

*1-29   If $a_1, a_2, \ldots, a_n$ are either all positive, or all negative and greater than -1, show that

$$\prod_{k=1}^{n} (1 + a_k) \geq 1 + \sum_{k=1}^{n} a_k$$

*1-30   If $a_1, a_2, \ldots, a_n$ are all greater than -1, prove that

$$\prod_{k=1}^{n} (1 + a_k) \leq \left[1 + \frac{1}{n} \sum_{k=1}^{n} a_k \right]^n$$

Chapter 2

SEQUENCES
AND THEIR
LIMITS

The subject matter of this chapter is usually not taken up until infinite series are studied—and even then it appears only as an adjunct. The reason we take it up this early is our belief that the concept of *limit*, which is one of the most basic ideas of analysis, is best introduced by the study of limits of sequences.

If we assume that we know what is meant by the terms *set* and *correspondence*, then we can formulate the following definition of a sequence. A *sequence* is a correspondence between the set of positive integers and a set S such that to each positive integer there corresponds exactly one member of S. The set S is usually a set of numbers, although the members of S may be any sort of objects. For example, in the last chapter we had occasion to deal with sequences of statements when we took up induction. We denote the object in S that corresponds to the integer n by a_n, b_n and so on; this object is called the nth *element* or nth *term* of the sequence. The sequence as an entity will be denoted by $\{a_n\}$, $\{b_n\}$, and so on.

EXAMPLE 2-1 Probably the most natural example of a sequence is the sequence of positive integers; that is, the sequence for which $a_n = n$ for $n = 1, 2, \ldots$. Other examples of sequences are $\{1/n\}$, $\{(-1)^n\}$, and *constant sequences* wherein $a_n = c$ for all n. Notice that such a constant sequence is denoted by $\{c\}$, and that this distinguishes the sequence from the number c.

EXAMPLE 2-2 A common way of specifying sequences is by means of a *recurrence relation* which defines a_n in terms of some (or all) of $a_1, a_2, \ldots,$

17

a_{n-1}. A famous example is *Fibonacci's sequence* which is defined by $a_1 = 1$, $a_2 = 1$, and $a_n = a_{n-1} + a_{n-2}$ for $n \geq 3$. The first few terms are 1, 1, 2, 3, 5, 8, 13, ... The terms of Fibonacci's sequence are called *Fibonacci numbers.*

As with numbers, an algebra can be developed for sequences whose terms are numbers. Two sequences $\{a_n\}$ and $\{b_n\}$ are *equal,* written $\{a_n\} = \{b_n\}$, if $a_n = b_n$ for all n. Notice that even though two sequences may have the same set of elements, they need not be equal. Thus, $\{(-1)^n\} \neq \{(-1)^{n+1}\}$. New sequences can be formed from two sequences $\{a_n\}$ and $\{b_n\}$ in various ways. We define their *sum* by $\{a_n\} + \{b_n\} = \{a_n + b_n\}$ and their *difference* by $\{a_n\} - \{b_n\} = \{a_n - b_n\}$. Similarly, their *product* and *quotient* are defined by $\{a_n\} \cdot \{b_n\} = \{a_n b_n\}$ and $\{a_n\}/\{b_n\} = \{a_n/b_n\}$ respectively. Finally, their *convolution* is defined by

$$\{a_n\} * \{b_n\} = \left\{ \sum_{k=1}^{n} a_k \, b_{n+1-k} \right\}$$

EXAMPLE 2-3 We will perform these operations on the two sequences $\{n\}$ and $\{1/n\}$. We have $\{n\} \pm \{1/n\} = \{(n^2 \pm 1)/n\}$, $\{n\} \cdot \{1/n\} = \{1\}$, $\{n\}/\{1/n\} = \{n^2\}$, and

$$\{n\} * \{1/n\} = \left\{ \sum_{k=1}^{n} \frac{k}{n+1-k} \right\}$$

EXERCISE 2-1 Write out the first five terms of each of the sequences obtained in Example 2-3.

Another notion connected with sequences is that of *boundedness.* A sequence $\{a_n\}$ is *bounded* if there is a number $B > 0$ such that $|a_n| < B$ for all n; a sequence is *unbounded* if it is not bounded.

EXAMPLE 2-4 The sequence $\{1/n\}$ is bounded since $0 < 1/n \leq 1$ for $n = 1, 2, \ldots$. The sequence $\{(-1)^n n\}$ is unbounded since $|(-1)^n n| = n$, and given any $B > 0$ we have $n > B$ for all sufficiently large n.

The following definition is a formulation of the most important notion connected with sequences.

Definition 2-1 *The sequence* $\{a_n\}$ *has limit L if given any* $\varepsilon > 0$ *there exists an integer N such that if* $n > N$, *then*

$$|a_n - L| < \varepsilon \qquad (2\text{-}1)$$

(Here ε is the lower case Greek letter episilon.) The notation

$$\{a_n\} \to L \qquad (2\text{-}2)$$

will be used to mean that the limit of $\{a_n\}$ is L.

This definition is simply a rigorous way of stating what one intuitively feels should be conveyed by the assertion $\{a_n\} \to L$. Namely, that the terms of $\{a_n\}$ become "close to" L and remain "close to" L for all "sufficiently large" n. The number ε provides a quantitative measure of "closeness," the number N a quantitative measure of "sufficient largeness."

Sequences that have limits are said to be *convergent*; sequences that have no limits are said to be *divergent*.

EXAMPLE 2-5 The sequences $\{n^2\}$ and $\{(-1)^n\}$ are both divergent, but in essentially different ways. Sequence $\{n^2\}$ is divergent because no matter what L and ε we choose, the inequality $L - \varepsilon < n^2 < L + \varepsilon$ can hold for only finitely many n. Sequence $\{(-1)^n\}$ is divergent because if L is any number, we can find a number ε such that at least one of the inequalities $L - \varepsilon < -1 < L + \varepsilon$ or $L - \varepsilon < 1 < L + \varepsilon$ is false. Hence the inequality $L - \varepsilon < (-1)^n < L + \varepsilon$ will be false for infinitely many n.

The type of divergence exemplified by $\{n^2\}$ is of sufficient interest to merit special consideration. Suppose that $\{a_n\}$ is a sequence such that $a_n > 0$ for all but a finite number of n, and let k be an integer such that if $m \geq k$, then $a_m > 0$. If $\{1/a_{k+n}\} \to 0$, we say that $\{a_n\}$ diverges to infinity and we denote this by $\{a_n\} \to \infty$. This is no more than a suggestive phrase for expressing the fact that the sequence $\{a_n\}$ is divergent in the very particular sense that its terms become larger and larger, and remain large. The symbol ∞ does not denote a number of any sort. Similarly, if $\{a_n\}$ is a sequence such that there is an integer k with the property that if $m > k$, then $a_m < 0$, we say that $\{a_n\}$ diverges to minus infinity, denoted by $\{a_n\} \to -\infty$, if $\{1/a_{k+n}\} \to 0$.

EXAMPLE 2-6 We illustrate the use of Definition 2-1 by showing that $\{1/n\} \to 0$ (and hence $\{n\} \to \infty$, as it should). Let $\varepsilon > 0$, and put $a_n = 1/n$, $L = 0$ in Eq. (2-1). We obtain $|1/n - 0| < \varepsilon$, or $1/n < \varepsilon$. Now this last inequality is clearly true if $n > 1/\varepsilon$. Thus if N is any integer $> 1/\varepsilon$, then $|1/n - 0| < \varepsilon$ for all $n > N$. In other words, we have proved that $\{1/n\} \to 0$.

Remark Notice here that the smaller ε is taken, the larger N must be taken This is generally the case in dealing with limits of sequences; but not always, as any constant sequence shows. Notice also that there are many choices for N (that this is always the case should be clear from the wording of Definition 2-1).

EXAMPLE 2-7 It is a direct consequence of Definition 2-1 that if $\{a_n\} \to L$, then $\{a_{n \pm k}\} \to L$ for any fixed integer k. Indeed, let $\varepsilon > 0$ be given. We know that there is an integer M such that if $n > M$, then $|a_n - L| < \varepsilon$. But if $n \pm k > M \pm k$, we also have $|a_{n \pm k} - L| < \varepsilon$. In other words, the N of Definition 2-1 for the sequence $\{a_{n+k}\}$ can be taken to be $M + k$; the N for the sequence $\{a_{n-k}\}$ can be taken to be $M - k$.

EXERCISES

2-2   Is the relation $\{(-1)^n n\} \to \infty$ true or false?   Why?

2-3   Prove that $\{(n^2 + 1)/(2n^2 - 1)\} \to \frac{1}{2}$.

2-4   Prove that a convergent sequence can have only one limit.

The following two theorems are very useful, both computationally and theoretically, in dealing with limits of sequences.

Theorem 2-1   *Let $\{a_n\} \to A$, and $\{b_n\} \to B$. Then*

(a) $\{a_n\} \pm \{b_n\} \to A \pm B$,

(b) $\{a_n\} \cdot \{b_n\} \to A \cdot B$,

(c) $\{a_n\}/\{b_n\} \to A/B$, *provided that $B \neq 0$.*

It can be shown by an easy induction that parts (a) and (b) of Theorem 2-1 hold for the sum (or difference) and product of a finite number of convergent sequences.

Theorem 2-2   *If $\{a_n\} \to L$, $\{b_n\} \to L$, and $a_n \leq c_n \leq b_n$ for all but a finite number of n, then $\{c_n\} \to L$.*

EXAMPLE 2-8   We use Theorem 2-1 to prove that $\{(n - 1)/(2n + 1)\} \to \frac{1}{2}$.   The theorem cannot be applied immediately because neither $\{n - 1\}$ nor $\{2n + 1\}$ have limits.   In order to be able to apply it, we use the trick of dividing the general term of $\{(n - 1)/(2n + 1)\}$ by n in both numerator and denominator.   We obtain in this way the relation $\{(n - 1)/(2n + 1)\} = \{1 - 1/n\}/\{2 + 1/n\}$.   Now we know from Example 2-6 that $\{1/n\} \to 0$. Therefore, Theorem 2-1a enables us to conclude that $\{1 - 1/n\} \to 1$, and $\{2 + 1/n\} \to 2$.   Since $2 \neq 0$, Theorem 2-1b can now be applied to yield $\{(n - 1)/(2n + 1)\} \to 1/2$.

EXAMPLE 2-9   We illustrate the use of Theorem 2-2 by showing that $\{(1 - 1/n)^{1/2}\} \to 1$.   Note first that inasmuch as $0 < 1 - 1/n < 1$, we have $1 - 1/n \leq (1 - 1/n)^{1/2} \leq 1$.   Next, we have $\{1\} \to 1$ and $\{1 - 1/n\} \to 1$. Thus, by Theorem 2-2, $\{(1 - 1/n)^{1/2}\} \to 1$ as well.

Remark   The case where one of $\{a_n\}$ or $\{b_n\}$ in Theorem 2-2 is a constant sequence, as in this example, is particularly important.   Indeed, this is the way that Theorem 2-2 is most often applied.

EXERCISES

2-5   Show that $\{(2n^2 - 5)/(5n^2 - 2)\} \to 2/5$.

2-6   Let k be a fixed integer.   Prove that

$$\left\{ \sum_{m=1}^{k} \frac{1}{n + m} \right\} \to 0$$

One serious difficulty in applying the definitions and theorems of this chapter is that the limit(s) involved must be known beforehand. While the problem of determining the limit of a sequence known (or suspected) to be convergent is often most difficult, it is frequently of secondary interest. The primary concern is most often whether the sequence converges or not. Of the several ways of showing a sequence to be convergent without having to find its limit, we give the most elementary and perhaps the most useful. The other methods are best considered in more advanced courses.

The method is based on a fundamental property of *monotone* sequences. A sequence $\{a_n\}$ is *monotone increasing* (increasing for short) if $a_n \leq a_{n+1}$ for all n. A sequence $\{a_n\}$ is *monotone decreasing* (decreasing for short) if $a_n \geq a_{n+1}$ for all n. The property possessed by monotone sequences that is of interest here is a version of the completeness property of the real number system that distinguishes the real numbers from the rational numbers. It is that *every bounded monotone sequence is convergent*. This is as intrinsic property of the real number system; it does not require proof. Indeed, it can be taken as an axiom of the real number system.

EXAMPLE 2-10 The sequence $\{n/(n + 1)\}$ is increasing because the inequality $n/(n + 1) < (n + 1)/(n + 2)$ is equivalent to $n(n + 2) < (n + 1)^2$, or $n^2 + 2n < n^2 + 2n + 1$. Since the last inequality is obviously true, the first must also be true. Further, it is clear that $n/(n + 1) < 1$. In other words, $\{n/(n + 1)\}$ is monotone and bounded. Consequently it converges.

Doubly monotone sequences will be encountered occasionally. A sequence $\{a_n\}$ is *doubly monotone* if $a_2 \leq a_4 \leq \cdots \leq a_{2n} \leq \cdots$, and $a_1 \geq a_3 \geq \cdots \geq a_{2n-1} \geq \cdots$, or vice versa. The procedure with such sequences is to determine first, using the basic property of monotone sequences cited above, whether $\{a_{2n}\} \to A$ and $\{a_{2n-1}\} \to B$. If this is the case, try to determine if $A = B$ holds.

EXAMPLE 2-11 Let $a_0 > 0$, $a_1 > 0$ and put $a_{2n} = (a_{2n-2} + a_{2n-1})/2$, $a_{2n+1} = 2a_{2n-2}a_{2n-1}/(a_{2n-2} + a_{2n-1})$. We will show that $\{a_n\}$ is doubly monotone, and that $\{a_n\} \to (a_0 a_1)^{1/2}$. To this end, notice two facts. First, a_{2n} is the arithmetic mean of a_{2n-2} and a_{2n-1}, and a_{2n+1} is the harmonic mean of these two numbers. Hence, by the remark following the solution of Problem 1-16, $a_{2n+1} \leq a_{2n}$ for all n. Secondly, we may rewrite the defining equation of a_{2n+1} as $[(a_{2n-2} + a_{2n-1})/2]a_{2n+1} = a_{2n-2}a_{2n-1}$. By the defining equation of a_{2n}, this identity is seen to be $a_{2n}a_{2n+1} = a_{2n-2}a_{2n-1}$, or

$$\frac{a_{2n}}{a_{2n-2}} = \frac{a_{2n-1}}{a_{2n+1}}$$

From this equation it can be seen that if $\{a_{2n}\}$ is decreasing, then $\{a_{2n-1}\}$ is increasing because it would be the case that $1 \geq a_{2n}/a_{2n-2} = a_{2n-1}/a_{2n+1}$.

We first show that $\{a_{2n}\}$ is indeed decreasing. Now $a_{2n+2} - a_{2n} = (a_{2n+1} + a_{2n})/2 - a_{2n} = (a_{2n+1} - a_{2n})/2$, but we have seen that $a_{2n+1} \leq a_{2n}$ and so $(a_{2n+1} - a_{2n})/2 \leq 0$. Thus $\{a_{2n}\}$ is decreasing and, as we have pointed out, this implies that $\{a_{2n-1}\}$ is increasing. We next prove that both of these sequences have limits. This is so for $\{a_{2n}\}$ because $a_{2n} > 0$ for all n. Let L_2 be the limit of $\{a_{2n}\}$. Since $a_{2n+1} \leq a_{2n} \leq a_2$, it can be seen that $\{a_{2n-2}\}$ is likewise bounded and hence has a limit L_1. Our final task is to show that $L_1 = L_2 = (a_0 a_1)^{1/2}$. To this end we apply Theorem 2-1 and Example 2-7 to the equation $a_{2n} = (a_{2n-1} + a_{2n-2})/2$, and conclude that $L_2 = (L_1 + L_2)/2$, or $L_1 = L_2$. Finally, it follows from the inequality of the harmonic-geometric-arithmetic means that

$$a_{2n+1} \leq (a_{2n-2} a_{2n-1})^{1/2} \leq \frac{a_{2n-2} + a_{2n-1}}{2} = a_{2n}$$

But $a_{2n-2} a_{2n-1} = a_{2n-4} a_{2n-3} = \cdots = a_0 a_1$ and so

$$a_{2n+1} \leq (a_0 a_1)^{1/2} \leq a_{2n}$$

This inequality implies that $L_1 \leq (a_0 a_1)^{1/2} \leq L_2$. Since $L_1 = L_2$, we have proved that $\{a_n\} \to (a_0 a_1)^{1/2}$.

We conclude this chapter by pointing out that if a sequence defined by a recurrence relation can be shown to have a limit, then there is a method that sometimes yields the actual value of the limit. It consists of using the result of Example 2-7 in the recurrence relation to obtain an equation for the limit. One then tries to solve this equation and thereby evaluate the limit. The next example will illustrate this point.

EXAMPLE 2-12   Let $a_1 = 2^{1/2}$, and $a_n = (2 + a_{n-1})^{1/2}$ for $n \geq 2$. We prove first that $a_n < 2$ for all n. This we do by induction. The statement is clearly true when $n = 1$. Now assume that $a_n < 2$. Then $a_{n+1} = (2 + a_n)^{1/2} < (2 + 2)^{1/2} = 2$. This completes the proof by induction. We next show that $\{a_n\}$ is increasing. Now $a_{n+1}/a_n = (2/a_n^2 + 1/a_n)^{1/2}$. From the arithmetic-geometric mean inequality we have $2/a_n^2 + 1/a_n \geq 2(2/a_n^3)^{1/2}$. Therefore,

$$\frac{a_{n+1}}{a_n} \geq \left[2\left(\frac{2}{a_n^3}\right)^{1/2} \right]^{1/2} = \left(\frac{2}{a_n}\right)^{3/4}$$

But $2/a_n > 1$ and so $a_{n+1}/a_n > 1^{3/4} = 1$. In other words, $\{a_n\}$ is increasing as well as bounded, and hence it must have a limit L. If we equate the limit of both sides of the recurrence relation $a_n = (2 + a_{n-1})^{1/2}$, we get $L = (2 + L)^{1/2}$, or $L^2 - L - 2 = 0$, or $(L - 2)(L + 1) = 0$. Inasmuch as L must be positive, it follows that $L = 2$.

EXERCISES

2-7 Show that $\{[(n + 1)^{1/2} - n^{1/2}]/n^{3/2}\}$ is decreasing. Deduce that it has a limit without finding it.

2-8 Show that

$$\left\{ \sum_{k=1}^{n} \frac{(-1)^{k+1}}{k} \right\}$$

is doubly monotone. Does it have a limit?

2-9 Put $a_1 = 1$, and $a_{n+1} = 2 - 1/a_n$ for $n \geq 1$. Show that $\{a_n\} \to 1$.

SOLUTIONS TO EXERCISES

2-1 The first five terms of $\{n\} + \{1/n\}$ are $2/1$, $5/2$, $10/3$, $17/4$, and $26/5$; the first five terms of $\{n\} - \{1/n\}$ are $0/1$, $3/2$, $9/3$, $15/4$, and $24/5$; the first five terms of $\{n\} \cdot \{1/n\}$ are 1, 1, 1, 1, and 1; the first five terms of $\{n\}/\{1/n\}$ are 1, 4, 9, 16, and 25; and finally, the first five terms of $\{n\} * \{1/n\}$ are $1/1$, $1/2 + 2/1$, $1/3 + 2/2 + 3/1$, $1/4 + 2/3 + 3/2 + 4/1$, and $1/5 + 2/4 + 3/3 + 4/2 + 5/1$, or 1, $5/2$, $26/6$, $77/12$, and $522/60$.

2-2 This relation is false because even though it is true that $\{1/(-1)^n n\} \to 0$, it is not true that $(-1)^n n > 0$ for all but a finite number of n. Since $(-1)^n n < 0$ is likewise not true for all but a finite number of n, it is also false that $\{(-1)^n n\} \to -\infty$.

2-3 Take $\varepsilon > 0$. We are interested in the equality $|(n^2 + 1)/(2n^2 - 1) - \frac{1}{2}| < \varepsilon$ which is the same as $|[2(n^2 + 1) - (2n^2 - 1)]/2(2n^2 - 1)| < \varepsilon$, or $|3/2(2n^2 - 1)| < \varepsilon$. Since $3/2(2n^2 - 1) > 0$ for all n, the last inequality is the same as $3/2(\varepsilon + 1) < \varepsilon$ which is equivalent to $3/2\varepsilon + 1 < 2n^2$, or $n > [(\frac{1}{2})(3/2n^2 - 1)]^{1/2}$. Thus, if N is any integer larger than $[(\frac{1}{2})(3/2\varepsilon + 1)]^{1/2}$, we see that if $n > N$, then $|(n^2 + 1)/(2n^2 - 1) - \frac{1}{2}| < \varepsilon$.

2-4 Suppose that $\{a_n\}$ has L_1 and L_2 as limits, and that $L_1 \neq L_2$. We may take $L_1 < L_2$. Let us choose ε so that $0 < \varepsilon < (L_2 - L_1)/2$. Since $\{a_n\} \to L_1$, we can find N_1 such that if $n > N_1$, then $|a_n - L_1| < \varepsilon$. And since $\{a_n\} \to L_2$, we can find N_2 so that if $n > N_2$, then $|a_n - L_2| < \varepsilon$. If we let N be the larger of N_1 and N_2, then both the inequalities $|a_n - L_1| < \varepsilon$ and $|a_n - L_2| < \varepsilon$ will be true for all $n > N$. In other words, $a_n < L_1 + \varepsilon$ and $a_n > L_2 - \varepsilon$ for all $n > N$. However, our choice of ε means that $L_1 + \varepsilon < L_1 + (L_2 - L_1)/2 = (L_2 + L_1)/2$, and $L_2 - \varepsilon > L_2 - (L_2 - L_1)/2 = (L_2 + L_1)/2$. Hence $L_1 + \varepsilon < L_2 - \varepsilon$. From this inequality it follows that if $n > N$, then $a_n < a_n$. We have thus arrived at a contradiction, and so our supposition that $L_1 \neq L_2$ must be false.

2-5 We have $\{(2n^2 - 5)/(5n^2 - 2)\} = \{2 - 5/n^2\}/\{5 - 2/n^2\}$. Now we know from Example 2-6 that $\{1/n\} \to 0$. Inasmuch as $\{1/n^2\} = \{1/n\}\{1/n\}$ it thus follows from Theorem 2-1b that $\{1/n^2\} \to 0$. Hence $\{2 - 5/n^2\} \to 2$ and $\{5 - 2/n^2\} \to 5$. Since $5 \neq 0$, we can conclude from Theorem 2-1c that $\{(2n^2 - 5)/(5n^2 - 2)\} \to \frac{2}{5}$.

2-6 Since $1/(n + m) \leq 1/(n + 1)$ for $m = 1, 2, \ldots, k$, we have

$$0 < \sum_{m=1}^{k} \frac{1}{n + m} \leq \frac{k}{n + 1}$$

But now Theorem 2-2 gives us the stated result because $\{k/(n + 1)\} \to 0$.

2-7 Notice that $\{[(n + 1)^{1/2} - n^{1/2}]/n^{3/2}\} = \{1/[(n + 1)^{1/2} + n^{1/2}]n^{3/2}\}$. Since $[(n + 2)^{1/2} + (n + 1)^{1/2}](n + 1)^{3/2} > [(n + 1)^{1/2} + n^{1/2}]n^{3/2}$, or $1/[(n + 2)^{1/2} + (n + 1)^{1/2}](n + 1)^{3/2} < 1/[(n + 1)^{1/2} + n^{1/2}]n^{3/2}$, the sequence is decreasing. Because $[(n + 1)^{1/2} - n^{1/2}]/n^{3/2} > 0$, the sequence is bounded. It therefore has a limit.

2-8 Denote the general term of this sequence by a_n. Then $a_{2n+2} - a_{2n} = 1/(2n + 1) - 1/(2n + 2) > 0$, while $a_{2n+1} - a_{2n-1} = -1/2n + 1/(2n + 1) < 0$. Hence $\{a_{2n}\}$ is increasing, and $\{a_{2n-1}\}$ is decreasing. Now

$$a_{2n} = \sum_{k=1}^{n} \left(\frac{1}{2k - 1} - \frac{1}{2k} \right)$$

and since each term of the sum is positive, we see that $a_{2n} > 0$. Also

$$a_{2n+1} = 1 + \sum_{k=1}^{n} \left(-\frac{1}{2k} + \frac{1}{2k + 1} \right)$$

and since each term under the $\sum$ symbol is negative, we see that $a_{2n+1} < 1$. But $a_{2n+1} - a_{2n} = 1/(2n + 1) > 0$. Therefore, $0 < a_{2n} < a_{2n+1} < 1$. This inequality shows that both $\{a_{2n}\}$ and $\{a_{2n-1}\}$ are bounded. We thus know that $\{a_{2n}\} \to L_2$, and that $\{a_{2n-1}\} \to L_1$. Since $a_{2n} < a_{2n+1}$, we can state further that $a_{2n} \leq L_2 \leq L_1 \leq a_{2n+1}$, or $0 \leq L_1 - L_2 \leq a_{2n+1} - a_{2n}$. Inasmuch as $a_{2n+1} - a_{2n} = 1/(2n + 1)$, we have $0 \leq L_1 - L_2 \leq 1/(2n + 1)$. Because $\{1/(2n + 1)\} \to 0$ it follows from the last inequality that $L_1 = L_2$, or that the sequence does have a limit.

2-9 We first show that $a_n \geq 1$ for all n; the proof is by induction. This inequality is true by definition when $n = 1$. If we assume that $a_n \geq 1$, then $-1/a_n \geq -1$ and $a_{n+1} = 2 - 1/a_n \geq 2 - 1 = 1$. The induction is complete. (We note in passing that we have also established that $a_n \neq 0$, or that the sequence in question is well defined.) We next have

$$a_n - a_{n+1} = a_n - 2 + \frac{1}{a_n} = \frac{a_n^2 - 2a_n + 1}{a_n} = \frac{(a_n - 1)^2}{a_n} > 0$$

In other words, $\{a_n\}$ is decreasing. Since $\{a_n\}$ is also bounded, it has a limit $L \geq 1$. But L must satisfy the equation $L = 2 - 1/L$; since this equation is the same as $(L-1)^2 = 0$, we see that $L = 1$.

PROBLEMS

2-1 Give the first five terms of each of the following sequences.
   (a) $\{n^2 - n\}$ (b) $\{n/(n+1)\}$
   (c) $\{n^2 - n\} + \{n/(n+1)\}$ (d) $\{n^2 - n\} - \{n/(n+1)\}$
   (e) $\{n^2 - n\} \cdot \{n/(n+1)\}$ (f) $\{n^2 - n\}/\{n/(n+1)\}$
   (g) $\{n^2 - n\} * \{n/(n+1)\}$

2-2 Prove each of the following limits using Definition 2-1:
   (a) $\{1/n^p\} \to 0$ for $p > 0$ (b) $\{(n^2 - 1)/(2n^3 + 3)\} \to \frac{1}{2}$
   (c) $\{\sin n/n\} \to 0$

2-3 Find the limits of the following sequences by using the appropriate limit theorem(s).

   (a) $\left\{\dfrac{n^2 + 2n - 3}{5n^2}\right\}$ (b) $\left\{\dfrac{2n}{n+1} - \dfrac{n+1}{2n}\right\}$

   (c) $\left\{\dfrac{1 + (-1)^n}{n^2}\right\}$ (d) $\{(n+1)^{1/2} - n^{1/2}\}$

   (e) $\left\{\dfrac{n+1}{n^2}\right\}$ (f) $\left\{\dfrac{1}{(n^2 + 1)^{1/2}}\right\}$

   (g) $\left\{\dfrac{n^2 + a}{n + a} \bigg/ \dfrac{n^2 + b}{n + b}\right\}$ (h) $\left\{\left(\dfrac{n^2 - 1}{2n^2 + 3}\right)^3\right\}$

2-4 if $\{a_n\} \to L$, show that
   (a) $\{|a_n|\} \to |L|$ (b) $\{a_n\}$ is bounded.
   Give examples to show that the converses to these statements are false.

2-5 Investigate the convergence or divergence of the following sequences. Find the limit of each convergent sequence.
   (a) $\{a^n\}$ where $a > 0$ (b) $\{a^{1/n}\}$ where $a > 0$
   *(c) $\{n^2/2^n\}$ *(d) $\{na^n\}$ where $0 < a < 1$
   *(e) $\{n^{1/n}\}$ (f) $\{(n^m)^{1/n}\}$, m a positive integer

2-6 Let $\{a_n\}$ be a sequence of positive terms.
   (a) If $a_{n+1} > ka_n$ and $k > 1$ show that $\{a_n\} \to \infty$.
   (b) If $a_{n+1} < ka_n$ and $0 < k < 1$, show that $\{a_n\} \to 0$.
   (c) If $\{a_{n+1}/a_n\} \to l$, show that $\{a_n\} \to \infty$ if $l > 1$, while $\{a_n\} \to 0$ if $0 < l < 1$.

2-7 Show that
   *(a) $\{a^n/n!\} \to 0$ for any $a > 0$ *(b) $\{(n!)^{1/n}\} \to \infty$

2-8   Examine the convergence or divergence of the following sequences.

(a) $\{n!/n^n\}$

(b) $\left\{\left(\dfrac{1}{n}\displaystyle\prod_{k=1}^{n}\dfrac{2k}{2k-1}\right)^2\right\}$

*(c) $\left\{\displaystyle\sum_{k=1}^{n}\dfrac{1}{n+k}\right\}$

*(d) $\left\{\left(1+\dfrac{1}{n}\right)^n\right\}$

2-9   Let $a_1 = 2^{1/2}$ and set $a_{n+1} = (2a_n)^{1/2}$ for $n > 1$. Show that $\{a_n\} \to 2$.

*2-10   Let $a_1 > 0$ and set $a_{n+1} = 1 + 1/a_n$ for $n \geq 1$. Find the limit of $\{a_n\}$.

*2-11   Let $0 < a_1 < b_1$ and set $a_{n+1} = (a_n b_n)^{1/2}$ and $b_{n+1} = \frac{1}{2}(a_n + b_n)$ for $n > 1$. Show that $\{a_n\}$ and $\{b_n\}$ converge and have the same limit.

*2-12   If a_1 and a_2 are arbitrary and $a_n = (a_{n-1} + a_{n-2})/2$ for $n \geq 3$, find the limit of $\{a_n\}$.

2-13   For a sequence $\{a_n\}$ put

$$A_n = \frac{1}{n}\sum_{k=1}^{n} a_k$$

(a) If $\{a_n\}$ is monotone, prove that $\{A_n\}$ is also monotone.
(b) If $\{a_n\} \to L$, prove that $\{A_n\} \to L$ as well.
(c) Let $a_n > 0$ for all n and suppose that $\{a_n\} \to L > 0$. If

$$H_n = \frac{n}{\displaystyle\sum_{k=1}^{n}\frac{1}{a_k}}$$

show that $\{H_n\} \to L$.
*(d) If $a_n > 0$ for all n, $\{a_n\} \to L > 0$ and

$$G_n = \left(\prod_{k=1}^{n} a_k\right)^{1/n}$$

prove that $\{G_n\} \to L$.

*2-14   If $a_n > 0$ for $n = 0, 1, 2, \ldots$, and $\{a_{n+1}/a_n\} \to L > 0$, show that $\{a_n^{1/n}\} \to L$.

Chapter 3

FUNCTIONS
AND THEIR
LIMITS

The concept of a function is very similar to the concept of a sequence, and the definition of a sequence given in Chapter 2 needs only one modification to serve as the definition of a function (provided that we again agree that we know the meanings of the terms "set" and "correspondence").

A *function* is a correspondence between two sets D and R such that to each member of D there corresponds exactly one member of R, and every member of R corresponds to at least one member of D. The set D is called the *domain* of the function; the set R is called the *range* of the function. If E is a subset of D, then the correspondence that specifies the function can be taken to hold only on E; the new function thereby determined is called the *restriction* of the original function to E. The sets D and R are usually sets of numbers in calculus, but there is nothing in our definition to prevent them from being sets of objects of any kind. Functions for which D and R are sets of numbers will be called *numerical*.

Observe that a sequence can now be regarded as a special function whose domain is the set of natural numbers. In other words, the one difference between the concept of a function and the concept of a sequence is that the domain of a function can be an arbitrary set while the domain of a sequence is the set of natural numbers.

The fact that sequences are at the same time functions means that we already have many examples of functions, but apart from these the reader should realize that he has encountered other examples of functions both in mathematics and everyday life. Thus a familiar function from mathematics is the

27

one that corresponds to each rectangle its area; while a familiar function from everyday life is the one that corresponds to each person who has at least one finger his set of fingerprints.

Functions will, for the most part, be denoted by lower case letters f, g, h, and so on.   If x is in the domain of f, we will denote by $f(x)$, read "f of x", that object in the range of f which corresponds to x.   When dealing with a numerical function, this notation furnishes an economical way of describing it.   One simply specifies its domain and then writes the *equation* of the function $y = f(x)$ (with the understanding that it holds only for x in the domain of f).

It should be kept in mind, however, that a function is not the same as its equation; it is a more abstract mathematical entity than its equation.   The blurring of this distinction between a function and its equation can lead to a certain amount of confusion.   It is a common mistake with beginners.   We will return to this point shortly.

The domains of functions are frequently sets called *intervals*.   We distinguish four types of intervals and we introduce a notation for each of them. Let $a < b$, and denote by $[a, b]$, (a, b), $[a, b)$, and $(a, b]$ all x satisfying the inequalities $a \leq x \leq b$, $a < x < b$, $a \leq x < b$, and $a < x \leq b$ respectively. The interval $[a, b]$ is called a *closed* interval, while (a, b) is called an *open* interval.   The notations $[a, \infty)$, (a, ∞), $(-\infty \ a]$, and $(-\infty, a)$ will be used to denote all x satisfying the inequalities $x \geq a$, $x > a$, $x \leq a$, and $x < a$ respectively.   The notation $(-\infty, \infty)$ will be used to refer to all real numbers. Needless to say, the same cautions about the use and interpretation of the symbol ∞ that were made in Chapter 2 apply here as well.

We now proceed to build up a similar sort of algebra for functions as we did for sequences.   Let f and g be functions.   We say that f and g are *equal*, written $f = g$, if f and g have the same domain and $f(x)$ is the same as $g(x)$ for all x in this common domain.   This definition of equality shows how important it is to keep in mind the distinction between a function and its equation.   Without this distinction, the simple fact that the function f with domain $[0, \infty)$ and equation $f(x) = x^2$ is not equal to the function g with domain $(-\infty, \infty)$ and equation $g(x) = x^2$ would make functions appear to be mysterious and nebulous.   [Incidentally, it is pointless to argue that the "real" domain of f is also $(-\infty, \infty)$ because the domain of the function is independent of the correspondence involved.   No domain is any more "real" or "natural" than any other.]

Now, let f and g be two numerical functions, and denote their domains by D_f and D_g respectively.   Let us further denote by $D_{f,g}$ all common members of D_f and D_g (in set theoretic terminology, $D_{f,g}$ is the intersection of D_f and D_g).   For f and g, $D_{f,g}$ will be the domain of their *sum* $f + g$, their *difference* $f - g$, and their *product* $f \cdot g$.   (We tacitly assume that there is at least one member in $D_{f,g}$; if $D_{f,g}$ has no members, none of these operations can be performed on f and g.)   The equations of each of these functions are defined as follows: $(f+g)(x) = f(x) + g(x)$, $(f-g)(x) = f(x) - g(x)$, and $(f \cdot g)(x) = f(x) \cdot g(x)$.   For the *quotient* f/g, the domain is all numbers x in $D_{f,g}$ such that $g(x) \neq 0$, and the equation is defined by $(f/g)(x) = f(x)/g(x)$.   Observe that these four operations are consistent with the algebraic operations between sequences defined in Chapter 2.

We conclude by defining an operation that does not depend on the elementary arithmetic operations.   Contrary to what might be expected because of the close parallel between functions and sequences up to now, this operation is not the analog of the convolution of two sequences.   (It is possible to define the convolution of two functions, but to do so it is necessary to employ the integral.)   The operation in question is called *composition*.   To form the composition of f by g, denoted by $f(g)$ and referred to as a *composite function*, it must be the case that the range of g and the domain of f have common members (in set theoretic terms, their intersection must be nonempty).   If this is the case, then $f(g)(x) = f(g(x))$.   The domain of $f(g)$ is that subset of D_g consisting of all numbers x such that $g(x)$ is a member of D_f.   If the function g is itself a composite function (the composite of p by q, for instance), then the composition of f by g can be regarded as the triple composition of f by p by q which would be written as $f(p(q))$.   It is clear that we can form the composition of any finite number of functions having the necessary relations between their domains and ranges.

Since the formation of a composition depends on rather special conditions, it should not be too surprising that the operation of composition has rather special properties.   In the first place, it may be possible to form $f(g)$ but not $g(f)$.   In the second place, even if both $f(g)$ and $g(f)$ can be formed, it need not be true that $f(g) = g(f)$.   Thus, it is important to make the order of formation clear.   It is for this reason that we speak of the composition of f by g, and not of the composition of f and g.   At the same time, it should not be inferred from the special conditions under which a composition can be formed that it is an operation of only limited importance.   Quite the contrary is true.   Composition is perhaps the most important of the functional operations because it depends most strongly on the concept of a function.   That is to say, while both functions and numbers can be added, subtracted, multiplied, and divided, only functions can be composed.

EXAMPLE 3-1   As with sequences, one of the most natural examples of a function is an *identity function* for which D and R are the same, and every x in D corresponds to itself.   Thus, if f is a numerical identity function, then $f(x) = x$ for all x in D.   Another simple numerical function is a *constant function* for which R consists of only one number c, and $f(x) = c$ for all x in D.

Other common numerical functions are the *power functions*, which have domain $(-\infty, \infty)$ and are defined by the equation $f(x) = x^n$ for some positive integer n.

The product of a constant function and a power function is called a *monomial*, and the sum of several monomials is called a *polynomial*. Thus, a polynomial f has the equation

$$f(x) = a_n x^n + a_{n-1} x^{n-1} + \cdots + a_1 x + a_0$$

where $a_0, a_1, \ldots, a_n$ are constants. The quotient of two polynomials is called a *rational function*.

Closely connected with the power functions are the *root functions*. Their definitions are similar, but require a bit more care. If n is an even integer, the domain of the root function is $[0, \infty)$, while if n is odd the domain is $(-\infty, \infty)$. In either case, the equation of the root function is $f(x) = x^{1/n}$.

EXAMPLE 3-2   A function we have already seen is the absolute value function defined by Eq. (1-20). Its domain is $(-\infty, \infty)$, and its range is $[0, \infty)$. Equation (1-21) shows that the absolute value can alternately be defined as the composite of the square-root function (the second-root function) by the square function (the second-power function). Thus, just as a number may have different representatives as a sum, product, and so on, we see that a function may also be represented in different ways.

EXAMPLE 3-3   In the previous example we saw that the absolute-value function could be defined as the composition of a root function by a power function. This method can be used to define power functions for any rational exponent r. First, suppose that $r > 0$ and that $r = m/n$, where m and n are positive integers. The rth power function is then defined to be the composition of the nth root function by the mth power functions. That is, $f(x) = (x^m)^{1/n}$. If $r < 0$, define the rth power function by $f(x) = 1/x^{-r}$.

Remark   This discussion leads naturally to the problem of how to define a yth power function, where y is any real number. While the complete answer to this question is too lengthy for a book of this nature, it seems appropriate to at least indicate one way of making such a definition.

In the first place it is clear that we have to impose some restrictions. On the basis of the definition of root functions, we should take $x \geq 0$. Further, we put $0^y = 0$ for all $y \neq 0$, $1^y = 1$ for all y, and $x^0 = 1$ for all $x > 0$.

Accordingly, take $x > 0$, $x \neq 1$, and $y \neq 0$. If y is rational, we know how to define x^y. We can thus concentrate on irrational y. Now if p and q are rational, and $p < q$, then

$$x^p \begin{cases} < x^q & \text{if} \quad x > 1 \\ > x^q & \text{if} \quad 0 < x < 1 \end{cases} \tag{3-1}$$

For y irrational pick a monotone sequence $\{r_n\}$ of rational numbers such that $\{r_n\} \to y$. It follows from Eq. (3-1) that $\{x^{r_n}\}$ is also monotone. Since $\{x^{r_n}\}$ is clearly bounded, it must have a limit λ. It can then be proved that λ does not depend on the choice of $\{r_n\}$, only on y. This fact can then be used to prove that if $\{p_n\}$ is any sequence of rational numbers such that $\{p_n\} \to y$, then $\{x^{p_n}\} \to \lambda$. We now define x^y to be the number λ.

Once x^y has been given meaning, it is apparent that two functions have been defined. On the one hand, we can keep y fixed, regard x as a variable and so obtain the yth power function. On the other hand, we can keep $x > 0$ fixed, regard y as a variable, and so obtain an *exponential function*.

It can be shown that an exponential function satisfies the relations

$$a^{x+z} = a^x a^z$$

and

$$(a^x)^z = a^{xz}$$

EXERCISES

3-1   If for all x, $f(x) = x + 2$, and $g(x) = x^2 - 4$, find
   (a) $f(2)$ (b) $f(98)$ (c) $g(4)$
   (d) $g(-2)$ (e) $(f+g)(1)$ (f) $(f \cdot g)(6)$
   (g) $(g/f)(3)$ (h) $(g-f)(-3)$ (i) $f(g(2))$
   (j) $g(f(2))$

3-2   Find functions f and g such that:
   (a) Neither $f(g)$ nor $g(f)$ can be formed.
   (b) $f(g)$ can be formed, but $g(f)$ cannot be formed.
   (c) Both $f(g)$ and $g(f)$ can be formed, but $f(g) \neq g(f)$.
   (d) Both $f(g)$ and $g(f)$ can be formed, and $f(g) = g(f)$.

Just as monotone sequences are easier to deal with than arbitrary sequences, some numerical functions have special characteristics that make them easier to deal with than arbitrary functions. A function whose domain is symmetric about zero (that is, if x belongs to D_f, then $-x$ also belongs to D_f) is *even* if $f(-x) = f(x)$, and it is *odd* if $f(-x) = -f(x)$. If the set S is a subset of D_f, then f is *strictly increasing* on S if whenever x_1 and x_2 are in S and $x_1 < x_2$, then $f(x_1) < f(x_2)$, and f is *increasing* on S if $f(x_1) \leq f(x_2)$. Similarly, f is *strictly decreasing* on S if whenever x_1 and x_2 are in S and $x_1 < x_2$, then $f(x_1) > f(x_2)$, and f is *decreasing* on S if $f(x_1) \geq f(x_2)$. Functions that are increasing or decreasing on S are collectively called *monotone* on S. Strictly increasing or strictly decreasing functions on S are collectively called *strictly monotone* on S. If a function is monotone or strictly monotone on its whole domain, we say simply that it is monotone or strictly monotone respectively. Finally, a function is *one-to-one* if $f(x_1) = f(x_2)$ can hold only for $x_1 = x_2$.

EXAMPLE 3-4   Typical even functions are the power functions whose exponents are even positive integers. That is, the functions whose equations are $f(x) = x^{2n}$ for $n = 1, 2, \ldots$. The power functions having odd positive integers for exponents are clearly examples of odd functions. There are, of course, odd and even functions that do not involve odd and even powers in any obvious way. Such are the familiar sine and cosine functions. Sine is an odd function, and cosine is an even function.

EXERCISES

3-3   Show that the sum and product of two even functions, both having the same domain, are also even. What can you say about the eveness or oddness of the sum and product of two odd functions?

3-4   Prove that a strictly monotone function is one-to-one. Is the converse true? (That is, is it true that if a function is one-to-one, then it is strictly monotone?)

Let us now turn from properties that involve a single function to an important relation that involves pairs of functions. Two functions f and g are *inverse functions* if the domain of f is the same as the range of g, the domain of g is the same as the range of f, and if $f(g)(y) = y$ for all y in D_g, $g(f)(x) = x$ for all x in D_f. [Notice that the conditions on the domains and ranges of f and g ensure that both $f(g)$ and $g(f)$ can be formed.] We refer to g as f *inverse* and denote it by f^{-1}. Symmetrically, we refer to f as g *inverse* and denote it by g^{-1}. The identities $(f^{-1})^{-1} = g$ and $(g^{-1})^{-1} = g$ are immediate consequences of these notations.

We now establish a simple criterion for determining whether a function has an inverse. First assume that f has an inverse, and put $y = f(x)$. Because $x = f^{-1}(f(x)) = f^{-1}(y)$, and f^{-1} is a function, it must be the case that x is uniquely determined by y. That is, if $f(x_1) = y = f(x_2)$, then $x_1 = x_2$. Thus if f^{-1} exists, then f is one-to-one. Next assume that f is one-to-one, and again put $y = f(x)$. Since f is one-to-one each such y uniquely determines an x in the domain of f. This means that there is a correspondence between members of the range of f and the domain of f such that to each y in the range of f there corresponds precisely one x in the domain of f. But this is exactly the notion of a function. In other words, a function g, whose domain is the range of f, is determined by the condition that for all y in the range of f the equation $x = g(y)$ holds if and only if $y = f(x)$. For g we have $f(g)(y) = f(g(y)) = f(x) = y$, and $g(f)(x) = g(f(x)) = g(y) = x$. Hence f and g are inverse functions, and we have shown that if f is one-to-one, then f^{-1} exists. To sum up, we have proved that *a function has an inverse if and only if it is one-to-one*.

EXAMPLE 3-5   The integer power functions provide clear examples of functions having inverses. If n is an odd integer, then the functions

$f(x) = x^n$ and $g(x) = x^{1/n}$ obviously satisfy the relations $f(g)(x) = x = g(f)(x)$, and are thus a pair of inverse functions. The situation is similar for even powers, except that we must consider the restriction of such a power function to $[0, \infty)$ because it is not one-to-one on $(-\infty, \infty)$. The introduction of an appropriate restriction of a function for the purpose of introducing an inverse function is a common technique. It is particularly evident in the definition of the inverse trigonometric functions (for which, consult any standard text).

EXAMPLE 3-6 A famous inverse function is the *logarithm function* to the base b. If $b \neq 1$ is a positive number, then we know, from the remark following Example 3-3, that an exponential function f with domain $(-\infty, \infty)$ can be defined by the equation $f(y) = b^y$. It is a consequence of the way in which b^y is defined that f is strictly increasing if $b > 1$, and strictly decreasing if $b < 1$. In either case, it follows from Exercise 3-4 that f is one-to-one, or that f has an inverse. This inverse function is called the logarithm to the base b, and it is denoted by $\log_b$. The domain of $\log_b$ is the range of f which is, as is easy to see, $(0, \infty)$. Otherwise stated, $\log_b$ is a function having domain $(0, \infty)$ which satisfies the condition that, for all $x > 0$, $y = \log_b x$ if and only if $x = b^y$. (The base b most frequently encountered in numerical calculations is $b = 10$; $\log_{10} x$ is called the *common* logarithm of x.) It can be shown that

$$\log_b xz = \log_b x + \log_b z$$

and

$$\log_b x^z = z \log_b x$$

for any base b.

EXERCISE 3-5 Show that the function f whose equation is $f(x) = (16 - 4x^2)^{1/2}$ on the domain $[0, 2]$ is strictly decreasing. Find f^{-1} and give its domain.

Up to now the functions we have been considering were defined by explicitly giving their equations or correspondences. However, there is another method of defining functions which results in *implicit functions*.

Consider a function F whose domain is a subset of the set of all ordered pairs of real numbers, and whose range is a subset of the set of real numbers. We use the notation (x, y) to indicate that the pair of numbers x and y is to be regarded as an ordered pair; that is, (x, y) and (u, v) are equal if and only if $x = u$ and $y = v$. Thus $(1, 2)$ is not the same as $(2, 1)$. [There is some risk of confusing the ordered pair (a, b) with the open interval (a, b), but it is usually clear from the context which meaning is intended.] Let c be any number in the range of F, and form the equation $F((x, y)) = c$.   It is then possible, at least in principle, to determine all ordered pairs satisfying this equation. Select any collection C of these ordered pairs subject only to the condition that if (x, y) and (x, z) are both in C, then $y = z$.   Because of this condition the collection C can be used to define a function f by setting $y = f(x)$ if and only if (x, y) is a member of C.   The domain of f is the set of all first members x of each ordered pair (x, y) of C, while the range of f is the set of all second members y.   This function f is called an *implicit function* determined by the equation $F((x, y)) = c$.   Observe that $F((x, f(x))) = c$. Since any such collection of ordered pairs determines a function, it follows that there are many (usually infinitely many) implicit functions determined by the same equation.   In other words, the equation $F((x, y)) = c$ actually determines a family of implicit functions.   However, it often happens that this family can be specified by listing only a few of its members—specified in the sense that the remaining members of the family can be obtained from appropriate restrictions of the functions in the list.   If $b \neq c$ is also in the range of F, then it is generally the case that the equation $F((x, y)) = b$ determines a completely different family than the one determined by $F((x, y)) = c$.

EXAMPLE 3-7   Let a, b, and c be three nonzero numbers, and put $F((x, y)) = ay^2 + bxy + cx^2$.   Since $F((0, 0)) = 0$, we see that 0 is in the range of F.   Consequently, we may expect that there is a family of implicit functions determined by the equation $F((x, y)) = 0$, or

$$ay^2 + bxy + cx^2 = 0 \tag{3-2}$$

To determine all ordered pairs (x, y) satisfying Eq. (3-2), we apply the quadratic formula, with x regarded as temporarily fixed, and find that

$$y = \frac{-bx \pm [(b^2 - 4ac)x^2]^{1/2}}{2a} \tag{3-3}$$

Now, if $b^2 - 4ac < 0$, then the only ordered pair satisfying Eq. (3-3) is $(0, 0)$. So in this case the family of implicit functions consists only of the function that has the single number 0 as both its domain and range.   If $b^2 - 4ac = 0$, then Eq. (3-3) becomes $y = -bx/2a$, and the example becomes more interesting because it can now be used to illustrate how a family of implicit functions can be specified by listing, in this case, only one function.   The function is of course the function f defined by the equation $f(x) = -bx/2a$ on the domain $(-\infty, \infty)$.   The family of implicit functions determined by Eq. (3-2) when $b^2 - 4ac = 0$ is then simply the family of all restrictions of f.   Finally, if $b^2 - 4ac > 0$, the situation is almost the same as the previous case except that now two functions f_1 and f_2 are necessary to specify the family of implicit functions.   Both f_1 and f_2 have $(-\infty, \infty)$ as domain.   The equation of f_1

is $f_1(x) = (-bx + [(b^2 - 4ac)x^2]^{1/2})/2$, and the equation of f_2 is $f_2(x) = (-bx - [(b^2 - 4ac)x^2]^{1/2})/2$. If S_1 is a subset of the domain of f_1, and S_2 is a subset of the domain of f_2, and if S_1 and S_2 have no common members, then a function g can be defined on the domain consisting of the set of all x that are in S_1 or in S_2 (in set theoretic terminology, this is the set $S_1 \cup S_2$) by taking $g(x)$ to be $f_1(x)$ if x is in S_1, and $g(x)$ to be $f_2(x)$ if x is in S_2. The family of all such g is the family of implicit functions determined by Eq. (3-2) when $b^2 - 4ac > 0$.

Remark   It should be pointed out that the above example is not typical of what happens in general.   Usually such equations as Eq. (3-2) cannot be solved completely.   This does not mean that there is no family of implicit functions, but rather that knowledge of this family must be gained by indirect means.   We shall encounter such an indirect method in the next chapter.

   EXERCISE 3-6   Specify the family of implicit functions determined by the equation $x^4 - y^4 = 0$.

   As with sequences, the most important notion that arises in connection with functions, at least as far as calculus is concerned, is that of the *limit* of a function.

Definition 3-1   *The function f has limit L at a, written*

$$\lim_{x \to a} f(x) = L \tag{3-4}$$

if given $\varepsilon > 0$, there exists $\delta > 0$ such that if

$$0 < |x - a| < \delta \tag{3-5}$$

then

$$|f(x) - L| < \varepsilon \tag{3-6}$$

(Here δ is the lower case Greek letter delta.)   It is tacitly assumed in this definition that the numbers x satisfying Eq. (3-5) are actually in the domain of f, but note that it is irrelevant whether or not a belongs to the domain of f. Definition 3-1 is no more than a precise way of stating that $f(x)$ is " arbitrarily close" to L when x is "sufficiently close" to a.   The number ε gives us a quantitative measure of "arbitrary closeness", and the number δ does the same for "sufficient nearness."

   It is clear that Definition 3-1 has much in common with the definition of the limit of a sequence; indeed, its very phraseology is almost identical with that of Definition 2-1.   But at the same time, there is no obvious way of relating one to the other.   For the sake of logical completeness, since sequences are, after all, only special functions, we should be able to say what connection there is between the two definitions.   We do this by defining *limits at infinity*.

Definition 3-1a   *The function f has limit L at infinity, written*

$$\lim_{x \to \infty} f(x) = L \tag{3-7}$$

if given any $\varepsilon > 0$, there exists $X > 0$ such that if $x > X$, then $|f(x) - L| < \varepsilon$.

EXERCISES

3-7   Verify that Definition 3-1a is the same as Definition 2-1 when the function f is a sequence (that is, when f is a function whose domain is the set of positive integers).

3-8   By mimicking Definition 3-1a almost word-for-word, formulate an analogous definition of the symbol $\lim_{x \to -\infty} f(x) = L$.

There is another variant of Definition 3-1 that likewise leads to a useful and interesting variant of the limit concept. As with Definition 3-1a, we will give one version of it, and leave it to the reader to formulate the companion definition (in Exercise 3-9).

Definition 3-1b   *The function f has right-hand limit L at a, written*

$$\lim_{x \to a+} f(x) = L \tag{3-8}$$

if given $\varepsilon > 0$, there exists $\delta > 0$ such that if

$$0 < x - a < \delta \tag{3-9}$$

then $|f(x) - L| < \varepsilon$.

The only difference between this definition and Definition 3-1 is that Eq. (3-5) has been replaced by Eq. (3-9). Now Eq. (3-5) is the same as $0 < x - a < \delta$ or $-\delta < x - a < 0$ by Theorem 1-1. Therefore, Eq. (3-9) is simply the "right-hand" part of Eq. (3-5). Right-hand limits arise in a natural way when a function has an interval of the form (a, b); $[a, b)$; $(a, b]$; $[a, b]$; (a, ∞); or $[a, \infty)$ as its domain, and it is of interest to examine the limiting behavior of f at a.

EXERCISES

3-9   Formulate a definition of the sentence "The left-hand limit of f at a is L," which is denoted by $\lim_{x \to a-} f(x) = L$, along the lines of Definition 3-1b.

3-10   Prove that $\lim_{x \to a} f(x) = L$ if and only if $\lim_{x \to a+} f(x) = L = \lim_{x \to a-} f(x)$.

The remaining variations of the limit concept can be defined by making use of the definitions we already have. They are the so-called *infinite limits*. Because there is little difference between the definition of limits that are "positive" infinite and those that are "negative" infinite, we are going to give one definition that will serve for both if the following convention is adhered to. In various parts of the upcoming Definition 3-1c we are going to insert the symbol $\left\{ \begin{matrix} \infty \\ -\infty \end{matrix} \right.$; whenever another $\left\{ \right.$ is encountered in the subsequent course of the definition, it will be understood that the top line to the right of the bracket is to be associated with ∞, while the bottom line is to be associated with $-\infty$. This is a common space-saving device that is used when two (or more) similar concepts are being defined or stated in a theorem. From now on we will make use of it without prior comment.

Definition 3-1c (i) *The function f has right-hand limit* $\left\{ \begin{matrix} \infty \\ -\infty \end{matrix} \right.$ *at a, written*

$$\lim_{x \to a+} f(x) = \left\{ \begin{matrix} \infty \\ -\infty \end{matrix} \right., \text{ if there is an interval of the form } (a, b) \text{ such that } f(x) \left\{ \begin{matrix} >0 \\ <0 \end{matrix} \right.$$

for all x in (a, b) and

$$\lim_{x \to a+} \frac{1}{f(x)} = 0$$

(ii) *The function f has left-hand limit* $\left\{ \begin{matrix} \infty \\ -\infty \end{matrix} \right.$ *at a, written* $\lim\limits_{x \to a+} f(x) = \left\{ \begin{matrix} \infty \\ -\infty \end{matrix} \right.$,

if there is an interval of the form (c, a) such that $f(x) \left\{ \begin{matrix} >0 \\ <0 \end{matrix} \right.$ *for all x in (c, a), and*

$$\lim_{x \to a-} \frac{1}{f(x)} = 0$$

(iii) *The function f has limit* $\left\{ \begin{matrix} \infty \\ -\infty \end{matrix} \right.$ *at a, written*

$$\lim_{x \to a} f(x) = \left\{ \begin{matrix} \infty \\ -\infty \end{matrix} \right., \text{ if } \lim_{x \to a+} f(x) = \lim_{x \to a-} f(x) = \left\{ \begin{matrix} \infty \\ -\infty \end{matrix} \right.$$

(iv) *The function f has limit* $\left\{ \begin{matrix} \infty \\ -\infty \end{matrix} \right.$ *at* ∞, *written* $\lim\limits_{x \to \infty} f(x) = \left\{ \begin{matrix} \infty \\ -\infty \end{matrix} \right.$, *if*

there is a number K > 0 such that $f(x) \left\{ \begin{matrix} >0 \\ <0 \end{matrix} \right.$ *for all x > K, and*

$$\lim_{x \to \infty} \frac{1}{f(x)} = 0$$

(v) *The function f has limit* $\left\{ \begin{matrix} \infty \\ -\infty \end{matrix} \right.$ *at* $-\infty$, *written* $\lim\limits_{x \to -\infty} f(x) = \left\{ \begin{matrix} \infty \\ -\infty \end{matrix} \right.$,

if there is a number M < 0 such that $f(x) \left\{ \begin{matrix} >0 \\ <0 \end{matrix} \right.$ *for all x < M, and*

$$\lim_{x \to -\infty} \frac{1}{f(x)} = 0$$

EXAMPLE 3-8   We illustrate the use of Definition 3-1 by proving that $\lim_{x \to 2} (2x - 4) = 0$. To this end, let $\varepsilon > 0$. We have to deal with the inequality $|(2x - 4) - 0| < \varepsilon$, or $|x - 2| < \frac{1}{2}\varepsilon$. From the second inequality we see that if δ is any number satisfying $0 < \delta < \frac{1}{2}\varepsilon$, then $0 < |x - 2| < \delta$ implies that $|(2x - 4) - 0| < \varepsilon$. In other words, we have proved that $\lim_{x \to 2} (2x - 4) = 0$.

Remark   This example also illustrates a point similar to that made in the remark following Example 2-6. Namely, that for a fixed $\varepsilon > 0$ there are infinitely many choices for δ in the sense that δ can be any number in some fixed interval, but that for smaller and smaller ε the choice of the corresponding δ must be made, in general, from a shorter and shorter interval.

EXAMPLE 3-9   Since $2x - 4 > 0$ for all $x > 2$, and $2x - 4 < 0$ for all $x < 2$, the previous example, in view of Definition (3-1c), also proves that $\lim_{x \to 2+} 1/(2x - 4) = \infty$, and $\lim_{x \to 2-} 1/(2x - 4) = -\infty$.

EXAMPLE 3-10   We will prove that $\lim_{x \to \infty} 1/x^2 = 0$ by applying Definition 3-1a. For $\varepsilon > 0$ we must examine the inequality $|1/x^2 - 0| < \varepsilon$ or $1/x^2 < \varepsilon$. But the second inequality is obviously satisfied whenever $x > 1/\varepsilon^{1/2}$. Thus the X we seek can be any number larger than $1/\varepsilon^{1/2}$. The proof is complete.

EXAMPLE 3-11   To show the necessity for the concepts of right-hand and left-hand limits, consider the function f defined by

$$f(x) = \begin{cases} 1 & \text{if } x \geq 0 \\ -1 & \text{if } x < 0 \end{cases}$$

It is clear that $\lim_{x \to 0+} f(x) = 1$, and that $\lim_{x \to 0-} f(x) = -1$.

Remark   The function f of this example is one of an important class of numerical functions known as *step functions*. A step function has for its domain two or more (possibly infinitely many) nonoverlapping and abutting intervals (the pairs of intervals $(a, b]$ and (b, c), or (a, b) and $[b, c)$ are nonoverlapping and abutting). Further, and this is really what characterizes a step function, it has the property that its restriction to any one of these intervals is a constant function. Step functions arise naturally in probability and statistics.

EXERCISES

3-11   Let f be defined as follows: The domain of f is the set of all nonzero real numbers, and the equation of f is

$$f(x) = \begin{cases} x^2 + 1 & \text{for } x < 0 \\ 1 - x & \text{for } x > 0 \end{cases}$$

Prove that $\lim\limits_{x\to 0} f(x) = 1$ by showing that $\lim\limits_{x\to 0+} f(x) = 1 = \lim\limits_{x\to 0-} f(x)$.

3-12 Prove that $\lim\limits_{x\to\infty} x/(2x+1) = 1/2$.

3-13 Prove that $\lim\limits_{x\to 1-} 1/(x-1) = -\infty$

3-14 Prove that $\begin{cases} \lim\limits_{x\to\infty} f(x) = L \\ \lim\limits_{x\to-\infty} f(x) = L \end{cases}$ if and only if $\begin{cases} \lim\limits_{y\to 0+} f(1/y) = L \\ \lim\limits_{y\to 0-} f(1/y) = L \end{cases}$

The same difficulty arises with these definitions of the various types of limits of functions that arose with the definition of a limit of a sequence. Namely, that the limit must, in some way or other, be known or guessed before it can be proved. There are two ways of getting around this difficulty.

The first way is to use limits that have been previously established, or are easier to prove, to calculate the limit in question. Thus the function involved may be expressible as the sum or product of simpler functions, or it may be possible to "squeeze" it between two simpler functions which have the same limit. Both of these alternatives are made precise by the next two theorems.

Theorem 3-1 *Let* $\lim\limits_{x\to a} f(x) = A$, *and* $\lim\limits_{x\to a} g(x) = B$. *Then*

(a) $\lim\limits_{x\to a} [f(x) \pm g(x)] = A \pm B$,

(b) $\lim\limits_{x\to a} f(x)g(x) = AB$,

(c) $\lim\limits_{x\to a} f(x)/g(x) = A/B$, *provided that* $B \neq 0$.

It can be proved easily by induction that parts (a) and (b) are true for any finite number of functions having limits at a.

Theorem 3-2 *If* $\lim\limits_{x\to a} f(x) = L = \lim\limits_{x\to a} g(x)$, *and* $f(x) \le h(x) \le g(x)$ *for* $0 < |x-a| < b$, *for some* $b > 0$, *then* $\lim\limits_{x\to a} h(x) = L$.

Theorem 3-1 continues to hold for right-hand and left-hand limits, as well as for limits at ∞ and $-\infty$, with no other modification of its hypotheses. Theorem 3-2 also remains true for these additional limits if its hypotheses are modified as follows:

For $\begin{cases} \text{right-hand} \\ \text{left-hand} \end{cases}$ limits, we need $f(x) \le h(x) \le g(x)$ only for $\begin{cases} a < x < b \\ b < x < a \end{cases}$; for

limits at $\begin{cases} \infty \\ -\infty \end{cases}$ it is sufficient to have $f(x) \le h(x) \le g(x)$ whenever $\begin{cases} x > K \\ x < K \end{cases}$

for some fixed $K \begin{cases} > 0 \\ < 0 \end{cases}$

The second way of avoiding the difficulties inherent in the limit definitions is to single out functions that are well-behaved with respect to limits. These are the *continuous functions*.

Definition 3-2   *The function f is continuous at a if a is in the domain of f,* $\lim\limits_{x \to a} f(x)$ *exists, and*

$$\lim_{x \to a} f(x) = f(a) \tag{3-10}$$

If a function is continuous at each number in a set S, then we say that f is *continuous on S*.

It follows from Theorem 3-1 that if f and g are continuous at a, then their sum, difference, product, and quotient are continuous at a as well (provided of course that the limit in the denominator of the quotient is not zero). This fact implies that if f and g are continuous on S, then so are their sum, difference, product, and (with the usual restriction that the denominator not be zero on S), their quotient.

The concept of continuity allows us to formulate a limit theorem for the limit of a composition.

Theorem 3-3   *If* $\lim\limits_{x \to a} g(x) = b$, *and if f is continuous at b then*

$$\lim_{x \to a} f(g(x)) = f(b) \tag{3-11}$$

[It is tacitly assumed that the composition $f(g)$ can be formed.]   The formal meaning of this theorem is that, with appropriate restrictions, the $\lim\limits_{x \to a}$ symbol can be moved inside the parentheses of the composition.   That is, Eq. (3-11) can be more suggestively written as $\lim\limits_{x \to a} f(g(x)) = f\left(\lim\limits_{x \to a} g(x)\right)$.   An immediate corollary of this theorem is the fact that the composition of two continuous functions is a continuous function.   Theorem 3-3 is also true for limits at $\pm\infty$.

By employing right-hand and left-hand limits, we can define what it means for a function to be *right-continuous* or *left-continuous*.

Definition 3-2a   *The function f is* $\begin{cases} right\text{-}continuous \\ left\text{-}continuous \end{cases}$ *at a if a is in the domain of f,*

$$\begin{cases} \lim\limits_{x \to a+} f(x) \\ \lim\limits_{x \to a-} f(x) \end{cases} exists, and$$

$$f(a) = \begin{cases} \lim\limits_{x \to a+} f(x) \\ \lim\limits_{x \to a-} f(x) \end{cases}$$

It is clear that a function is continuous at a if and only if it is both right-continuous and left-continuous at a. If a function is $\begin{cases} \textit{right-continuous} \\ \textit{left- continuous} \end{cases}$ at each number in a set S, then we say that f is $\begin{cases} \textit{right-continuous on } S \\ \textit{left-continuous on } S \end{cases}$. Because Theorem 3-1 is true for right-hand and left-hand limits, it follows that right-hand and left-hand continuity, whether at a number a or on a set S, is preserved by addition, subtraction, multiplication, and, again with the restriction that the denominator is not zero, division. Theorem 3-3 continues to hold for right-hand and left-hand limits if the condition on f in its hypothesis is changed to right-continuity or left-continuity.

If a function is not continuous at a, then f is said to be *discontinuous at a*, or f has a *discontinuity at a*. (Right-hand and left-hand discontinuities can also be defined in an obvious way.) We give a classification of the various ways that it is possible for a function to be discontinuous at a. To facilitate in this, we introduce the following notation: $\lim_{x \to a+} f(x)$ will be denoted by $f(a+)$, and $\lim_{x \to a-} f(x)$ will be denoted by $f(a-)$. If $\lim_{x \to a} f(x)$ exists but does not equal $f(a)$, or if $\lim_{x \to a} f(x)$ exists and a is not in the domain of f, we say that f has a *removable discontinuity at a*. [The reason for this terminology is that f may be made continuous at a by setting $f(a) = \lim_{x \to a} f(x)$.] If $f(a+) = \pm\infty$ and $f(a-) = \pm\infty$, or if $f(a+) = \pm\infty$ and $f(a-)$ exists, or if $f(a+)$ exists and $f(a-) = \pm\infty$, we say that f has an *infinite discontinuity at a*. If $f(a+)$ and $f(a-)$ exist but $f(a+) \neq f(a-)$, we say that f has a *simple discontinuity at a*. Finally, if $f(a+)$ or $f(a-)$ do not exist and neither is $\pm\infty$, we say that f has an *oscillatory discontinuity at a*.

EXAMPLE 3-12 The limit Theorems 3-1, 3-2, and 3-3 are not only useful computationally but have theoretical applications as well. As an illustration, we will use Theorem 3-1a to prove that $\lim_{x \to a} f(x) = L$ if and only if $\lim_{x \to a} [f(x) - L] = 0$. (Despite its obvious nature, this simple little result can be useful on occasion.) Suppose first that $\lim_{x \to a} f(x) = L$. It is clear that $\lim_{x \to a} L = L$ (here L is being regarded as a constant function). Thus, $\lim_{x \to 0} [f(x) - L] = L - L = 0$. If, conversely, $\lim_{x \to a} [f(x) - L] = 0$, then, since $f(x) = f(x) + L - L$, we have

$$\lim_{x \to a} f(x) = \lim_{x \to a} [(f(x) - L) + L] = \lim_{x \to a} [f(x) - L] + L = 0 + L = L$$

EXAMPLE 3-13 To show how Theorem 3-1 is used for limits at ∞, we prove that $\lim_{x \to \infty} 1/x^4 = 0$. This follows easily from the relation $\lim_{x \to \infty} 1/x^2 = 0$, proved in Example 3-10, the identity $1/x^4 = (1/x^2)(1/x^2)$, and Theorem 3-1b. Interestingly enough, we can use the fact that $\lim_{x \to \infty} 1/x^2 = 0$ to prove that $\lim_{x \to \infty} 1/x^4 = 0$ by applying Theorem 3-2 instead of Theorem 3-1. This is because of the inequality $0 < 1/x^4 \le 1/x^2$ which holds for all $x \ge 1$.

EXAMPLE 3-14   We know from Example 3-8 that $\lim_{x \to 2} (2x - 4) = 0$. By the result of Example 3-12, this means that $\lim_{x \to 2} 2x = 4$; but $\lim_{x \to 2} \frac{1}{2} = \frac{1}{2}$ and so, by Theorem 3-1b, we have $\lim_{x \to 2} x = \lim_{x \to 2} (\frac{1}{2})2x = (\frac{1}{2}) \lim_{x \to 2} 2x = 2$. In other words, the function $f(x) = x$ is continuous at 2. (Of course, the natural way to prove this somewhat unsurprising result is to make direct use of Definition 3-1. See Problem 3-13b.)

With this fact as a starting point, we can deduce the continuity at 2 of a large class of functions by systematic applications of the various parts of Theorem 3-1. First, we can obtain from part (b) the fact that if n is a positive integer, then the monomial $g(x) = x^n$ is continuous at 2. It next follows, from part (b) again, that if a_n is a constant, then $a_n g(x) = a_n x^n$ is also continuous at 2. But now we can apply part (a) and conclude that the sum of a finite number of monomials is continuous at 2; that is, any polynomial

$$p(x) = \sum_{k=0}^{n} a_k x^k$$

is continuous at 2. Finally, it now follows from part (c) that any rational function is continuous at 2 provided that 2 is not a root of its denominator.

EXAMPLE 3-15   Now that we know, from the previous example, that the monomial $f(x) = x^n$, where n is a positive integer, is continuous at 2, we can give an illustration of Theorem 3-3. All we need to do is to pick a function g such that $\lim_{x \to b} g(x) = 2$. Then $\lim_{x \to b} [g(x)]^n = 2^n$. More generally, if f is any of the functions of the previous example, then $\lim_{x \to b} f(g(x)) = f(2)$.

EXAMPLE 3-16   We conclude this group of examples by giving an illustration of each of the four types of discontinuities. Three of these have already been encountered in previous examples or exercises, and need only be pointed out. Thus the function f of Exercise 3-11 provides an example of a removable discontinuity at 0 because $f(0+) = f(0-)$, but 0 is not in the domain of f. The step function f of Example 3-11 is an example of a simple discontinuity at 0 because $f(0+) = 1 \ne -1 = f(0-)$. Next, the function $f(x) = 1/(2x - 4)$ of Examples 3-8 and 3-9 furnishes an instance of an infinite discontinuity at 2 because $f(2+) = \infty$ [or $f(2-) = -\infty$].

All that remains is to give an example of an oscillatory discontinuity. The function f having domain $[0, 1]$ and equation

$$f(x) = \begin{cases} 1 \text{ if } x \text{ is a rational number} \\ 0 \text{ if } x \text{ is an irrational number} \end{cases}$$

is such an example because, as we shall show, neither $f(\frac{1}{2} +)$ nor $f(\frac{1}{2} -)$ exist. To see that $f(\frac{1}{2} +)$ does not exist, consider an interval of the form $(\frac{1}{2}, \frac{1}{2} + \delta)$. This interval will always contain both rational numbers and irrational numbers. Consequently, there will always be numbers x and y in $(\frac{1}{2}, \frac{1}{2} + \delta)$ such that $f(x) = 1$ and $f(y) = 0$. Hence, no inequality of the form $|f(x) - L| < \varepsilon$ can be satisfied for any L and all $\varepsilon > 0$, and for all x in $(\frac{1}{2}, \frac{1}{2} + \delta)$. In other words, $f(\frac{1}{2} +)$ cannot exist. In the same way, $f(\frac{1}{2} -)$ cannot exist either. Further, since $|f(x)| \leq 1$, we cannot have $f(\frac{1}{2} +) = \pm\infty$, or $f(\frac{1}{2} -) = \pm\infty$. Therefore, f has an oscillatory discontinuity at $\frac{1}{2}$. It is not hard to see that the same argument would hold for any number in $(0, 1)$, and that f, therefore, actually has an oscillatory discontinuity at all x in $(0, 1)$.

EXERCISES

3-15   Use Theorem 3-1 and the result of Example 3-10 to prove that $\lim\limits_{x \to \infty} 6x^2/(3x^2 + 7) = 2$.

3-16   Use the results of the previous exercise and Example 3-14 to prove that

$$\lim_{x \to \infty} \left[a\left(\frac{6x^2}{3x^2 + 7}\right)^2 + b\left(\frac{6x^2}{3x^2 + 7}\right) + c \right] = 4a + 2b + c$$

3-17   Prove that the absolute value function is continuous on $(-\infty, \infty)$.

3-18   Why is the rational function $f(x) = 1/(x - 1)$ discontinuous at 1? What is the nature of its discontinuity?

SOLUTIONS TO EXERCISES

3-1   (a) $f(2) = 2 + 2 = 4$
(b) $f(98) = 98 + 2 = 100$
(c) $g(4) = 4^2 - 4 = 16 - 4 = 12$
(d) $g(-2) = (-2)^2 - 4 = 4 - 4 = 0$
(e) $(f + g)(1) = (1 + 2) + (1^2 - 4) = 3 - 3 = 0$
(f) $(f \cdot g)(6) = (6 + 2) \cdot (6^2 - 4) = 8 \cdot 32 = 256$
(g) $(g/f)(3) = (3^2 - 4)/(3 + 2) = 5/5 = 1$
(h) $(g - f)(-3) = [(-3)^2 - 4] - (-3 + 2) = 5 - (-1) = 6$
(i) $f(g(2)) = (2^2 - 4) + 2 = 0 + 2 = 2$
(j) $g(f(2)) = (2 + 2)^2 - 4 = 4^2 - 4 = 12$

3-2   There are of course many answers to this exercise. We will give only one example in each part.

(a)   Define f on the domain $(0, 1)$ by the equation $f(x) = 1$, and define g on the domain $(-1, 0)$ by the equation $g(x) = -1$.   Then the range of f is the number 1, and the range of g is the number -1.   Clearly the range of f has no member in common with the domain of g, and vice versa.   Thus neither $f(g)$ nor $g(f)$ can be formed.

(b)   Let f have domain $(0, 1)$ and put $f(x) = -x$.   Let g have domain $(0, 1)$ and put $g(x) = x^2$.   Then the range of f is $(-1, 0)$, while the range of g is $(0, 1)$.   Because the range of g is the same as the domain of f, $f(g)$ can be formed.   But since the range of f has no member in common with the domain of g, $g(f)$ cannot be formed.

(c)   If f has equation $f(x) = -x$ on the domain $[0, 1]$, and g has equation $g(x) = x^2$ on the domain $[-1, 0]$, then the range of f is $[-1, 0]$, and the range of g is $[0, 1]$.   Hence both $f(g)$ and $g(f)$ may be formed.   But since the domain of $f(g)$ is $[-1, 0]$, and the domain of $g(f)$ is $[0, 1]$, we have $f(g) \neq g(f)$.

(d)   Let f have equation $f(x) = x^2$ on $[0, 1]$, and g have equation $g(x) = x^3$ on $[0, 1]$.   Then their ranges are $[0, 1]$, and hence both $f(g)$ and $g(f)$ may be formed.   Note that $f(g)$ and $g(f)$ have $[0, 1]$ as a common domain.   Inasmuch as $f(g)(x) = (x^3)^2 = x^6$, and $g[f(x)] = (x^2)^3 = x^6$, it follows that $f(g) = g(f)$.

3-3   Let f and g be even functions.   Then

$$(f + g)(-x) = f(-x) + g(-x) = f(x) + g(x) = (f + g)(x)$$

and

$$(f \cdot g)(-x) = f(-x) \cdot g(-x) = f(x) g(x) = (f \cdot g)(x)$$

Now let k and l be odd functions.   Then

$$(k + l)(-x) = k(-x) + l(-x) = -k(x) - l(x)$$
$$= -[k(x) + l(x)] = -(k + l)(x)$$

Thus the sum of two odd functions is odd.   Further,

$$(k \cdot l)(-x) = k(-x) l(-x) = [-k(x)][-l(x)] = k(x) \cdot l(x) = (k \cdot l)(x)$$

and so the product of two odd functions is even.

3-4   Suppose for definiteness that f is strictly increasing and that $f(x_1) = f(x_2)$ for some x_1 and x_2 in the domain of f.   For x_1 and x_2 exactly one of the relations $x_1 < x_2$, $x_1 = x_2$, or $x_1 > x_2$ must hold.   But if $x_1 < x_2$, then $f(x_1) < f(x_2)$, and if $x_1 > x_2$, then $f(x_1) > f(x_2)$.   It must thus be the case that $x_1 = x_2$, or that f is one-to-one.   The proof when f is strictly decreasing is similar.

The converse is false. A simple counterexample is the function f defined on the domain consisting of the set of numbers 1, 2, and 3 by the equations $f(1) = 2$, $f(2) = 3$, and $f(3) = 1$. Clearly f is one-to-one. But $1 < 3$ and $f(1) > f(3)$; consequently, f is not strictly increasing. At the same time, $1 < 2$ but $f(1) < f(2)$, and so f is not strictly decreasing.

3-5 Let x_1 and x_2 be such that $0 \le x_1 < x_2 \le 2$. We have to show that $f(x_1) > f(x_2)$, or $(16 - 4x_1{}^2)^{1/2} > (16 - 4x_2{}^2)^{1/2}$. The last inequality is equivalent to $16 - 4x_1{}^2 > 16 - 4x_2{}^2$ or $x_2{}^2 > x_1{}^2$. Since $x_2{}^2 > x_1{}^2$ is a true inequality and the intervening inequalities are equivalent, we have established that $f(x_1) > f(x_2)$.

As for f^{-1}, we first determine the range of f. If $0 \le x \le 2$, then $0 \le x^2 \le 4$, and $0 \ge -4x^2 \ge -16$. Hence $16 \ge 16 - 4x^2 \ge 0$, or $4 \ge (16 - 4x^2)^{1/2} \ge 0$. In other words, the range of f is a subset of $[0, 4]$. But if y is in $[0, 4]$, the equation $(16 - 4x^2)^{1/2} = y$ can be solved for x in $[0, 2]$ by means of the quadratic formula; we omit the details. Thus the range of y is all of $[0, 4]$. Now let y be in $[0, 4]$, and consider the equation $y = f[f^{-1}(y)]$, or $y = [16 - 4[f^{-1}(y)]^2]^{1/2}$. Solving the last equation for $f^{-1}(y)$ yields $f^{-1}(y) = (4 - \frac{1}{4}y^2)^{1/2}$ (the positive square root being taken because the range of f^{-1} consists of the nonnegative numbers in $[0, 2]$).

3-6 We can write $x^4 - y^4 = 0$ either as $(x^2 + y^2)(x^2 - y^2) = 0$, or as $(x^2 + y^2)(x + y)(x - y) = 0$. Now the product of these three factors will be zero only if at least one of them is zero. But $x^2 + y^2$ is zero only if $x = 0$ and $y = 0$; $x + y$ is zero if $y = -x$; and $x - y$ is zero if $y = x$. Hence the family of implicit functions can be determined, as in Example 3-7, from the two functions f_1 and f_2, both having domain $(-\infty, \infty)$, defined by the equations $f_1(x) = x$ and $f_2(x) = -x$.

3-7 If f is a function whose domain is the set of natural numbers, denote $f(n)$ by a_n for any natural number n. If $\lim_{x \to \infty} f(x) = L$ and $\varepsilon > 0$, then there is a number $X > 0$ such that if $x > X$, then $|f(x) - L| < \varepsilon$. Pick any integer $N > X$. Then if $n > N$, we have $|f(n) - L| < \varepsilon$, or $|a_n - L| < \varepsilon$. In other words, we have turned Definition (3-1a) into Definition (2-1) when f is a sequence.

3-8 The definition of $\lim_{x \to -\infty} f(x) = L$ is: Given $\varepsilon > 0$, there exists $X < 0$ such that if $x < X$, then $|f(x) - L| < \varepsilon$.

3-9 The definition of $\lim_{x \to a-} f(x) = L$ is: Given $\varepsilon > 0$, there exists $\delta > 0$ such that if $-\delta < x - a < 0$, then $|f(x) - L| < \varepsilon$.

3-10 Assume first that $\lim_{x \to a} f(x) = L$. Given $\varepsilon > 0$, we can find $\delta > 0$ such that if $0 < |x - a| < \delta$, then $|f(x) - L| < \varepsilon$. But the inequality $0 < |x - a| < \delta$

is equivalent to the two inequalities $0 < x - a < \delta$ and $-\delta < x - a < 0$. Consequently, if $0 < x - a < \delta$, then $|f(x) - L| < \varepsilon$, and if $-\delta < x - a < 0$, then $|f(x) - L| < \varepsilon$. In other words, $\lim\limits_{x \to a+} f(x) = L = \lim\limits_{x \to a-} f(x)$.

Suppose now that $\lim\limits_{x \to a+} f(x) = L = \lim\limits_{x \to a-} f(x)$. Take $\varepsilon > 0$. We can find numbers $\delta_1 > 0$ and $\delta_2 > 0$ such that if $0 < x - a < \delta_1$, then $|f(x) - L| < \varepsilon$ and if $-\delta_2 < x - a < 0$, then $|f(x) - L| < \varepsilon$. Let δ be the smaller of δ_1 and δ_2. Then if $0 < |x - a| < \delta$, the inequalities $0 < x - a < \delta_1$, and $-\delta_2 < x - a < 0$ are true simultaneously. Thus if $0 < |x - a| < \delta$, then $|f(x) - L| < \varepsilon$, or $\lim\limits_{x \to a} f(x) = L$.

3-11   Let us first show that $\lim\limits_{x \to 0+} f(x) = 1$. Take $\varepsilon > 0$. We have to find $\delta > 0$ such that if $0 < x < \delta$, then $|f(x) - 1| < \varepsilon$, or $|(1 - x) - 1| < \varepsilon$ or $|-x| < \varepsilon$, or $|x| < \varepsilon$. It is clear that δ can be any positive number less than ε. Next, let us show that $\lim\limits_{x \to 0-} f(x) = 1$. If $\varepsilon > 0$, we have to produce a $\delta > 0$ such that if $-\delta < x < 0$, then $|f(x) - 1| < \varepsilon$, or $|x^2 + 1 - 1| < \varepsilon$, or $x^2 < \varepsilon$. But the last inequality is the same as $-\varepsilon^{1/2} < x < \varepsilon^{1/2}$. We may thus let δ be any positive number small than $\varepsilon^{1/2}$.

3-12   For $\varepsilon > 0$ we have to consider the inequality $|x/(2x + 1) - \frac{1}{2}| < \varepsilon$. But $x/(2x + 1) - \frac{1}{2} = -1/2(2x + 1)$. Consequently the inequality is the same as $|-1/2(2x + 1)| < \varepsilon$. If we solve the second inequality for x, we get $x > (\frac{1}{2})(1/2\varepsilon - 1)$. Choose X to be any positive number if $1/2\varepsilon - 1 \le 0$; otherwise choose X to be $(\frac{1}{2})(1/2\varepsilon - 1)$. Then, because the above inequalities are all equivalent, it follows that if $x > X$, then $|x/(2x + 1) - \frac{1}{2}| < \varepsilon$.

3-13   According to Definition 3-1c we have to show that $x - 1 < 0$ in some interval of the form $(b, 1)$, and that $\lim\limits_{x \to 1-} (x - 1) = 0$. But $x - 1 < 0$ in all intervals of the form $(b, 1)$, and $\lim\limits_{x \to 1-} (x - 1) = 0$ is obvious. [It can be proved similarly that $\lim\limits_{x \to 1+} 1/(x - 1) = \infty$.]

3-14   We shall prove this only for limits at ∞, the proof at $-\infty$ being almost identical. Accordingly, assume that $\lim\limits_{x \to \infty} f(x) = L$. Then for $\varepsilon > 0$ we can find $X > 0$ such that if $x > X$, then $|f(x) - L| < \varepsilon$. Let us replace x by $1/y$. Then if $1/y > X$, we have $|f(1/y) - L| < \varepsilon$. But since $X > 0$, we can rewrite $1/y > X$ as $1/X > y > 0$. Put $\delta = 1/X$. It is now clear that if $0 < y < \delta$, then $|f(1/y) - L| < \varepsilon$. In other words, we have proved that $\lim\limits_{y \to 0+} f(1/y) = L$. It is easy to see that the above steps can be reversed to prove the converse.

3-15   First notice the identity $6x^2/(3x^2 + 7) = 6/(3 + 7/x^2)$. By Example 3-10, $\lim\limits_{x \to \infty} 1/x^2 = 0$. Since $\lim\limits_{x \to \infty} 7 = 7$, Theorem 3-1b means that $\lim\limits_{x \to \infty} 7/x^2 =$

$7 \cdot 0 = 0$. Next, Theorem 3-1a implies that $\lim_{x \to \infty} (3 + 7/x^2) = 3 + 0 = 3$.
Because $3 \neq 0$, we can apply part (c) of Theorem 3-1 and conclude that $\lim_{x \to \infty} 6/(3 + 7/x^2) = 6/3 = 2$.

3-16 Since $f(x) = ax^2 + bx + c$ is a polynomial, we know from Example 3-14 that it is continuous at 2. By the previous exercise $\lim_{x \to \infty} 6x^2/(3x^2 + 7) = 2$.
Hence Theorem 3-3 can be applied to give us

$$\lim_{x \to \infty} \left[a \left(\frac{6x^2}{3x^2 + 7} \right)^2 + b \left(\frac{6x^2}{3x^2 + 7} \right) + c \right] = a(2)^2 + b \cdot 2 + c$$

3-17 Let a be any real number. We have to show that $\lim_{x \to a} |x| = |a|$. To this end, take $\varepsilon > 0$ and consider the inequality $||x| - |a||$. By Problem 1-11f we know that $||x| - |a|| \leq |x - a|$. Let δ satisfy $0 < \delta < \varepsilon$. Then if $|x - a| < \delta$, it follows that $||x| - |a|| \leq \delta < \varepsilon$. This completes the proof.

3-18 This function is discontinuous at 1 because 1 is not in its domain. As to the nature of its discontinuity, we know from Exercise 3-13 that $\lim_{x \to 1-} 1/(x - 1) = -\infty$ and $\lim_{x \to 1+} 1/(x - 1) = \infty$. Thus the function has an infinite discontinuity at 1.

PROBLEMS

3-1 Find the equation of the function f such that:
    (a) $f(A)$ is the radius of a circle of area A.
    (b) $f(s)$ is the area of an equilateral triangle of side s.
    *(c) $f(x)$ is the first number to the right of the decimal point in the decimal expansion of x where $0 \leq x \leq 1$.

3-2 Let $f(x) = x^2/(x^2 + 1)$ for all x, and $g(x) = x^{1/2}$ for $x \geq 0$. Find the equation of each of the following functions and give their domains.
    (a) $f + g$ (b) $f - g$ (c) $f \cdot g$
    (d) g/f (e) $f(g)$ (f) $g(f)$

3-3 Determine which of the following functions are even and which are odd. The domain of each is $(-\infty, \infty)$.
    (a) $f(x) = 1/(1 + x^2)$ (b) $f(x) = x/(1 + x^2)$
    (c) $f(x) = \sin x$ (d) $f(x) = \cos x$

3-4 Assume that all the compositions in this problem can be formed.
    (a) Prove that the composition of an odd function by an odd function is odd.
    (b) Prove that the composition of an even function by an odd function is even.
    (c) Prove that the composition of an arbitrary function by an even function is always even.

3-5 Let f have domain $(-\infty, \infty)$ or $(-a, a)$ where $a > 0$.
  (a) Show that $\frac{1}{2}[f(x) + f(-x)]$ is even.
  *(b) Show that f can be expressed as the sum of an even function and an odd function.

3-6 Determine which of the following functions are strictly increasing and which are strictly decreasing. Find the inverse of each function and give its domain.
  (a) $f(x) = x^{1/2} + 3$ for $0 \leq x \leq 1$
  (b) $f(x) = 1/(x^2 - 1)$ for $2 \leq x \leq 5$
  (c) $f(x) = 2x^2 + x - 3$ for $1 \leq x \leq 9$

3-7 If $f(x) = (ax + b)/(cx + d)$ for $x > -d/c$ with $a \neq 0$ and $c \neq 0$, prove that f is strictly increasing if $ad - bc > 0$, strictly decreasing if $ad - bc < 0$. What can you say about f if $ad - bc = 0$? Find f^{-1} if $ad - bc \neq 0$. Is it ever the case that $f = f^{-1}$?

3-8 If f is strictly increasing on $[a, b]$, prove that f^{-1} is strictly increasing on $[f(a), f(b)]$.

3-9 Let f and g be increasing and have the same domain. Show that $f + g$ is increasing. Give an example to show that $f \cdot g$ need not be increasing. Find a condition on f and g that is sufficient to ensure that $f \cdot g$ is increasing.

3-10 Let g be increasing on $[a, b]$, and let f be increasing on $[g(a), g(b)]$. Prove that $f(g)$ is increasing on $[a, b]$.

3-11 If f is defined on $(0, \infty)$, has an inverse f^{-1}, and satisfies the relation $f(x) + f(y) = f(xy)$ for all $x > 0$ and $y > 0$, find a corresponding relation satisfied by f^{-1}.

3-12 Find the family of implicit functions determined by each of the given equations.
  (a) $x^2 + xy = 1$ (b) $x^2 + y^2 = 1$
  (c) $|x| + |y| = 1$ (d) $x^{2/3} + y^{2/3} = 1$

3-13 Prove each of the following limits by using the appropriate definition.
  (a) $\lim_{x \to 0+} x^p = 0$ for $p > 0$ (b) $\lim_{x \to c} = (ax + b) = ac + b$
  *(c) $\lim_{x \to a} x^2 = a^2$ *(d) $\lim_{x \to a} 1/x = 1/a$ for $a \neq 0$
  *(e) $\lim_{x \to a} x^{1/2} = a^{1/2}$ for $a \geq 0$

3-14 Find each of the following limits by using the appropriate limit theorem(s). (Here n, k, m, and j are positive integers.)
  (a) $\lim_{x \to \infty} 1/x^p$ for $p > 0$ (b) $\lim_{x \to a} x^n$
  (c) $\lim_{x \to a} 1/x^n$, $a \neq 0$ (d) $\lim_{x \to a} \sum_{k=0}^{n} c_k x^k$
  (e) $\lim_{x \to a} \left[\sum_{k=0}^{n} c_k x^k \Big/ \sum_{k=0}^{m} b_k x^k \right]$, where $\sum_{k=0}^{m} b_k a^k \neq 0$

(f) $\lim\limits_{x\to\infty}\left[\sum\limits_{k=0}^{n} a_k x^k \bigg/ \sum\limits_{k=0}^{n} b_k x^k\right]$ (g) $\lim\limits_{x\to\infty}\left[\sum\limits_{k=0}^{n} a_k x^k \bigg/ \sum\limits_{k=0}^{n} b_k x^k\right]^m$

(h) $\lim\limits_{x\to a}\sum\limits_{k=0}^{n} a_k\left[\sum\limits_{j=0}^{m} b_j x^j\right]^k$ (i) $\lim\limits_{x\to a}\left[\left|\sum\limits_{k=0}^{n} c_k x^k\right|\right]^{1/2}$

3-15 Find each of the following one-sided limits
   (a) $\lim\limits_{x\to 0+} x\sin(1/x^{1/2})$
   (b) $\lim\limits_{x\to\frac{1}{2}-} f(x)$ where f is the function of Problem 3-1c.
   (c) $\lim\limits_{x\to a+} (x^2 - a^2)^{1/2}/(x-a)^{1/2}$ for $a>0$

3-16 Determine where each of the following functions are continuous and where they are discontinuous. Classify each discontinuity and remove any removable discontinuities.
   (a) $f(x) = x^{1/4}$ for $x>0$

   (b) $f(x) = \begin{cases} x & \text{for } x\le 0 \\ 1/x & \text{for } x>0 \end{cases}$

   (c) $f(x) = \begin{cases} |x|/x & \text{for } x\ne 0 \\ 1 & \text{for } x=0 \end{cases}$

   (d) $f(x) = \begin{cases} \frac{1}{2}x^2 - 2 & \text{for } x<2 \\ 1 & \text{for } x=2 \\ 2 - 8/x^2 & \text{for } x>2 \end{cases}$

*3-17 Let f be defined by $f(x) = \sin(1/x)$ for $x\ne 0$ and $f(0)=0$. Show that f has an oscillatory discontinuity at zero.

*3-18 If f is monotone on $[a, b]$, show that $f(c+)$ exists for $a\le c<b$ and that $f(c-)$ exists for $a<c\le b$. What does this imply about the possible discontinuities of f?

*3-19 Let f be continuous on $(-\infty, \infty)$. If $f(x+y) = f(x) + f(y)$ for all x and y, prove that there exists a constant c such that $f(x) = cx$ for all x.

Chapter 4

THE DERIVATIVE AND ITS APPLICATIONS

The concern of elementary algebra or trigonometry is primarily with the study of functions that are obtained from other functions by means of the four algebraic operations or the operation of composition. Calculus, while also involving itself with such matters, is primarily concerned with the study of functions that are obtained from other functions by means of two limiting operations. One of these limiting operations results in the integral, which we will take up in Chapter 6; the other results in the *derivative*.

Definition 4-1 *The function f is differentiable at a if*

$$\lim_{h \to 0} \frac{f(a + h) - f(a)}{h} \tag{4-1}$$

exists; otherwise f is nondifferentiable at a. If the limit of Eq. (4-1) exists, it is called the derivative of f at a.

The derivative of f at a will be denoted by any one of the symbols $f'(a)$, $df(a)/dx$, or $Df(a)$. Thus, $f'(a) = \lim_{h \to 0} [f(a + h) - f(a)]/h$.

In Definition 4-1 it is tacitly assumed that a is in the domain of f, and that there is also an open interval including a that is in the domain of f (see the discussion following Definition 3-1). In other words, if a is not in the domain of f, then f is nondifferentiable at a, or if a is in the domain of f but there are numbers not in the domain that are arbitrarily close to a, then f is likewise nondifferentiable at a.

EXERCISE 4-1 Prove that $f'(a)$ could have been defined by

$$f'(a) = \lim_{x \to a} \frac{f(x) - f(a)}{x - a} \tag{4-2}$$

As with the property of continuity, the property of differentiability can be extended from a single number to a set of numbers S. If f is differentiable at each member of S, then f is said to be *differentiable on S*. With the aid of this notion we can obtain a new function from a given function. This new function is called the *derivative* of f, and it is denoted by any one of the symbols f', df/dx, or Df. The domain of f' is the largest set on which f is differentiable, and the equation of f' is $f'(x) = \lim_{h \to 0} [f(x + h) - f(x)]/h$.

It may happen that a derivative f' is itself a differentiable function. If so, we can obtain its derivative. This new function is called the *second derivative* of f, and it is denoted by any of the symbols f'', d^2f/dx^2, or D^2f. More generally, if the $(n - 1)$st derivative of f is differentiable, then its derivative is called the nth *derivative*, and it is denoted by any of the symbols $f^{(n)}$, d^nf/dx^n, or D^nf. The second, third, fourth, ... derivatives of f are referred to as the *higher derivatives* of f.

EXAMPLE 4-1 Let $f(x) = x^2$ for all x; we will show that $f'(x) = 2x$ for all x. To this end,

$$\frac{f(x + h) - f(x)}{h} = \frac{(x + h)^2 - x^2}{h} = 2x + h$$

Consequently, $f'(x) = \lim_{h \to 0} (2x + h) = 2x$.

It is not hard to see that $f''(x) = 2$, and that $f^{(n)}(x) = 0$ for $n \geq 3$.

EXERCISE 4-2 Calculate the derivative of the polymoninal $f(x) = 2x^3 - 5x^2 - 7x + 1$.

Two areas of investigation arise in connection with the derivative. One is the practical problem of how to compute derivatives without having to make use of the computationally awkward Definition 4-1. The other is the more theoretical question of what differentiability of f on S implies about the behavior of f on S. We shall take up first the fairly complete solution that can be given to the problem of calculating derivatives.

The idea is to reduce the computation of "complicated" derivatives to the computation of "simpler" derivatives by examining the structure of the function involved with respect to the functional operations. In this way whole families of functions can be differentiated in a more or less mechanical way once the derivatives of a few functions have been determined. The precise meaning of these somewhat vague statements is contained in the next two theorems.

Theorem 4-1   Let f and g be differentiable on S. Then, on S,
  (a) $D[f \pm g] = Df \pm Dg$
  (b) $D[f \cdot g] = f Dg + g Df$
  (c) $D[f/g] = (g Df - f Dg)/g^2$
Note that if f is a constant function, say $f(x) = k$, then part (b) becomes, since the derivative of a constant function is clearly zero, $D[kg] = k Dg$.

Theorem 4-2   (*The Chain Rule*). Let f be differentiable on T, and g be differentiable on S. Suppose further that the composition $f(g)$ can be formed with domain S. Then, on S,

$$D[f(g)] = Df(g) \cdot Dg \qquad (4\text{-}3)$$

Equation (4-3) requires some amplification. It states that the derivative of the composite of f by g is equal to the composite of the derivative of f by g multiplied by the derivative of g. In the prime notation for derivatives Eq. (4-3) can be written as

$$[f(g)]' = f'(g)g' \qquad (4\text{-}4)$$

EXAMPLE 4-2   Here is an interesting and useful application of Theorem 4-2. If f is differentiable, and if f has an inverse function, then it can be shown that f^{-1} is also differentiable provided that $f' \neq 0$.
  Once this is known, Df^{-1} can be calculated by applying the chain rule to the equation $x = f(f^{-1}(x))$. Thus, $1 = Df(f^{-1}(x)) = Df(f^{-1}(x))Df^{-1}(x)$, or

$$Df^{-1} = \frac{1}{Df(f^{-1})} \qquad (4\text{-}5)$$

To make effective use of these theorems, it is necessary to know the derivatives of certain functions. The following list seems to be adequate.

$$Da^x = a^x \log_e a \qquad (4\text{-}6)$$

where e is the limit of the sequence of Problem 2-8d,

$$D \sin x = \cos x \qquad (4\text{-}7)$$

and

$$D \cos x = -\sin x \qquad (4\text{-}8)$$

With the aid of these three equations, together with Eq. (4-5) and Theorems 4-1 and 4-2, all the differentiation formulas of elementary calculus can be obtained. We illustrate how in the following group of examples and exercises.

EXAMPLE 4-3 Let us begin by calculating $D \log_b x$. As we know $\log_b x$ is the inverse of the exponential b^x. We can thus apply Eqs. (4-5) and (4-6) to obtain $D \log_b x = 1/b^{\log_b x} \log_e b = 1/x \log_e b$. We display this result for future reference.

$$D \log_b x = \frac{1}{x \log_e b} \tag{4-9}$$

Remark It is mainly because of the simplicity of Eq. (4-9) when $b = e$ that logarithms to the base e are called *natural logarithms*. In this connection, we make the following notational convention. *Whenever the expressions* $\log x$ *or* $\log f(x)$ *appear, it will be understood that they are natural logarithms.*

EXAMPLE 4-4 One of the most important differentiation formulas is

$$Dx^p = px^{p-1} \tag{4-10}$$

where p is any real number. To derive this, take the derivative of each side of the equation $x^p = e^{\log x^p} = e^{p \log x}$. The derivative of the right side can be calculated by the Chain Rule and Eqs. (4-6) and (4-9). Thus,

$$De^{p \log x} = e^{p \log x} D[p \log x] = x^p \cdot p/x = px^{p-1}$$

EXAMPLE 4-5 Equation 4-10 has many consequences. For one, when used in conjunction with Theorem 4-1, it implies that

$$D\left[\sum_{k=0}^{n} a_k x^k \right] = \sum_{k=1}^{n} k a_k x^{k-1}$$

That is, we can easily calculate the derivative of any polynomial. This result can now be used in conjunction with Theorem 4-1c to find the derivative of any rational function. More generally, by using Eq. (4-10) together with Theorems 4-1 and 4-2, we can find the derivative of any *algebraic function*, where an algebraic function is any function whose equation can be obtained by a finite number of additions, subtractions, multiplications, divisions, and exponentiations applied to the symbol x. That is, the equation of an algebraic function is obtained from a finite number of algebraic operations on x. The function f whose equation is

$$f(x) = \left[\frac{x^{1/2} + (x + x^{1/2})^{1/2}}{2x^{2/3}(1 + x)^{1/3}} \right]^{2/5}$$

is an example of an algebraic function.

EXAMPLE 4-6 Since the other trigonometric functions are algebraic combinations of sines or cosines, it follows that Eqs. (4-7) and (4-8) are sufficient to calculate the derivative of any trigonometric function. For instance, $\tan x = \sin x/\cos x$. Consequently, we can apply Theorem 4-1c and obtain

$$D \tan x = [\cos x \cdot \cos x - \sin x \, (-\sin x)]/\cos^2 x = 1/\cos^2 x = \sec^2 x$$

We thus have

$$D \tan x = \sec^2 x \qquad \qquad (4\text{-}11)$$

The following formulas can be obtained similarly.

$$D \sec x = \sec x \tan x \qquad \qquad (4\text{-}12)$$

$$D \operatorname{cosec} x = - \operatorname{cosec} x \cot x \qquad \qquad (4\text{-}13)$$

$$D \cot x = - \operatorname{cosec}^2 x \qquad \qquad (4\text{-}14)$$

EXAMPLE 4-7   Once the derivatives of the trigonometric functions are known, Eq. (4-5) can be employed to find the derivatives of the inverse trigonometric functions. (Since the trigonometric functions are not one-to-one, the inverse trigonometric functions are actually the inverses of appropriate restrictions of the trigonometric functions. Although there is no complete agreement, it is customary to take $\sin^{-1}$ to be the inverse of the restriction of sine to $[-\tfrac{1}{2}\pi, \tfrac{1}{2}\pi]$; $\tan^{-1}$ to be the inverse of the restriction of tangent to $(-\tfrac{1}{2}\pi, \tfrac{1}{2}\pi)$; and $\sec^{-1}$ to be the inverse of the restriction of secant to the two nonoverlapping intervals $[0, \tfrac{1}{2}\pi)$ and $[\pi, \tfrac{3}{2}\pi)$. (Inverses of cosine, cotangent, and cosecant can be defined similarly, but they are not usually discussed). For $\sin^{-1}$, we have $D \sin^{-1} x = 1/\cos(\sin^{-1} x)$. This result can be expressed in a more convenient form. Indeed, we have $\sin^2 y + \cos^2 y = 1$; here let $y = \sin^{-1} x$. Then we have $x^2 + \cos^2(\sin^{-1} x) = 1$, or $\cos(\sin^{-1} x) = \pm (1 - x^2)^{1/2}$. All that remains is to remove the ambiguity of the $\pm$ sign. This we can do because sine is a strictly increasing function on $[-\tfrac{1}{2}\pi, \tfrac{1}{2}\pi]$ and hence, as we know from Problem 3-8, $\sin^{-1}$ must also be strictly increasing. As we shall see presently, a strictly increasing differentiable function cannot have a negative derivative. This fact means that we must choose the $+$ sign. To sum up, we have proved that

$$D \sin^{-1} x = \frac{1}{(1 - x^2)^{1/2}} \qquad \qquad (4\text{-}15)$$

In the same way it can be proved that

$$D \tan^{-1} x = \frac{1}{1 + x^2} \qquad \qquad (4\text{-}16)$$

and

$$D \sec^{-1} x = \frac{1}{x(x^2 - 1)^{1/2}} \qquad \qquad (4\text{-}17)$$

EXERCISES

4-3 Find the higher derivatives of the function of Exercise 4-2.

4-4 Calculate the derivative of the function appearing in the last sentence of Example 4-5.

4-5 Prove Eq. (4-12).

4-6 Prove Eq. (4-16).

One consequence of the rules for differentiation, and especially the Chain Rule, is the notion of *implicit differentiation*. As we know from the previous chapter, an equation of the form $F((x, y)) = c$ generally determines a family of implicit functions. Some members of this family may be differentiable. If any of them are, and the equation $F((x, y)) = c$ has a continuous derivative with respect to x when y is held fixed, and vice versa, then we can find their derivatives by regarding y as a differentiable function of x, differentiating $F((x, y)) = c$ with respect to x, and solving for Dy. Notice that the chain rule necessarily plays a key role in this procedure. Notice also that the family of implicit functions does not have to be specified; the method of implicit differentiation indirectly gives us information about some of its members.

EXAMPLE 4-8 Let us apply implicit differentiation to the equation $ay^2 + bxy + cx^2 = 0$ of Example 3-7. We have

$$0 = D0 = D[ay^2 + bxy + cx^2] = aDy^2 + bD[xy] + cDx^2 =$$
$$2ayDy + b[xDy + y] + 2cx$$

From this equation we find that

$$Dy = -(by + 2cx)/(2ay + bx)$$

EXERCISE 4-7 Apply the method of implicit differentiation to the equation of Exercise 3-6.

We now turn to some of the theoretical aspects of differentiation. They are contained in the next five theorems.

Theorem 4-3 *If f is differentiable on a set S, then f is continuous on S.*

Since it is frequently a matter of routine to decide if a given function is differentiable, this theorem provides a fairly effective test for continuity. Thus, it is now an immediate consequence of Example 4-5 that every polynomial is continuous on $(-\infty, \infty)$.

Theorem 4-4 *If f is differentiable on S, and if $f'(x) \begin{cases} > 0 \\ < 0 \end{cases}$ for all x in S, then f is strictly $\begin{cases} increasing \\ decreasing \end{cases}$ on S.*

We thus have another fairly effective test; in this case for monotonicity.

Theorem 4-5 (*The Mean Value Theorem*) *If f is continuous on* $[a, b]$, *and differentiable on* (a, b), *then there exists a number z such that* $a < z < b$ *and*

$$f(b) - f(a) = f'(z)(b - a) \qquad (4\text{-}18)$$

Notice that it is possible for there to be more than one such z, and that no mention is made of the location of z in (a, b).

Despite this vagueness, the Mean Value Theorem is one of the most, if not the most, important results that arises in connection with differentiability. We shall see that its applications are both many and varied.

There is a corollary to the Mean Value Theorem that is important in its own right. It is known as Rolle's Theorem. (Rolle's Theorem is actually equivalent to the mean value theorem in the sense that it can be deduced from the Mean Value Theorem, and conversely.) To state it we need a preliminary definition. The number c is a *critical number* of f if c is in the domain of f and either $f'(c) = 0$, or $f'(c)$ fails to exist.

Theorem 4-6 (*Rolle's Theorem*) *If f is continuous on* $[a, b]$, *and if* $f(a) = f(b)$, *then there exists a critical number c of f such that* $a < c < b$.

The next theorem generalizes the Mean Value Theorem; it is known as Cauchy's mean value theorem.

Theorem 4-7 *If f and g are continuous on* $[a, b]$ *and differentiable on* (a, b), *if* $g(a) \neq g(b)$, *and if not both* $f'(x) = 0$ *and* $g'(x) = 0$ *for any x in* (a, b), *then there exists a number z such that* $a < z < b$ *and*

$$\frac{f(b) - f(a)}{g(b) - g(a)} = \frac{f'(z)}{g'(z)} \qquad (4\text{-}19)$$

Notice that Eq. (4-19) reduces to Eq. (4-18) when $g(x) = x$.

EXAMPLE 4-9   We know that $D \sin x = \cos x$, and we also know that $\cos x > 0$ for all x in $(-\tfrac{1}{2}\pi, \tfrac{1}{2}\pi)$. It thus follows from Theorem 4-4 that sine is strictly increasing on $(-\tfrac{1}{2}\pi, \tfrac{1}{2}\pi)$.

EXAMPLE 4-10   As an illustration of the great utility of the Mean Value Theorem, we will use it to prove Theorem 4-4 when S is an open interval (a, b). Accordingly, take $f'(x) \begin{cases} > 0 \\ < 0 \end{cases}$ on (a, b), suppose that both x_1 and x_2 belong to (a, b), and that $x_1 < x_2$. Since f is differentiable on (a, b), it must be continuous on (a, b). Inasmuch as $[x_1, x_2]$ is a subset of (a, b), it follows that f is differentiable on (x_1, x_2) and continuous on $[x_1, x_2]$.

By the Mean Value Theorem, there is a z between x_1 and x_2 such that $f(x_2) - f(x_1) = f'(z)(x_2 - x_1)$. In view of the fact that $x_2 - x_1 > 0$, this equation implies that $f(x_2) - f(x_1) \begin{cases} > 0 \\ < 0 \end{cases}$ on (a, b) because, in particular, $f'(z) \begin{cases} > 0 \\ < 0 \end{cases}$.

EXERCISE 4-8 Determine the set S on which the polynomial $f(x) = 2x^3 + 3x^2 - 12x + 5$ is strictly increasing, and the set T on which it is strictly decreasing.

The applications of the derivative can be grouped into four rough classifications: (i) geometrical applications, (ii) the solution of maximum and minimum problems, (iii) the derivative as a rate of change, and (iv) computational applications. We will discuss them in that order.

The first geometrical application of the derivative is the definition of the *tangent line* to the graph of a function at a point $(a, f(a))$. If $f'(a)$ exists, this line is defined to be the graph of the equation,

$$y = f'(a)(x - a) + f(a) \qquad (4\text{-}20)$$

The rationale behind this definition lies in Eq. (4-2). There the quantity $(f(x) - f(a))/(x - a)$ is the slope of a secant line of the graph of f. If these slopes have a limit, it is natural to interpret it as the slope of the tangent line. It is in this capacity that $f'(a)$ appears in Eq. (4-20).

Closely connected with tangent lines are normal lines. A line is the *normal line* to the graph of f at $(a, f(a))$ if it contains the point $(a, f(a))$ and is perpendicular to the tangent line of the graph of f at $(a, f(a))$. It follows that this normal line is the graph of the equation

$$y = -\frac{1}{f'(a)}(x - a) + f(a) \qquad (4\text{-}21)$$

provided that $f'(a) \neq 0$.

To give the second geometrical application of differentiation, we need some preliminary definitions.

Definition 4-2 *The graph of f is concave* $\begin{cases} \textit{upward} \\ \textit{downward} \end{cases}$ *at the point $(a, f(a))$ if there is an open interval I containing a such that the graph of f on I is* $\begin{cases} \textit{above} \\ \textit{below} \end{cases}$ *the tangent line to the graph of f at $(a, f(a))$.*

As with continuity and differentiability, we say that the graph of f is concave $\begin{cases} \text{upward} \\ \text{downward} \end{cases}$ on a set S if it is concave $\begin{cases} \text{upward} \\ \text{downward} \end{cases}$ at each point

$(x, f(x))$ for all x in S. We can now define a *point of inflection*. The point $(a, f(a))$ is a point of inflection of the graph of f if there is an open interval (b, c) containing a such that the graph of f is concave upward on (b, a) and concave downward on (a, c), or vice versa. In other words, at a point of inflection the graph changes concavity.

The connection between concavity and differentiation is given by the next theorem.

Theorem 4-8   *If f' is differentiable on some open interval containing a, then the graph of f is concave $\begin{cases} upward \\ downward \end{cases}$ at $(a, f(a))$ if $f''(a) \begin{cases} > 0 \\ < 0 \end{cases}$.*

It is an immediate consequence of this theorem that if f'' changes sign at a, then $(a, f(a))$ is a point of inflection of the graph of f.

EXAMPLE 4-11   Tangent lines and/or normal lines often have interesting geometrical properties. As an illustration, we show that if A denotes the point of intersection with the x axis of the tangent line to the graph of the function $f(x) = 1/x$, for $x > 0$, at an arbitary point $P = (x_0, y_0)$, and if O denotes the point (o, o), then the triangle OPA is isosceles. Since $f'(x_0) = -1/x^2_0$, the equation of the tangent line is $y = (-1/x_0^2)(x - x_0) + 1/x_0 = (-y_0/x_0)(x - x_0) + y_0$. We can rewrite this equation as $x_0 y = -y_0 x + 2x_0 y_0$, or $x_0 y + y_0 x = 2x_0 y_0$. The tangent line intersects the x axis when $y = 0$, or $x = 2/y_0 = 2x_0$. Thus the side PA has length

$$[(2x_0 - x_0)^2 + y_0^2]^{1/2} = (x_0^2 + y_0^2)^{1/2}$$

But OP also has length $(x_0^2 + y_0^2)^{1/2}$. Therefore, triangle OPA is indeed isosceles.

EXAMPLE 4-12   We have $D^2 \sin x = D \cos x = -\sin x$. Now, $-\sin x < 0$ for x in $(0, \pi)$, and $-\sin x > 0$ for x in $(-\pi, 0)$. Therefore, the graph of sine is concave downward on $(0, \pi)$; concave upward on $(-\pi, 0)$. It follows that the point $(0, 0)$ is a point of inflection of the graph of sine.

EXERCISES

4-9   A right triangle with its base on the x axis is formed by the x axis, a normal line to the graph of $f(x) = 2x^{1/2}$, and the line perpendicular to the x axis and through the point where the normal line intersects the graph of f. Show that the base of this triangle has the same length regardless of which point is chosen on the graph of f.

4-10   Find the set on which the graph of $f(x) = \tan^{-1} x$ is concave upward and the set on which it is concave downward. Also find the inflection points of this graph.

We turn now to the second application of differentiation. The number $f(a)$ is a $\begin{cases} maximum \\ minimum \end{cases}$ of the function f if there is an open interval I containing a such that $f(a) \begin{cases} >f(x) \\ <f(x) \end{cases}$ for all $x \neq a$ belonging to I. A variant of this is the notion of an *endpoint maximum* or *minimum*. The number $f(a)$ is an endpoint $\begin{cases} maximum \\ minimum \end{cases}$ of f if either (i) there is an interval (b, a) that is not in the domain of f and another interval (a, c) such that $f(a) \begin{cases} >f(x) \\ <f(x) \end{cases}$ for all x in (a, c) or (ii) there is an interval (a, c) that is not in the domain of f and another interval (b, a) such that $f(a) \begin{cases} >f(x) \\ <f(x) \end{cases}$ for all x in (b, a). For example, the number 0 is an endpoint minimum of the root function $f(x) = x^{1/2}$. More generally, if f is strictly monotone on the domain $[a, b]$, then one of $f(a)$ and $f(b)$ is an endpoint maximum, the other an endpoint minimum.

Finally, $f(a)$ is the *absolute* $\begin{cases} maximum \\ minimum \end{cases}$ of the function f if $f(a) \begin{cases} \geq f(x) \\ \leq f(x) \end{cases}$ for all x in the domain of f.

A basic theorem of analysis states that *any function that is continuous on a closed interval I must have both an absolute maximum and an absolute minimum on I*. The existence of an absolute maximum and an absolute minimum under the stated conditions naturally raises the question of whether these values can be determined. The following discussion will show that we can give a very satisfactory answer to this question by applying the derivative.

The main result is given by the following theorem.

Theorem 4-9 *If f is continuous on $[a, b]$, and if $f(c)$ is a maximum or a minimum of f for $a < c < b$, then c is a critical number of f.*

Notice that the theorem gives only a necessary condition for a maximum or minimum, not a sufficient condition. That is, it is possible for c to be a critical number of f and for $f(c)$ to be neither a maximum nor a minimum of f. An example of such behavior is given by the power function $f(x) = x^3$. Since $f'(x) = 3x^2$, it is clear that 0 is a critical number of f. But if $x > 0$, then $x^3 > 0$, while if $x < 0$, then $x^3 < 0$. In view of the fact that $f(0) = 0$, this means that $f(0)$ cannot be a maximum or minimum of f.

Such examples mean that we cannot simply find all critical numbers of f and assert that f has a maximum or minimum at each one. Rather, we must have a test that will determine when a critical number actually gives rise to a maximum or minimum. There are two such tests.

One is known as the *first derivative test*. Suppose that c is a critical number of f and that there is an interval (a, b) containing c in which f has no other critical numbers. Then

(a) $f(c)$ is a maximum if $f'(x) > 0$ for x in (a, c) and $f'(x) < 0$ for x in (c, b),

(b) $f(c)$ is a minimum if $f'(x) < 0$ for x in (a, c) and $f'(x) > 0$ for x in (c, b),

(c) $f(c)$ is neither a maximum nor a minimum if $f'(x)$ is either >0 or <0 for all x in (a, c) and (c, b).

[The function $f(x) = x^3$ that was considered above is an instance of case (c).] The reason why this test works will be clear if we examine (a). In this case, according to Theorem 4-4, f must be increasing on (a, c), and decreasing on (c, b). Obviously, $f(c)$ must be a maximum. Parts (b) and (c) can be justified similarly.

The other test is known as the *second derivative test*. Suppose that $f'(c) = 0, f''(c) \neq 0$, and f is differentiable on some open interval I containing c. Then

$$f(c) \text{ is a maximum if } f''(c) < 0, \text{ a minimum if } f''(c) > 0$$

This test can be established geometrically. Because of Theorem 4-8, the graph of f is either concave upward or concave downward at c. But, in view of Eq. (4-20), the tangent line to the graph of f has equation $y = f(c)$; that is, it is a horizontal line. Consequently, if the graph of f is concave $\begin{cases} \text{upward} \\ \text{downward} \end{cases}$ at c, then the graph of f lies $\begin{cases} \text{above} \\ \text{below} \end{cases}$ this horizontal tangent line on some open interval J containing c. In other words, $f(x) \begin{cases} >f(c) \\ <f(c) \end{cases}$ for all $x \neq c$ in J.

Theorem 4-9 and these two tests give us a systematic procedure for determining the absolute maximum and the absolute minimum of a function that is continuous on a closed interval $[a, b]$ provided that it has only a finite number of critical numbers on $[a, b]$. It is as follows: Find the critical numbers of f—this will isolate all possible maximums and minimums; next, determine which of these critical numbers give rise to maximums and minimums; next, form the set of maximums and minimums and add to it the numbers $f(a)$ and $f(b)$; finally, determine the largest and smallest numbers in this augmented set. The largest one will be the absolute maximum of f, and the smallest one will be the absolute minimum of f.

Although we have concentrated only on functions that are continuous on a closed interval, this method extends to functions that are continuous on infinite intervals provided that their behavior can be determined outside some closed interval. The next example will illustrate this.

EXAMPLE 4-13 Consider the function f whose domain is $[-1, \infty)$ and whose equation is $f(x) = (1 + x)^\alpha - (1 + \alpha x)$, where $\alpha > 1$. It is clear that $\lim_{x \to \infty} f(x) = \infty$. Thus, f cannot have an absolute maximum. But since f is continuous on any interval of the form $[-1, a]$, it must have an absolute minimum. We will determine this minimum. We first find the critical numbers of f. Now $f'(x) = \alpha(1 + x)^{\alpha-1} - \alpha$. Since f' exists for all $x > -1$, the only critical numbers of f are the roots of the equation $\alpha(1 + x)^{\alpha-1} - \alpha = 0$, or $(1 + x)^{\alpha-1} = 1$. Thus, 0 is the only critical number of f. Since $f''(x) = \alpha(\alpha - 1)(1 + x)^{\alpha-2}$, and $f''(0) = \alpha(\alpha - 1) > 0$, we see that $f(0) = 0$ is a minimum of f. Inasmuch as $f(-1) = \alpha - 1 > 0$, 0 must be the absolute minimum of f.

Remark We can state this result as $(1 + x)^{\alpha} - (1 + \alpha x) \geq 0$, or $(1 + x)^{\alpha} \geq 1 + \alpha x$, when $\alpha > 1$ and $x \geq -1$. If α is a positive integer, then this inequality becomes Bernoulli's inequality [see the solution of Problem 1-29].

EXAMPLE 4-14 Maximum and minimum problems can also arise in everyday situations. For instance, suppose that you want to mail a box, and that its cross section must be a square. Suppose in addition that postal regulations specify that the sum of the length and the girth of the box cannot exceed 60 inches. How should you choose the dimensions of the box so as to maximize its volume?

In the first place, it is clear that the combined length and girth should be exactly 60 inches. If x denotes the length and y the height of the box in inches, then $x + 4y = 60$ and you want to maximize $V = xy$. Now $y = (60 - x)/4$ and so, $V = 15x - x^2/4$, $dV/dx = 15 - x/2$, and $dV/dx = 0$ when $x = 30$. Since $d^2V/dx^2 = \frac{1}{2} < 0$, $x = 30$ will indeed give rise to the maximum volume. When $x = 30$, $y = 7.5$; these are the dimensions you want.

EXERCISES

4-11 Let $x \geq 0$, $y \geq 0$, and $x + y = k$, where k is a constant. Find the maximum value of $x \cdot y$.

4-12 Of all right-circular cylinders with a given volume, find the one with the smallest surface area.

The third application of differentiation is the interpretation of the derivative as a rate of change. This interpretation can be motivated by considering a special case. Suppose that we examine a moving point P (we will call such moving points *particles*) which is constrained to move back and forth on a coordinate line. A function s is called a *position function* (it is also called an equation of motion) if $s(t)$ is the coordinate of P at time t. The quotient

$$\frac{s(t_1) - s(t_0)}{t_1 - t_0} \tag{4-22}$$

is the average velocity of P on the time interval with endpoints t_0 and t_1. By taking t_1 close to t_0, we obtain a more accurate estimate of the velocity of P at t_0. But this is just the process of taking the limit, and it would result in $s'(t_0)$. For this reason the derivative of the position is interpreted as the (*instantaneous*) *velocity* of P. It is customary to denote s' by v. In the same way the acceleration of P is defined to be v', or s''; the acceleration is customarily denoted by a. If $v(t) > 0$, that is $s'(t) > 0$, on an interval I, then s is increasing on I. Consequently P moves to the right as t increases on I. Similarly if $v(t) < 0$ on I, then P moves to the left as t increases on I. It is clear that P can change direction at t_0 only if $v(t_0) = 0$. Similar considerations, together with Newton's second law (which states that the force acting on a particle is proportional to the product of its mass and its acceleration), yield the results that if $a(t) > 0$ on I, then P is being pushed to the right by some force (even though P may actually be moving to the left), and if $a(t) < 0$, then P is being pushed to the left.

More generally, if g is some "physical" quantity such as distance, area, mass, and so on; if r is some other "physical" quantity; and if there is a function f such that $g = f(r)$, then the (*instantaneous*) *rate of change* of g with respect to r is defined to be df/dr. It is thus possible to speak of the rate of change of the area of a circle with respect to its radius, or the rate of change of the mass of a body with respect to its velocity.

EXAMPLE 4-15   The requirement that P moves on a straight line is more flexible than it first appears. For instance, suppose that a particle P' moves around the circle $x^2 + y^2 = r^2$ in a counterclockwise direction with constant angular velocity ω starting at $(r, 0)$. If P is the projection of P' on the x axis, then P is constrained to move on a coordinate line, and we can examine the motion of P.

We have (see Figure 4-1) $x = OP = r \cos \theta$, where $\theta = \angle P'OP$. Thus $v = dx/dt = Dr \cos \theta = -r \sin \theta \, d\theta/dt = -r\omega \sin \theta$, and $a = d^2x/dt^2 = -r\omega \cos \theta \, d\theta/dt = -r\omega^2 \cos \theta$. It can be seen that P oscillates between $(-r, 0)$

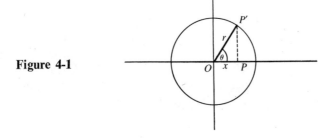

Figure 4-1

and $(r, 0)$, and that its speed (that is, $|v|$) is greatest at $(0, 0)$ and zero at $(-r, 0)$ and $(r, 0)$; while its acceleration is zero at $(0, 0)$ and greatest at $(-r, 0)$ and $(r, 0)$, being positive at $(-r, 0)$ and negative at $(r, 0)$. This regular oscillation of P is called *simple harmonic motion*.

EXAMPLE 4-16 Let r denote the radius of a sphere, S its surface area, and V its volume. Suppose that r varies with respect to time t. Then, since $V = \frac{4}{3}\pi r^3$ and $S = 4\pi r^2$, we have $dV/dt = 4\pi r^2\, dr/dt$ and $dS/dt = 8\pi r\, dr/dt$. We have thus calculated the rate of change of the volume and surface area with respect to time. We can combine our results into the single formula $dV/dt = \frac{1}{2}r\, dS/dt$.

EXERCISES

4-13 Discuss the motion of the particle whose position function s is given by $s(t) = e^{-t} \cos t$ for $t \geq 0$.

4-14 A spherical raindrop gathers moisture in such a way that its radius is increasing at a rate equal to its diameter. Find the rate of increase of its volume.

We will take up two computational applications of the derivative. Although they involve different problems, both are consequences of the Cauchy mean value theorem.

The first of these is a method for evaluating the limits of *indeterminate forms*. There are several such indeterminate forms, but they can usually be reduced to either a so-called 0/0 form, or an ∞/∞ form. A quotient f/g of two functions is said to be of type 0/0 at c if $\lim_{x \to c} f(x) = 0 = \lim_{x \to c} g(x)$; it is of type ∞/∞ at c if $\lim_{x \to c} f(x) = \infty = \lim_{x \to c} g(x)$. In either of these forms the limits could be right-handed or left-handed, or the limits could be at infinity. Such quotients are called indeterminate because they can exhibit any sort of limiting behavior at c. For example, the quotients x^2/x^4, x^2/x, and $x \sin (1/x)/x$ are all of type 0/0 at zero, yet the first has limit ∞ at zero, the second has limit 0 at zero, and the third has no limit at zero.

A difference $f - g$ of two functions is an $\infty - \infty$ form at c if $\lim_{x \to c} f(x) = \infty = \lim_{x \to c} g(x)$. Such a form can frequently be reduced to a 0/0 form or an ∞/∞ form by means of algebraic manipulation, or by replacing the variable x by a suitable function $h(y)$. A product $f \cdot g$ of two functions is a $0 \cdot \infty$ form at c if $\lim_{x \to c} f(x) = 0$, and $\lim_{x \to c} f(x) = \infty$. The identity $f \cdot g = f/(1/g)$ reduces such a form to a 0/0 form, while the identity $f \cdot g = g/(1/f)$ reduces it to an ∞/∞ form. Finally, f^g is a 1^∞, 0^0, or ∞^0 form at c if $\lim_{x \to c} f(x) = 1$ and $\lim_{x \to c} g(x) = \infty$, or $\lim_{x \to c} f(x) = 0 = \lim_{x \to c} g(x)$, or $\lim_{x \to c} f(x) = \infty$ and $\lim_{x \to c} g(x) = 0$

respectively.   The identity $\log f^g = g \log f$ reduces such forms to the $0 \cdot \infty$ type.   Just as with $0/0$ or ∞/∞ forms, in each of these the limits could be right-handed, left-handed, or at infinity.

L'Hospital's rule is a method by means of which the limits of $0/0$ or ∞/∞ forms can sometimes be evaluated.   For $0/0$ forms at least, it is a consequence of Eq. (4-19).   Indeed, if f/g is a $0/0$ form at c, if $f(c) = 0 = g(c)$, and if f and g are differentiable on some interval I including c with not both $f'(x) = 0$ and $g'(x) = 0$ for x in I, then

$$\frac{f(x)}{g(x)} = \frac{f'(z)}{g'(z)}$$

where z is between c and x.   Since z is between c and x, it follows that if $\lim_{x \to c} f'(x)/g'(x) = L$, then $\lim_{x \to c} f'(z)/g'(z) = L$.   Consequently,

$$\lim_{x \to c} \frac{f(x)}{g(x)} = \lim_{x \to c} \frac{f'(x)}{g'(x)} \qquad (4\text{-}23)$$

Equation (4-23) is L'Hospital's rule for $0/0$ forms.   It also holds when $\lim_{x \to c} f'(x)/g'(x) = \pm \infty$.   It can be shown, with some difficulty, that Eq. (4-23) is also true for ∞/∞ forms.   For both $0/0$ and ∞/∞ forms L'Hospital's rule also holds for limits at infinity.

Since a $0 \cdot \infty$ form can be reduced to a $0/0$ form or an ∞/∞ form, it is clear that L'Hospital's rule can be of use in evaluating its limit.   The same is true for an $\infty - \infty$ form that reduces to a $0/0$ form or an ∞/∞ form. Finally, a form arising from an exponential f^g can be written $f^g = e^{g \log f}$. Since the exponential function is continuous, we have $\lim_{x \to c} f(x)^{g(x)} =$ $e^{[\lim_{x \to c} g(x) \log f(x)]}$; further, $g \log f$ is a $0 \cdot \infty$ form.   We can thus try to apply L'Hospital's rule to $\lim_{x \to c} g(x) \log f(x)$.   If it can be shown that $\lim_{x \to c} g(x) \log f(x) = L$, then $\lim_{x \to c} f(x)^{g(x)} = e^L$.   To sum up, it can be seen that L'Hospital's rule may be of use with forms other than the $0/0$ or ∞/∞ types.

It sometimes happens that an application of Eq. (4-23) results in another indeterminate form.   In such cases another application is frequently successful.

EXAMPLE 4-17   The expression $\sec x - \tan x$ is an $\infty - \infty$ form at $\pi/2$ as a left-hand limit.   We will reduce it to a $0/0$ form and then apply L'Hospital's rule.   Now, $\sec x - \tan x = (1 - \sin x)/\cos x$ and the last expression $0/0$ form at $\pi/2$.   Since $\lim_{x \to \pi/2} D(1 - \sin x)/D \cos x = \lim_{x \to \pi/2} - \cos x/ - \sin x = 0$, it follows that $\lim_{x \to \pi/2-} (\sec x - \tan x) = 0$.

EXAMPLE 4-18   The expression $x^{1/x}$ is an ∞^0 form at infinity. Now $x^{1/x} = e^{(1/x)\,\log x}$, and $\log x/x$ is an ∞/∞ form at ∞. Since $\lim\limits_{x \to \infty} D \log x/Dx = \lim\limits_{x \to \infty} (1/x)/1 = 0$, we have that $\lim\limits_{x \to \infty} x^{1/x} = e^0 = 1$. Notice that Problem 2-5e is a special case of this result.

EXERCISES

4-15   Find $\lim\limits_{x \to 0} (\sin x - \tan x)/x^2$.

4-16   Find $\lim\limits_{x \to 1-} x^{1/(1-x)}$.

The second computational consequence of differentiation is the use of the Mean Value Theorem to prove inequalities. When it is necessary to estimate $f(b) - f(a)$, and it is possible to obtain an inequality of the form $f'(z) < g(z)$, or $f'(z) > h(z)$ for all z between a and b, the inequalities $f(b) - f(a) < g(z)(b - a)$, or $f(b) - f(a) > h(z)(b - a)$ follow immediately from the Mean Value Theorem if $b > a$. Such inequalities are particularly useful when the function g is increasing or the function h is decreasing. Then they become $f(b) - f(a) < g(b)(b - a)$, or $f(b) - f(a) > h(a)(b - a)$.

EXAMPLE 4-19   We illustrate this technique by proving the useful inequality $py^{p-1}(x - y) < x^p - y^p < px^{p-1}(x - y)$ if $p > 1$ and $x > y > 0$. We have $x^p - y^p = pz^{p-1}(x - y)$ where $y < z < x$. But the function $h(z) = z^{p-1}$ is increasing for $z > 0$ because $p - 1 > 0$. Therefore, $y^{p-1} < z^{p-1} < x^{p-1}$. But this means that $py^{p-1}(x - y) < pz^{p-1}(x - y) < px^{p-1}(x - y)$, or $py^{p-1}(x - y) < x^p - y^p < px^{p-1}(x - y)$.

EXERCISE 4-18   If $x > 0$, prove that $1/(x + 1) < \log(x + 1) - \log x < 1/x$.

SOLUTIONS TO EXERCISES

4-1   The substitution $h = x - a$ changes Eq. (4-1) into Eq. (4-2), and conversely, because $h \to 0$ if and only if $x \to a$. Therefore, the two formulations are equivalent.

4-2   We have

$f(x + h) - f(x)$
$$= 2(x + h)^3 + 5(x + h)^2 - 7(x + h) + 1 - (2x^3 + 5x^2 - 7x + 1)$$
$$= 6x^2h + 6xh^2 + 2h^3 + 10xh + 5h^2 - 7h$$

Consequently, $[f(x + h) - f(x)]/h = 6x^2 + 10x - 7 + h(6x + 5 + 2h)$ and so, $f'(x) = \lim\limits_{h \to 0} [6x^2 + 10x - 7 + h(6x + 5 + 2h)] = 6x^2 + 10x - 7$

4-3 For the solution of this exercise we use Theorem 4-1a and Eq. (4-10). We already know that $f'(x) = 6x^2 + 10x - 7$. It thus follows that $f''(x) = 12x + 10$, $f^{(3)}(x) = 12$, and $f^{(4)}(x) = 0$. It is now clear that if $n \geq 4$, then $f^{(n)}(x) = 0$ for all x.

4-4 For this function

$$f'(x) = \frac{2}{5}\left[\frac{x^{1/2} + (x + x^{1/2})^{1/2}}{2x^{2/3}(1 + x)^{1/3}}\right]^{-3/5}$$

$$\times \left[\frac{2x^{2/3}(1 + x)^{1/3}\left\{\frac{1}{2}x^{-1/2} + \frac{1}{2}(x + x^{1/2})^{-1/2}\left(1 + \frac{1}{2}x^{-1/2}\right)\right\}}{(2x^{2/3}(1 + x)^{1/3})^2}\right.$$

$$\left. - \frac{[x^{1/2} + (x + x^{1/2})^{1/2}]\left(\frac{4}{3}x^{-1/3}(1 + x)^{1/3} + 2x^{2/3}\frac{1}{3}(1 + x)^{-2/3}\right)}{(2x^{2/3}(1 + x)^{1/3})^2}\right]$$

In addition to the chain rule, this calculation required the use of every part of Theorem 4-1.

4-5 Since $\sec x = 1/\cos x$, we have

$$D \sec x = \frac{-(-\sin x)}{\cos^2 x} = \left(\frac{1}{\cos x}\right)\left(\frac{\sin x}{\cos x}\right) = \sec x \tan x$$

4-6 It follows from Eq. (4-5) and Eq. (4-11) that $D \tan^{-1} x = 1/\sec^{-1}(\tan^{-1} x)$. Now $\sec^2 y = 1 + \tan^2 y$, so that if we put $y = \tan^{-1} x$, then $\sec^2(\tan^{-1} x) = 1 + \tan^2(\tan^{-1} x) = 1 + x^2$. Therefore, $D \tan^{-1} x = 1/(1 + x^2)$.

4-7 The equation in question is $x^4 - y^4 = 0$. If y is a differentiable function, then $4x^3 + 4y^3(dy/dx) = 0$, or $dy/dx = -x^3/y^3$.

4-8 Now $f'(x) = 6x^2 + 6x - 12 = 6(x + 2)(x - 1)$. From this we see that $f'(x) > 0$ if either $x < -2$, or $x > 1$, and that $f'(x) < 0$ if $-2 < x < 1$. Hence S is the union of the intervals $(-\infty, -2)$ and $(1, \infty)$, while T is the interval $(-2, 1)$.

4-9 Let (x_0, y_0) be an arbitrary point on the curve. Then $y_0 = 2x_0^{1/2}$, $dy/dx = x^{-1/2}$, the slope of the normal line is $-x_0^{1/2}$, and the equation of the normal line is

$$y = -x_0^{1/2}(x - x_0) + y_0 = -x_0^{1/2}x + x_0^{1/2}(x_0 + 2)$$

We see that the normal line intersects the x axis at $(x_1, 0)$ where $x_1 = x_0 + 2$. The perpendicular through (x_0, y_0) intersects the x axis at $(x_0, 0)$. The base of the triangle thus has length $x_1 - x_0 = 2$.

4-10 We know that $f'(x) = 1/(1 + x^2)$, and so $f''(x) = -2x/(1 + x^2)^2$. Since $(1 + x^2)^2 > 0$. We see that $f''(x) > 0$ if $x < 0$, and that $f''(x) < 0$ if $x > 0$. Therefore, the graph of $\tan^{-1}$ is concave upward on $(-\infty, 0)$ and concave downward on $(0, \infty)$. It now follows that $(0, 0)$ is the only point of inflection.

4-11 If we let $f(x) = kx - x^2$ for $0 \le x \le k$, then $f(x) = xy$ and $f'(x) = k - 2x$. Thus $f'(x) = 0$ only if $x = k/2$ and since $f''(x) = 2 < 0$, $f(k/2)$ is a maximum. It is the absolute maximum since $f(0) = 0 = f(k)$. Therefore $f(x) \le k^2/2 - k^2/4 = k^2/4$, or $xy \le [(x + y)/2]^2$. This shows again that the geometric mean of two nonnegative numbers never exceeds their arithmetic mean.

4-12 If V denotes the volume, S the surface area, r the radius of base, and h the height of the cylinder, then $V = \pi r^2 h$ and $S = 2\pi r^2 + 2\pi rh$. Since V is constant, $0 = dV/dr = 2\pi rh + \pi r^2\, dh/dr$, or $dh/dr = -2h/r$. Thus $dS/dr = 4\pi r + 2\pi h + 2\pi r\, dh/dr = 4\pi r - 2\pi h$, and $dS/dr = 0$ when $h = 2r$. Inasmuch as $dS/dr < 0$ when $2r < h$ and $dS/dr > 0$ when $2r > h$, S is a minimum when $h = 2r$.

Remark Our solution of this exercise illustrates the usefulness of implicit differentiation when it is not necessary to find the absolute maximum or absolute minimum, but only to indicate the circumstances under which it is attained.

4-13 We have $v(t) = -e^{-t}\sin t - e^{-t}\cos t = -e^{-t}(\sin t + \cos t)$, and $a(t) = e^{-t}(\sin t + \cos t) - e^{-t}(\cos t - \sin t) = 2e^{-t}\sin t$. Now $v(t) = 0$ only if $\sin t + \cos t = 0$, or $\tan t = -1$ and this happens only if $t = \frac{3}{4}\pi + k\pi$, where $k = 0, 1, 2 \ldots$. Thus, the particle reverses direction infinitely many times. At the same time, since $|\cos t| \le 1$, it can be seen that $\lim_{t \to \infty} s(t) = 0$. Therefore, the particle oscillates back and forth about 0 with ever decreasing amplitude. This motion is called *exponentially damped vibration*.

4-14 If r denotes the radius and V the volume of the raindrop, we have $V = \frac{4}{3}\pi r^3$ and $dr/dt = 2r$. Thus $dV/dt = 4\pi r^2\, dr/dt = 8\pi r^3$.

4-15 By L'Hospital's rule $\lim_{x \to 0}(\sin x - \tan x)/x^2 = \lim_{x \to 0}(\cos x - \sec^2 x)/2x$. The last expression is another $0/0$ form so that we can apply L'Hospital's rule again: $\lim_{x \to 0}(\cos x - \sec^2 x)/2x = \lim_{x \to 0}(-\sin x - 2\sec^2 x \tan x)/2 = 0$. It now follows that $\lim_{x \to 0}(\sin x - \tan x)/x^2 = 0$.

4-16 We first consider $\log x^{1/(1-x)} = \log x/(1 - x)$. As $x \to 1-$ the last is a $0/0$ form. Thus, $\lim_{x \to 1-}\log x/(1 - x) = \lim_{x \to 1-}(1/x)/(-1) = -1$. Consequently $\lim_{x \to 1-} x^{1/(1-x)} = 1/e$.

4-17 From the Mean Value Theorem, we have $\log(x+1) - \log x = 1/z$, where $x < z < x + 1$. Since $1/(x+1) < 1/z < 1/x$, it follows that $1/(x+1) < \log(x+1) - \log x < 1/x$.

PROBLEMS

4-1 Use the definition of the derivative to find:
   (a) dx^n/dx, n a positive integer
   (b) $dx^{1/2}/dx$
   (c) $dx^{-1/2}/dx$

4-2 Differentiate each of the given functions. (Here $p > 0$.)
   (a) $f(x) = (ax + b)^p$ (b) $f(x) = (ax^2 + bx + c)^{-p}$
   (c) $f(x) = \left(\sum\limits_{k=0}^{n} c_k x^k \right)^p$ (d) $f(x) = (\tan x)^p$
   (e) $f(x) = (\sin^{-1} x)^p$ (f) $f(x) = (e^x + \log x)^p$
   (g) $f(x) = (\sin x^p)^p$ (h) $f(x) = \tan^{-1} x^p$

4-3 If $y = a \cos mx + b \sin mx$, show that $d^2y/dx^2 + m^2 y = 0$.

4-4 Show that the derivative of $y = (\tan x + \sec x)^m$ is $my \sec x$.

4-5 If u and v are differentiable functions, show that

$$D \tan^{-1}(u/v) = (v\, Du - uD\, v)/(u^2 + v^2)$$

4-6 Find $D\, e^{e^{e^x}}$ and $D \log (\log (\log x))$.

4-7 If f and h are differentiable, $f(x) > 0$ and $h(x) > 0$, find $D \log_{(hx)} f(x)$.

4-8 If f and g are differentiable and $f(x) > 0$, find $D\, f(x)^{g(x)}$.

4-9 Prove that the derivative of an even function is an odd function. What can you say about the derivative of an odd function?

4-10 *(a) Prove *Leibniz' rule*: If f and g have nth order derivatives, then

$$D^n[f \cdot g] = \sum\limits_{k=0}^{n} \binom{n}{k} D^k f\, D^{n-k} g$$

where $D^0 f$ and $D^0 g$ are interpreted as f and g respectively.
   (b) Find $D^n x^3 \cos x$.
   (c) Prove that if f has an nth derivative, and m is a positive integer, then

$$D^n x^m f(x) = \begin{cases} \sum\limits_{k=0}^{n} \binom{n}{k} \dfrac{m!}{(m-k)!} x^{m-k}\, D^{n-k} f(x) & \text{if } m \geq n \\[3mm] \sum\limits_{k=0}^{m} \binom{n}{k} \dfrac{m!}{(m-k)!} x^{m-k}\, D^{n-k} f(x) & \text{if } m < n \end{cases}$$

4-11 If $x^p y^q = (x + y)^{p+q}$, show that $x\, dy/dx = y$.

4-12 If $\tan^{-1}(x/y) + \log(x^2 + y^2)^{1/2} = 0$, prove that $dy/dx = (x + y)/(x - y)$ when $x \neq y$.

4-13 If $e^{x+y} = e^x + e^y$, show that $dy/dx = -e^{y-x}$.

4-14 Show that if $f(x) = |x|$, then f is not differentiable at zero.

*4-15 If $f(x) = x \sin(1/x)$ for $x \neq 0$, and $f(0) = 0$, investigate the continuity and differentiability of f at zero.

*4-16 Let $a > 0$ and define a function f by putting $f(x) = x^2 \sin (1/x) + ax$ for $x \neq 0$, and $f(0) = 0$. Find f' and show that f' is not continuous at zero.

4-17 If $f(x) = x - \sin x$, show that f is strictly increasing. For which values of α is $g(x) = \alpha x - \sin x$ a strictly increasing or strictly decreasing function?

4-18 Show that if $f(x) = x/\sin x$ for $0 < x \leq \pi/2$, and $f(0) = 1$, then f is strictly increasing on $[0, \pi/2]$.

4-19 *(a) If f and g are continuous on $[a, b]$ and differentiable on $(a, b]$; if $f(a) = 0 = g(a)$; if $g'(x) > 0$; if f'/g' is increasing on $(a, b]$, prove that f/g is increasing on $[a, b]$ provided that $(f/g)(a) = \lim_{x \to a} f(x)/g(x)$ and the limit exists.

   (b) What does (a) become when $g(x) = x$?

   (c) Use (a) to prove Problem 4-18.

   (d) Prove that $3 < \sin \pi x / x (1 - x) \leq 4$ if $0 < x < 1$.

4-20 Find the values (if such exist) where $f(x) = (x^2 - 2x - 8)^{1/2}$ has a horizontal tangent line.

4-21 Determine where the graphs of each of the following functions are concave upward, and where they are concave downward. Find all points of inflection of each of these graphs.

   (a) $f(x) = x(x - 1)^{1/2}$ for $x > 1$

   (b) $f(x) = (x^2 - 4)/(x^2 - 9)$ for all x such that $x^2 \neq 9$

   (c) $f(x) = e^{-x^2}$ for all x

4-22 A function is *convex* on an interval I if

$$f\left(\frac{x + y}{2}\right) \leq \frac{f(x) + f(y)}{2} \tag{4-24}$$

for all x and y in I. Observe that the midpoint of the line segment connecting $(x, f(x))$ and $(y, f(y))$ is $((x + y)/2, [f(x) + f(y)]/2)$. Therefore, the geometric meaning of Eq. (4-24) is that the midpoint of any chord of the graph of a convex function lies above or on the graph.

   *(a) Prove that if $f''(x) > 0$ for all x on an interval I, then f is convex on I, and Eq. (4-24) becomes $[f(x + y)]/2) < [f(x) + f(y)]/2$ if $x \neq y$.

   (b) Prove the inequalities

     (i) $\left(\dfrac{x + y}{2}\right)^p < \dfrac{x^p + y^p}{2}$ if $x > 0, y > 0, x \neq y$, and $p > 1$

     (ii) $e\left(\dfrac{x + y}{2}\right) < (e^x + e^y)/2$ if $x \neq y$

     (iii) $(x + y) \log \left(\dfrac{x + y}{2}\right) < x \log x + y \log y$ if $x > 0, y > 0$, and $x \neq y$

4-23 Determine $k > 0$ so that the normals at the points of inflection of the curve $y = k(x^2 - 3)^2$ will pass through the origin.

4-24 Find the equation of the tangent line to the curve $x^m y^n = a^{m+n}$, $a > 0$, at any point. Show that the portion of it included between the coordinate axes is divided in the ratio $|m|/|n|$ at the point of tangency.

4-25 Show that for the hypocycloid $x^{2/3} + y^{2/3} = a^{2/3}$, $a > 0$, the portion of the tangent line, at any point, that is included between the coordinate axes has constant length a.

4-26 Find the absolute maximum and absolute minimum of each of the following functions on the indicated interval.
   (a) $f(x) = x^3 - 9x^2 + 24x$ on $[0, 6]$
   (b) $f(x) = x^{2/3}$ on $[-1, 1]$
   (c) $f(x) = x \log x$ on $(0, 1)$

4-27 Determine the maximums and minimums (if any) of the function

$$f(x) = (x - a)^m (x - b)^n$$

where $a < b$ and m and n are positive integers in the cases:
   (a) m and n are both even
   (b) m and n are both odd
   (c) m is odd, and n is even
   (d) m is even, and n is odd

4-28 Establish each of the following inequalities.
   (a) $x < \tan x$ for $0 < x < \pi/2$
   (b) $e^x > \sum_{k=0}^{n} \dfrac{x^k}{k!}$ for all $x > 0$, and any integer n
   (c) $x^p(a - x)^q \leq p^p q^q a^{p+q}/(p + q)^{p+q}$ for $0 \leq x \leq a$ and where $p > 0$, $q > 0$
   (d) $a^2 \sec^2 x + b^2 \csc^2 x \geq (a + b)^2$ for $0 < x < \pi/2$ and where $a > 0$, $b > 0$.

4-29 Show that the function $f(x) = x^3 + ax^2 + bx + c$ has no maximums or minimums if and only if $a^2 \leq 3b$.

4-30 Given n fixed numbers $a_1, a_2, \ldots, a_n$, find x so that $f(x) = \sum_{k=1}^{n} (a_k - x)^2$ is a minimum.

4-31 If $p > 1$, show that $x^p - px + p - 1 \geq 0$ for all $x \geq 0$.

4-32 A piece of wire k units long is cut into two parts. One part is bent into a circle, the other into a square. Show that the sum of their areas is a minimum when the wire is cut in such a way that the diameter of the circle equals the side of the square. When is the sum a maximum?

4-33 A hall runs out of a long room m feet wide at a right angle. How wide is the hall if it is just possible to carry a piece of rigid pipe n feet long where $n > m$, from the room into the hall, keeping it horizontal?

4-34 If a and b are two sides of a triangle, determine the third side so that the area is a maximum.

4-35 Find the length of the shortest beam that can be used to brace a vertical wall if the beam must pass over another wall that is a feet high and b feet from the first wall.

4-36 A sheet of paper $ABCD$ has width $AB = a$. A triangle BEF is formed by folding the sheet in such a way that the corner B falls on the edge AD. Find the least length of EF.

4-37 A wheel of radius 1 foot rolls in the positive direction along the x axis without slipping. It makes 1 revolution per second. If Q is that point on the rim of the wheel that was at the origin at time $t = 0$, and P is the projection of Q on the x axis, find the acceleration of P.

4-38   If the equation of a rectilinear motion is $s(t) = (t+1)^{1/2}$, show that the acceleration is negative and proportional to the cube of the velocity.

4-39   The velocity of a point moving along a straight line is given by $v^2 = a + b/s$, where a and b are constants and s is the distance traveled. Show that the acceleration is $-b/s^2$.

4-40   A ladder 10 feet long is leaning against a fence 8 feet high, its upper end projecting over the fence. If the lower end of the ladder slides away from the fence at the rate of 2 ft/sec, at what rate is the upper end of the ladder approaching the ground? What is this rate when the upper end of the ladder reaches the top of the fence?

4-41   A line segment 10 units long moves so that its ends A and B remain on the x and y axes respectively in the first quadrant. If A moves away from the origin O at a constant rate k units/min, at what rate is the area formed by the axes and AB changing. If P is the midpoint of AB, at what rate is OP changing?

4-42   A vertical cylindrical tank of radius 10 inches has a hole of radius 1 inch in its base. The velocity with which the water contained runs out of the tank is given by the formula $v^2 = 2gh$, where h is the depth of the water, and g is the acceleration of gravity. How rapidly is the velocity changing?

4-43   Find:

(a) $\displaystyle\lim_{x \to 0} \frac{a^x - b^x}{x}$, $a > 0$ and $b > 0$

(b) $\displaystyle\lim_{x \to 1} \frac{1 - 4 \sin^2 \frac{\pi}{6} x}{1 - x^2}$

(c) $\displaystyle\lim_{x \to 0} \frac{e^x - e^{-x} - 2x}{x - \sin x}$

4-44   Find:

(a) $\displaystyle\lim_{x \to \infty} \frac{\log x}{x^\alpha}$, $\alpha > 0$

(b) $\displaystyle\lim_{x \to 0+} \frac{\log x}{\operatorname{cosec} x}$

(c) $\displaystyle\lim_{x \to 0+} \frac{\log (\sin x)}{\log (\tan x)}$

4-45   Find:

(a) $\displaystyle\lim_{x \to 1} \left[\frac{x}{x - 1} - \frac{1}{\log x} \right]$

(b) $\displaystyle\lim_{x \to \infty} [x(x^2 + a^2)^{1/2} - x^2]$

4-46   Find:
  (a) $\lim\limits_{x\to 0+} x^{\alpha} \log x, \; \alpha > 0$
  (b) $\lim\limits_{x\to\infty} x \sin (a/x)$

4-47   Find:
  (a) $\lim\limits_{y\to\infty} (1 + x/y)^y$
  (b) $\lim\limits_{x\to 0+} (\sin x)^x$
  (c) $\lim\limits_{x\to 0+} (1/x)^{\sin x}$

4-48   Find $\lim\limits_{x\to 0+} \dfrac{e^{-1/x}}{x}$

4-49   Show that $\lim\limits_{x\to 0} x^2 \sin (1/x)/\sin x = 0$, but that L'Hopital's rule does not apply.

4-50   Criticize the following:

$$\lim_{x\to 2} \frac{3x^2 - 4x - 4}{x^2 - 2x} = \lim_{x\to 2} \frac{6x - 4}{2x - 2} = \lim_{x\to 2} \frac{6}{2} = 3$$

4-51   If, for a function f, $\lim\limits_{x\to\infty} f(x) = \lim\limits_{x\to\infty} f'(x) = \lim\limits_{x\to\infty} f''(x) = \infty$, and $\lim\limits_{x\to\infty} \dot{x} f^{(3)}(x)/f''(x) = k$, find $\lim\limits_{x\to\infty} x f'(x)/f(x)$.

4-52   Given a circle with center at O, radius r and a tangent line AT. In the figure, AM equals arc AP and B is the intersection of the line through M and P with the line through A and O. Find the limiting position of B as P approaches A as a limiting position.

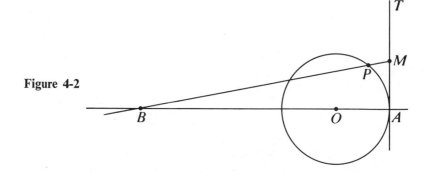

Figure 4-2

4-53   Show that if $x > 0$, then $x/(x^2 + 1) < \tan^{-1} x < x$.
4-54   Show that $e^a(x - a) < e^x - e^a < e^x(x - a)$ if $a < x$.
4-55   Prove that $|\sin x - \sin y| \le |x - y|$
4-56   Show that if $x > -1$ and $x \ne 0$, then $x/(x + 1) < \log (x + 1) < x$.
4-57   If $a_n/(n + 1) + a_{n-1}/n + \cdots + a_1/2 + a_0 = 0$, prove that the equation $a_n x^n + a_{n-1}x^{n-1} + \cdots + a_1 x + a_0 = 0$ has at least one root between 0 and 1.

Chapter 5

THE ANTIDERIVATIVE AND ITS APPLICATIONS

In the last chapter we examined the problem of finding the derivative of a given function. In this chapter we shall be concerned with the opposite problem; that is, given a function f, find all functions g such that $g' = f$. This reversal of the process of differentiation is called *antidifferentiation*.

Definition 5-1 *The function g is an antiderivative of f if*

$$g'(x) = f(x) \tag{5-1}$$

for all x in some interval I

Notice that we say *an* antiderivative rather than *the* antiderivative. The reason for this is that if a given function has one antiderivative, then it will have infinitely many antiderivatives. Indeed, if g is an antiderivative of f, and C is any constant, then $g + C$ is also an antiderivative of f. Fortunately, all antiderivatives of f are essentially of the form $g + C$. The precise meaning of this statement is contained in the following theorem.

Theorem 5-1 *Any two antiderivatives of f differ by a constant; in other words, if g and h are antiderivatives of f on an interval I, then there exists a constant C such that*

$$h(x) - g(x) = C \tag{5-2}$$

for all x in I.

This theorem characterizes all the antiderivatives of f. Let us denote this family of functions by $\int f(x)\, dx$. (From a logical point of view we should

73

denote this family by $\int f$ as the letter x and the symbol dx play no essential role, but the notation we use is hallowed by tradition, and, more importantly, it has many computational advantages.) If g is any antiderivative of f, then we can restate the theorem as

$$\int f(x)\, dx = g(x) + C \tag{5-3}$$

provided that we interpret C as the family of *all* real numbers. Since $\int f(x)\, dx$ does represent a family of functions, it is not mere pedantry to insist that the omission of C from Eq. (5-3) is incorrect.

As with the derivative, the areas of investigation of the antiderivative are the problem of how antiderivatives are calculated, and the determination of the theoretical consequences of antidifferentiation. Unlike the derivative, the theoretical ramifications of the antiderivative are relatively unimportant as long as it is considered in isolation (the true theoretical significance of the antiderivative will become apparent in the next chapter when it is con-sidered in connection with the integral). In fact, Theorem 5-1 is essentially the beginning and the end of the theoretical development of the antiderivative in this chapter. This is not to say that there are no other general properties of the antiderivative. Indeed, there are several. But they stem so naturally from the computation of antiderivatives that it seems more proper to deem them "computational" rather than "theoretical".

We are thus left with the problem of developing techniques for the calcula-tion of antiderivatives. The starting point of our development is the fact that every differentiation formula can also be viewed as an antidifferentiation formula. For instance, we can write Eq. (4-16), $D \tan^{-1} x = 1/(1 + x^2)$, as $\int [1/(1 + x^2)]\, dx = \tan^{-1} x + C$. In this way, from the corresponding differentiation formulas of Chapter 4, formulas can be obtained for the anti-derivatives of power functions, exponentials, logarithms, and so forth. These formulas follow.

$$\int x^p\, dx = \frac{1}{p+1}\, x^{p+1} + C \tag{5-4}$$

if $p \neq -1$.

$$\int a^x\, dx = \frac{1}{\log a}\, a^x + C \tag{5-5}$$

when $a = e$, Eq. (5-5) becomes

$$\int e^x\, dx = e^x + C \tag{5-5a}$$

$$\int \frac{1}{x} \, dx = \log b \, \log_b |x| + C \tag{5-6}$$

for any $b > 0$. When $b = e$, Eq. (5-6) becomes

$$\int \frac{1}{x} \, dx = \log |x| + C \tag{5-6a}$$

$$\int \sin x \, dx = -\cos x + C \tag{5-7}$$

$$\int \cos x \, dx = \sin x + C \tag{5-8}$$

$$\int \sec^2 x \, dx = \tan x + C \tag{5-9}$$

$$\int \operatorname{cosec}^2 x \, dx = -\cot x + C \tag{5-10}$$

$$\int \sec x \tan x \, dx = \sec x + C \tag{5-11}$$

$$\int \operatorname{cosec} x \cot x \, dx = -\operatorname{cosec} x + C \tag{5-12}$$

$$\int \frac{1}{(a^2 - x^2)^{1/2}} \, dx = \sin^{-1} \frac{x}{a} + C \tag{5-13}$$

$$\int \frac{1}{a^2 + x^2} \, dx = \frac{1}{a} \tan^{-1} \frac{x}{a} + C \tag{5-14}$$

$$\int \frac{1}{x(x^2 - a^2)^{1/2}} \, dx = \frac{1}{a} \sec^{-1} \frac{x}{a} + C \tag{5-15}$$

In the last three equations we assume that $a > 0$.

Equation (5-6) requires some additional explanation since Eq. (4-9) does not involve the absolute value. To establish Eq. (5-6), note that $D \log_b |x| = (1/|x| \log b) D|x|$. Now $D|x| = \begin{cases} 1 & \text{if } x > 0 \\ -1 & \text{if } x < 0 \end{cases}$, and so $(1/|x|)D|x| = \begin{cases} 1/|x| & \text{if } x > 0 \\ -1/|x| & \text{if } x < 0 \end{cases}$. But if $x > 0$, then $|x| = x$, and if $x < 0$, then $|x| = -x$. In either case we see that $D \log_b |x| = 1/x \log b$. Rather than Eq. (4-9) Eq. (5-6) is the counterpart of the last relation.

Equations (5-13), (5-14), and (5-15) are also not quite the same as the corresponding differentiation formulas of the last chapter. They are also established by showing that the derivative of the right-hand side is the function inside the antiderivative symbol.

Remark   An examination of Eqs. (5-5)–(5-15) gives some indication of why antidifferentiation is more difficult than differentiation. Consider the classification of functions into polynomials, rational functions, algebraic functions, and transcendental functions. If we think of polynomials as being the "simplest" type, and transcendental functions as being the most "complicated", with the other two types ranged in between in the order they are given above, then differentiation of a function does not result in a more complicated function, but the same is not true for antidifferentiation, as Eq. (5-6) or any of Eqs. (5-13)–(5-15) demonstrate.

EXAMPLE 5-1   Equations 5-6 and 5-6a, together with Theorem 5-1, enable us to obtain in a very simple way an identity we could have obtained directly, but not as neatly, when logarithms were introduced. It follows from Eqs. (5-6) and (5-6a) that $\log b \log_b x$ and $\log x$ are both antiderivatives of $1/x$. By Theorem 5-1 it must then be the case that $\log b \log_b x - \log x = C$ for all $x > 0$. If we let $x = b$ in this equation we find, since $\log_b b = 1$, that $C = \log b - \log b = 0$. We thus have the identity

$$\log b \log_b x = \log (x) \tag{5-16}$$

for converting logarithms to the base b into natural logarithms. Many other identities can be obtained by this method.

All but one of our techniques of antidifferentiation depend on reversing properties of the derivative. One of these stems from the formula $Dkf(x) = k Df(x)$, where k is some constant. In terms of antiderivatives, this can be written as

$$k \int f(x)\, dx = \int k f(x)\, dx \tag{5-17}$$

Another stems from the formula $D[f(x) \pm g(x)] = Df(x) \pm Dg(x)$, which has the relation

$$\int [f(x) \pm g(x)]\, dx = \int f(x)\, dx \pm \int g(x)\, dx \tag{5-18}$$

as its antidifferentiation counterpart. It is clear that Eq. (5-18) is true for the sum or difference of any finite number of functions having antiderivatives.

EXAMPLE 5-2   We can use Eqs. (5-4), (5-17), and (5-18) to find the antiderivative of any polynomial

$$f(x) = \sum_{k=0}^{n} a_k x^k$$

Indeed,

$$\int f(x)\,dx = \sum_{k=0}^{n} \int a_k x^k \, dx = \sum_{k=0}^{n} a_k \int x^k \, dx$$

$$= \sum_{k=0}^{n} \frac{a_k}{k+1} x^{k+1} + C$$

EXERCISE 5-1 Let f be a polynomial, and p be any real number. Find $\int [f(x)/x^p]\,dx$.

Our next antidifferentiation method is the chain rule stated in terms of antiderivatives. Suppose that F is an antiderivative of f, and that g is differentiable. The chain rule states that $DF(g(x)) = F'(g(x))g'(x) = f(g(x))g'(x)$. In terms of antiderivatives, we may state this result as $\int f(g(x))g'(x)\,dx = F(g(x)) + C$. On the right-hand side we may replace $g(x)$ by u, and obtain $F(u) + C = \int f(u)\,du$. In other words, if $u = g(x)$, then

$$\int f(g(x))g'(x)\,dx = \int f(u)\,du \tag{5-19}$$

Equation (5-19) is our method. We will call it the *change of variable method* because the right-hand side of Eq. (5-19) is obtained from the left-hand side by replacing $g(x)$ by a new variable u. In doing so we replace the symbol $g'(x)\,dx$ by du [it is for reasons such as this that we denote antiderivatives by $\int f(x)\,dx$ instead of $\int f$].

EXAMPLE 5-3 We use this method to find $\int[(x+1)/(x^2+2x-5)]dx$. If we put $u = x^2 + 2x - 5$, then $du = 2(x+1)\,dx$. Now we do not quite have the necessary form to apply Eq. (5-19), but we lack only a constant factor. Equation (5-17) permits us to make up this defect by multiplying by 2 inside the $\int$ symbol, and compensating by dividing by 2 outside the $\int$ symbol. Thus

$$\int \frac{x+1}{x^2+2x-5}\,dx = \frac{1}{2}\int \frac{2x+2}{x^2+2x-5}\,dx = \frac{1}{2}\int \frac{1}{u}\,du$$

$$= \frac{1}{2}\log|u| + C$$

$$= \frac{1}{2}\log|x^2+2x-5| + C$$

Remark   There is a common error involved in the use of this method. It lies in the fact that it is sometimes necessary, as in this example, to multiply by a constant inside the antiderivative, and divide by it outside.   Beginners often forget that Eq. (5-17), which is the justification of this procedure, is true only for *constant* factors.   This can only lead to erroneous results.   As an illustration, consider the following false argument.   In $\int (2/x^2)\, dx$, make the change of variable $u = x^2$; then $2x\, dx = du$.   Since $2/x^2\, dx$ does not contain $2x\, dx$, we write it as $(1/x) \int (2x/x^2)\, dx$.   Now $\int (2x/x^2)\, dx = \int (1/u)\, du = \log |u| + C = \log x^2 + C$.   However, it is not true that $\int (2/x^2)\, dx = (1/x) \log|x^2| + C/x$ since $\int (2/x^2)\, dx$ is, in fact, equal to $-2/x + C$.

EXAMPLE 5-4   One of the more important applications of the change of variable method is the calculation of the antiderivatives of various combinations of trigonometric functions.   We will examine such antiderivatives at some length in this example.

Consider   first $\int \tan x\, dx$.   Now $\int \tan x\, dx = \int (\sin x/\cos x)\, dx$;   put $u = \cos x$, then $du = -\sin x\, dx$,

$$\int \left(\frac{\sin x}{\cos x}\right) dx = -\int \left(-\frac{\sin x}{\cos x}\right) dx = -\int (1/u)\, du = -\log |u| + C =$$

$$-\log |\cos x| + C$$

Therefore,

$$\int \tan x\, dx = \log |\sec x| + C \tag{5-20}$$

The formula

$$\int \cot x\, dx = \log |\sin x| + C \tag{5-21}$$

can be obtained similarly.   As for $\int \sec x\, dx$, we have

$$\int \sec x\, dx = \int \frac{\sec x(\sec x + \tan x)}{\sec x + \tan x}\, dx = \int \frac{\sec^2 x + \sec x \tan x}{\sec x + \tan x}\, dx$$

Make the change of variable

$$u = \sec x + \tan x, \quad du = (\sec^2 x + \sec x \tan x)\, dx$$

Then

$$\int \frac{\sec^2 x + \sec x \tan x}{\sec x + \tan x}\, dx = \int \frac{1}{u}\, du = \log |u| + C$$

$$= \log |\sec x + \tan x| + C$$

In other words,

$$\int \sec x \, dx = \log |\sec x + \tan x| + C \tag{5-22}$$

in the same way it can be shown that

$$\int \operatorname{cosec} x \, dx = \log |\operatorname{cosec} x - \cot x| + C \tag{5-23}$$

Certain products of trigonometric functions can be dealt with by the change of variable method in conjunction with some trigonometric identities. Thus, if n is a nonnegative integer, and p is any real number, then $\int \sin^p x \cos^{2n+1} dx$ can be attacked by using the identity $1 - \sin^2 x = \cos^2 x$ to write $\cos^{2n} x$ as $(1 - \sin^2 x)^n$. Since $D \sin x = \cos x$, the substitution $u = \sin x$ will then change $\int \sin^p x \cos^{2n+1} x \, dx$ into $\int u^p (1 - u^2)^n \, du$. Now, using the binomial theorem,

$$\int u^p (1 - u^2)^n \, du = \int u^p \sum_{k=0}^{n} \binom{n}{k} (-1)^k u^{2k} \, du$$

$$= \sum_{k=0}^{n} (-1)^k \binom{n}{k} \int u^{2k+p} \, du$$

and the last expression may be evaluated by means of Eq. (5-4) and, if necessary, Eq. (5-6a). The antiderivative $\int \cos^p x \sin^{2n+1} x \, dx$ can be evaluated in a similar way by means of the identity $\sin^2 x = 1 - \cos^2 x$, and the substitution $u = \cos x$.

If n and m are nonnegative integers, then $\int \sin^{2n} x \cos^{2m} x \, dx$ can be reduced to a sum of antiderivatives of the form $\cos^{2k+1} y \, dy$, where k is a nonnegative integer, by means of the trigonometric identities $\sin^2 x = (1 - \cos 2x)/2$, and $\cos^2 x = (1 + \cos 2x)/2$ (possibly used several times). Inasmuch as antiderivatives of the form $\int \cos^{2k+1} y \, dy$ can be evaluated by the procedure of the previous paragraph $\left(\int \cos^{2k+1} y \, dy = \int \sin^0 y \cos^{2k+1} y \, dy \right)$, it follows that we can evaluate $\int \sin^{2n} x \cos^{2m} x \, dx$; at least in principle.

Finally, if p is arbitrary and n is a nonnegative integer, then the expression $\int \tan^p x \sec^{2n+2} x \, dx$ can be evaluated by writing $\sec^{2n} x = (1 + \tan^2 x)^n$, and substituting $u = \tan x$, $du = \sec^2 x \, dx$. Thus

$$\int \tan^p x \sec^{2n+2} x \, dx = \int \tan^p x (1 + \tan^2 x)^n \sec^2 x \, dx$$

$$= \int u^p (1 + u^2)^n \, du = \sum_{k=0}^{n} \binom{n}{k} \int u^{2k+p} \, du$$

We can find $\int \tan^{2n+1} x \sec^{p+1} x \, dx$ in a similar way by putting $\tan^{2n} x = (\sec^2 x - 1)^n$, and substituting $u = \sec x$, $du = \sec x \tan x \, dx$. This results in

$$\int \tan^{2n+1} x \sec^{p+1} x \, dx$$

$$= \int (\sec^2 x - 1)^n \sec^p x \sec x \tan x \, dx = \int (u^2 - 1)^n u^p \, du$$

$$= \sum_{k=0}^{n} (-1)^{n-k} \binom{n}{k} \int u^{2k+p} \, du$$

EXERCISES

5-2   Find $\int (\log x / x) \, dx$.

5-3   Establish Eqs. (5-21) and (5-23).

5-4   Find $\int \sin^3 x \, dx$.

We now come to one of the most important techniques for finding antiderivatives, It is known as *integration by parts*. This method is nothing more than the rule for calculating the derivative of a product expressed in terms of antiderivatives. If we take the antiderivative of each side of the equation $D[f(x)g(x)] = f(x)g'(x) + g(x)f'(x)$, we get

$$f(x)g(x) + C = \int [f(x)g'(x) + g(x)f'(x)] \, dx =$$
$$\int f(x)g'(x) \, dx + \int g(x)f'(x) \, dx$$

We may lump the family of constants C into either of the two antiderivatives on the right-hand side, transpose one of them, and obtain the integration by parts formula

$$\int f(x)g'(x) \, dx = f(x)g(x) - \int g(x)f'(x) \, dx \qquad (5\text{-}24)$$

Equation (5-24) is customarily rewritten. We replace $f(x)$ by u, $g(x)$ by v, $g'(x) \, dx$ by dv, and $f'(x) \, dx$ by du to attain the more compact formula

$$\int u \, dv = uv - \int v \, du \qquad (5\text{-}25)$$

EXAMPLE 5-5   Consider the problem of finding $\int \log x \, dx$. At first sight it looks as though integration by parts would be of no use because Eq. (5-24) seems to imply that integration by parts is concerned with finding antiderivatives of products, and we do not have a product here. But any function can be regarded as the product of itself and the constant function 1. This is what we do here. We put $u = \log x$ and $dv = dx$; then $du = (1/x)\, dx$, $v = x$, and $\int \log x \, dx = x \log x - \int (x/x)\, dx = x \log x - x + C$.

EXAMPLE 5-6 Occasionally it is necessary to integrate by parts several times to evaluate an antiderivative. Such is the case with $\int x^2 \sin x \, dx$. We first put $u = x^2$ and $\sin x \, dx = dv$. Since $du = 2\, x \, dx$ and $v = -\cos x$, we have $\int x^2 \sin x \, dx = -x^2 \cos x + 2 \int x \cos x \, dx$. We are not yet done, but it is clear we have made progress. In $\int x \cos x \, dx$ let $u = x$ and $\cos x \, dx = dv$; then $du = dx$, $v = \sin x$, and $\int x \cos x \, dx = x \sin x - \int \sin x \, dx = x \sin x + \cos x + C$. Consequently, $\int x^2 \sin x \, dx = -x^2 \cos x + 2\, x \sin x + 2 \cos x + C$.

EXERCISES

5-5 Find $\int x \sec^2 x \, dx$.

5-6 Find $\int x(1 - x^2)^{1/2} \sin^{-1} x \, dx$.

The key to the next method of antidifferentiation is already in our possession; all we have to do is point it out. Like the change of variable method, it is contained in Eq. (5-19)—we simply change its emphasis. Since the symbols x and u have no intrinsic meaning in Eq. (5-19), we may write it as

$$\int f(x) \, dx = \int f(g(u))g'(u) \, du \qquad (5\text{-}26)$$

with $x = g(u)$. If g has an inverse function, we can just as well say that $u = g^{-1}(x)$. When this condition is satisfied, Eq. (5-26) gives us another method for changing the form of an antiderivative; we will call it the *method of substituion*. With this terminology we are trying to indicate the difference between the two related methods of change of variable and substitution. In the former we start with a function and "change" it into a variable, whereas with the latter we start with a variable and "substitute" for it a function. Aside from this semantic difference, the main difference between the two methods is that with the method of substitution care must be taken that the function being substituted has an inverse.

EXAMPLE 5-7 We will find $\int [x/(x + 1)^{1/2}]dx$ by this method. If we put $x + 1 = u^2$, then $u = (x + 1)^{1/2}$, $x = u^2 - 1$, $dx = 2u \, du$, and $\int [x/(x + 1)^{1/2}] \, dx = \int [(u^2 - 1)/u]2u \, du = 2 \int (u^2 - 1) \, du = $
$$(2/3)u^3 - 2u + C.$$

Therefore, $\int [x/(x+1)^{1/2}]\,dx = (2/3)(x+1)^{3/2} - 2(x+1)^{1/2} + C$.

EXAMPLE 5-8 In addition to isolated applications of the method of substitution, as in the previous example, there are systematic applications of it as well. One of these stems from the fundamental trigonometric identity $\sin^2\theta + \cos^2\theta = 1$, or, as we shall be using it, $\cos\theta = (1 - \sin^2\theta)^{1/2}$ (at least for $-\pi/2 \le \theta \le \pi/2$). The idea is to substitute $a\sin\theta$ for x in the expression $(a^2 - x^2)^{1/2}$ where $a > 0$, to turn it into $a\cos\theta$. Inasmuch as $\sin^{-1}$ is available, the effect of this substitution is to change an antiderivative involving $(a^2 - x^2)^{1/2}$ into one involving trigonometric functions. In view of the fact that we can handle such antiderivatives in a fairly satisfactory way, there is some hope that the method of substitution will succeed in such cases.

As an illustration, let $a > 0$ and consider $\int (a^2 - x^2)^{1/2}\,dx$. If $x = a\sin\theta$, then $dx = a\cos\theta\,d\theta$, and $\theta = \sin^{-1}(x/a)$. By Eq. (5-26) we have

$$\int (a^2 - x^2)^{1/2}\,dx = \int (a^2 - a^2\sin^2\theta)^{1/2}\theta\cos\theta\,d\theta$$

$$= a^2 \int \cos^2\theta\,d\theta = a^2 \int [(1 + \cos 2\theta)/2]\,d\theta$$

$$= (a^2/2)\theta + (a^2/4)\sin 2\theta + C = (a^2/2)\theta + (a^2/2)\sin\theta\cos\theta + C$$

$$= (a^2/2)\sin^{-1}(x/a) + (a^2/2)\sin[\sin^{-1}(x/a)]\cos[\sin^{-1}(x/a)] + C$$

$$= (a^2/2)\sin^{-1}(x/a) + (a^2/2)(x/a)(a^2 - x^2)^{1/2}/a + C$$

$$= (a^2/2)\sin^{-1}(x/a) + (x/2)(a^2 - x^2)^{1/2} + C$$

Notice how the techniques outlined in Example 5-6 come into play as soon as the substitution was made. It should be clear that this is generally the case.

We can also write the identity $\sin^2\theta + \cos^2\theta = 1$ as $\tan^2\theta + 1 = \sec^2\theta$. Two versions of this are of interest. First, $\tan\theta = (\sec^2\theta - 1)^{1/2}$; second $\sec\theta = (\tan^2\theta + 1)^{1/2}$. The first of these reduces the expression $(x^2 - a^2)^{1/2}$ to $a\tan\theta$ when the substitution $x = a\sec\theta$ is made, and the second reduces $(x^2 + a^2)^{1/2}$ to $a\sec\theta$ when x is replaced by $a\tan\theta$. Consequently, antiderivatives involving either of the terms $(x^2 - a^2)^{1/2}$ or $(x^2 + a^2)^{1/2}$ can be changed into trigonometric antiderivatives.

EXERCISES

5-7 Find $\int [x^2/(x+1)^{1/2}]\,dx$.

5-8 Find $\int [x^3/(x^2 - 1)^{1/2}]\,dx$.

The final antidifferentiation method we take up is the only one that is not based on reversing a property of the derivative. Instead, it depends on a result from elementary algebra concerning rational functions. Now, a rational function is the quotient of two polynomials, and the result we need depends very strongly on properties of polynomials. Accordingly, we begin by recalling some definitions and stating, without proof, some theorems about polynomials.

We know that a polynomial f is a function whose equation is, for all x, $f(x) = a_n x^n + a_{n-1} x^{n-1} + \cdots + a_1 x + a_0$. The numbers $a_n, a_{n-1}, \ldots, a_1, a_0$, which we will always take to be real numbers, are called the *coefficients* of f, and the integer n is called the *degree* of f (that is, the degree is the largest power appearing in the equation of f). A *root* of the polynomial f is any number r such that $f(r) = 0$. Even though we are only considering polynomials with real coefficients, it is possible that some roots may be complex numbers. [Such, for example, is the case with the simple polynomial $f(x) = x^2 + 1$ which has for roots the complex numbers i and $-i$ where $i^2 = -1$.] It is thus necessary to give a brief account of complex numbers.

Let a and b be real numbers, and let i be such that $i^2 = -1$. The *complex number system* is the set of all numbers of the form $a + bi$ in which equality is defined by $a + bi = c + di$ if and only if $a = c$ and $b = d$, and addition and multiplication are defined by $(a + bi) + (c + di) = (a + c) + (b + d)i$ and $(a + bi)(c + di) = (ac - bd) + (ad + bc)i$. If $z = a + bi$, then its *conjugate* $\bar{z}$ is the complex number $a - bi[=a + (-b)i]$.

Let us now consider a polynomial f of degree n. The *Fundamental Theorem of Algebra* states that f has a root r. It can also be shown that f can be written as $f(x) = (x - r)g(x)$, where g is a polynomial of degree $n - 1$. By repeating this procedure we can write f as

$$f(x) = K \prod_{k=1}^{n} (x - r_k) \qquad (5\text{-}27)$$

where K is some constant (that is, a polynomial of degree zero). In other words, the polynomial f of degree n has n roots $r_1, r_2, \ldots, r_n$, and it can be expressed as a product of n first degree polynomials and a constant. Now in Eq. (5-27) not all of $r_1, r_2, \ldots, r_n$ have to be distinct. Let us suppose that among the numbers $r_1, r_2, \ldots, r_n$ there are m_1 equal to q_1, m_2 equal to q_2, $\ldots$, m_l equal to q_l. Clearly $l \leq n$, $1 \leq m_j \leq n$, and $m_1 + m_2 + \cdots + m_l = n$. We can now write Eq. (5-27) as

$$f(x) = K \prod_{j=1}^{l} (x - q_j)^{m_j} \qquad (5\text{-}28)$$

If $m_j > 1$ in Eq. (5-28), then the root q_j is called a *repeated root*. The exponent m_j associated with any root q_j, repeated or not, is called the *order* of q_j. We next have to realize that some (or all) of the numbers $q_1, q_2, \ldots, q_l$ may be complex. But a theorem from algebra states that if, as here, a polynomial f has real coefficients, and z is a complex root of f, then $\bar{z}$ is also a root of f. This means that if in Eq. (5-28) q_j is a complex root, then its conjugate is also a root, and they both have the same order. Let us denote the real numbers among $q_1, q_2, \ldots, q_l$ by $a_1, a_2, \ldots, a^s$, and the complex numbers among them by $c_1, \bar{c}_1, c_2, \bar{c}_2, \ldots, c_t, \bar{c}_t$. Clearly $2t + s = l$. Let n_j be the order of c_j and $\bar{c}_j$, and m_k be the order of a_k. We can now put Eq. (5-28) in the form

$$f(x) = K \prod_{k=1}^{s} (x - a_k)^{m_k} \prod_{j=1}^{t} [(x - c_j)(x - \bar{c}_j)]^{n_j} \qquad (5\text{-}29)$$

Finally, $(x - c_j)(x - \bar{c}_j) = x^2 - (c_j + \bar{c}_j)x + c_j \bar{c}_j$. But if $c_j = u_j + v_j i$, then $c_j + \bar{c}_j = 2u_j$ and $c_j \bar{c}_j = u_j^2 + v_j^2$. Equation (5-29) can thus be written as

$$f(x) = K \prod_{k=1}^{s} (x - a_k)^{m_k} \prod_{j=1}^{t} [x^2 - 2u_j x + (u_j^2 + v_j^2)]^{n_j} \qquad (5\text{-}30)$$

A polynomial with real coefficients is said to be *irreducible* if it has no real roots. With this terminology we can put Eq. (5-30) into words: *Every polynomial having real coefficients can be expressed as the product of first degree polynomials (linear factors) having real roots and irreducible second degree polynomials (irreducible quadratics) together with a real constant.*

After these preliminaries we are ready for the theorem from algebra that concerns rational functions.

Theorem 5-2 *Let f and g be polynomials with real coefficients, and suppose that the degree of g is less than the degree of f. If*

$$f(x) = K \prod_{j=1}^{s} (x - a_j)^{m_j} \prod_{k=1}^{t} (x^2 + 2b_k x + c_k)^{m_k}$$

where $a_1, a_2, \ldots, a_s$ are real, and $x^2 + 2b_k x + c_k$ is an irreducible quadratic for $k = 1, 2, \ldots, t$, then

$$\frac{g(x)}{f(x)} = \sum_{j=1}^{m_1} \frac{A_{1,j}}{(x - a_1)^j} + \sum_{j=1}^{m_2} \frac{A_{2,j}}{(x - a_2)^j} + \cdots$$

$$+ \sum_{j=1}^{m_s} \frac{A_{s,j}}{(x - a_s)^j} + \sum_{k=1}^{n_1} \frac{B_{1,k} x + C_{1,k}}{(x^2 + 2b_1 x + c_1)^k}$$

$$+ \sum_{k=1}^{n_2} \frac{B_{2,k} x + C_{2,k}}{(x^2 + 2b_2 x + c_2)^k} + \cdots \qquad (5\text{-}31)$$

$$+ \sum_{k=1}^{n_t} \frac{B_{t,k} x + C_{t,k}}{(x^2 + 2b_t x + c_t)^k}$$

where the A's, B's and C's are fixed real numbers.

Equation (5-31) is called the *partial fraction decomposition* of the rational function g/f; it requires some explanation. We know from Eq. (5-30) that every polynomial can be written as a product of linear factors and irreducible quadratics, and that each linear factor appears as many times as the order of the root it is associated with, while each irreducible quadratic appears as many times as the order of the conjugate pair of complex roots with which it is associated. In Eq. (5-31) it can be seen that a real root r of the denominator, having order p, gives rise to a sum in the partial fraction decomposition of the form $D_1/(x - r) + D_2/(x - r)^2 + \cdots + D_p/(x - r)^p$, where $D_1, D_2, \ldots, D_p$ are real constants. It is also clear that a complex root $c = u + vi$ of order q, together with its conjugate $\bar{c}$, gives rise to a sum of the form

$$\frac{(E_1 x + F_1)}{[x^2 - 2ux + (u^2 + v^2)]} + \frac{(E_2 x + F_2)}{[x^2 - 2ux + (u^2 + v^2)]^2} + \cdots$$

$$+ \frac{(E_q x + F_q)}{[x^2 - 2ux + (u^2 + v^2)]^q}$$

where $E_1, E_2, \ldots, E_q$, and $F_1, F_2, \ldots, F_q$ are real constants. In other words, the roots of the denominator of a rational function determine the form of its partial fraction expansion.

Even though the degree of g cannot exceed the degree of f in order to apply Eq. (5-31), we can still make use of it with a rational function h/f in which the degree of h is greater than or equal to the degree of f. This is so because we have

$$\frac{h(x)}{f(x)} = q(x) + \frac{g(x)}{f(x)} \tag{5-32}$$

where q is a polynomial, g is a polynomial, and the degree of g is smaller than the degree of f. Thus the partial fraction expansion of g/f can be found.

We now see that any rational function can be expressed as a sum of relatively simple terms. Since the antiderivative of a sum is the sum of the antiderivatives, it follows that we can find the antiderivative of any rational function if we can find the antiderivative of a few special rational functions.

According to Eq. (5-32) and Theorem 5-2, any rational function can be written as the sum of a polynomial and a partial fraction expansion. We know how to antidifferentiate the polynomial. The antiderivative of the partial fraction expansion reduces to three antiderivatives: $\int [1/(x + r)^n] \, dx$, $\int [x/(x^2 + bx + c)^n] \, dx$, and $\int [1/(x^2 + bx + c)^n] \, dx$, where n is a positive integer and, since $x^2 + bx + c$ is irreducible, $b^2 - 4c < 0$. For the first antiderivative we have

$$\int \frac{dx}{(x+r)^n} = \begin{cases} \log|x+r| + C & \text{if } n = 1 \\ \dfrac{1}{(1-n)(x+r)^{n-1}} + C & \text{if } n \neq 1 \end{cases}$$

As for the other two, we first "complete the square" to get $x^2 + bx + c = (x + b/2)^2 + (c - b^2/4)$. Since $4c - b^2 > 0$, we can let $a = (4c - b^2)^{1/2}/2$. The change of variable $y = x + b/2$ then reduces the problem to finding $\int [y/(y^2 + a^2)^n] \, dy$ and $\int [1/(y^2 + a^2)^n] \, dy$. Now

$$\int \frac{y}{(y^2 + a^2)^n} \, dy = \begin{cases} \dfrac{1}{2} \log(y^2 + a^2) + C & \text{if } n = 1 \\ \dfrac{1}{2} \dfrac{1}{(1-n)(y^2 + a^2)^{n-1}} + C & \text{if } n \neq 1 \end{cases}$$

and $\int [1/(y^2 + a^2)] \, dy = (1/a) \tan^{-1}(y/a) + C$. We are thus left with the task of evaluating $\int [1/(y^2 + a^2)^n] \, dy$ when $n > 1$. In $\int [1/(y^2 + a^2)^{n-1}] \, dy$ put $u = 1/(y^2 + a^2)^{n-1}$ and $dv = dy$. Then $du = [(1 - n)2y/(y^2 + a^2)^n] \, dy$, $v = y$, and

$$\int \frac{1}{(y^2 + a^2)^{n-1}} \, dy = \frac{y}{(y^2 + a^2)^{n-1}} + 2(n-1) \int \frac{y^2}{(y^2 + a^2)^n} \, dy$$

$$= \frac{y}{(y^2 + a^2)^{n-1}} + 2(n-1) \int \frac{(y^2 + a^2)}{(y^2 + a^2)^n} \, dy$$

$$- 2(n-1)a^2 \int \frac{1}{(y^2 + a^2)^n} \, dy$$

The foregoing may be written as

$$\int \frac{dy}{(y^2 + a^2)^n} = \frac{1}{2(n-1)a^2} \left[\frac{y}{(y^2 + a^2)^{n-1}} \right.$$

$$\left. + (2n-3) \int \frac{dy}{(y^2 + a^2)^{n-1}} \right] \qquad (5\text{-}33)$$

Repeated applications of Eq. (5-33) will reduce the problem of finding $\int [1/(y^2 + a^2)^n] \, dy$ to the known relation $\int [1/(y^2 + a^2)] \, dy = (1/a) \tan^{-1}(y/a) + C$. To sum up, we have shown how to find the antiderivative of any rational function. In fact, we have shown that the antiderivative is the sum of a polynomial, a rational function, a sum of logarithms, and a sum of inverse tangents. Because this method of antidifferentiating rational functions depends on the partial fraction expansion, it is called the *method of partial fractions*.

Remark Unfortunately, there is one serious drawback to the method of partial fractions. It lies in the fact that it is necessary to know the roots of the denominator of a rational function before it can be expanded into partial fractions. For second degree polynomials the roots are given by the familiar quadratic formula. Further, there are (less familiar) formulas that give the roots of any third or fourth degree polynomial algebraically in terms of their coefficients. (We use the term "algebraically" here in the sense of a finite number of additions, subtractions, multiplications, divisions, and extraction of roots.) But it can be proved, and this is one of the landmarks in mathematical history, that *if n is a positive integer such that n ≥ 5, then there does not exist a formula giving the roots of an arbitrary polynomial of degree n algebraically in terms of its coefficients.* This does not mean that the roots of a polynomial of degree 5 or higher cannot be found; but it does mean that the problem of finding the roots of such a polynomial must be attacked by largely *ad hoc* methods.

In view of this fact, we must qualify our assessment of the method of partial fractions. We can only say that it enables us to compute the antiderivative of any rational function "in principle", or "in theory".

EXAMPLE 5-9 Consider the problem of finding

$$\int [(x^4 - x^3 + 4x^2 - 2x + 1)/(x^3 + x)] \, dx$$

by the method of partial fractions. Since the degree of the numerator exceeds the degree of the denominator, we must first carry out the division indicated in Eq. (5-32). This results in

$$\frac{x^4 - x^3 + 4x^2 - 2x + 1}{x^3 + x} = x - 1 + \frac{3x^2 - x + 1}{x^3 + x}$$

Now $x^3 + x = x(x^2 + 1)$, and $x^2 + 1$ is an irreducible quadratic. Therefore, we must have

$$\frac{3x^2 - x + 1}{x^3 + x} = \frac{A}{x} + \frac{Bx + C}{x^2 + 1}$$

To determine A, B, and C we carry out the addition on the right-hand side to get

$$\frac{3x^2 - x + 1}{x^3 + x} = \frac{(A + B)x^2 + Cx + A}{x^3 + x}$$

Now this equality can hold only if $3x^2 - x + 1 = (A + B)x^2 + Cx + A$, and two polynomials can be equal only if their coefficients are equal. Thus $C = -1$, $A = 1$, and $A + B = 3$; and so $B = 2$.

Therefore

$$\int \frac{x^4 - x^3 + 4x^2 - 2x + 1}{x^3 + x} \, dx = \int (x - 1) \, dx + \int \frac{1}{x} \, dx$$

$$+ 2 \int \frac{x}{x^2 + 1} \, dx - \int \frac{dx}{x^2 + 1}$$

$$= \frac{x^2}{2} - x + \log |x| + \log (x^2 + 1)$$

$$- \tan^{-1} x + C$$

EXAMPLE 5-10   We will now calculate

$$\int \frac{2x^5 - 7x^4 + 7x^3 - 19x^2 + 7x - 6}{(x - 1)^2 (x^2 + 1)^2} \, dx$$

Since the degree of the denominator exceeds the degree of the numerator, we can concentrate on the partial fraction expansion. Notice that 1 is a double root, and that $x^2 + 1$ is an irreducible quadratic appearing with exponent 2. It follows that

$$\frac{2x^5 - 7x^4 + 7x^3 - 19x^2 + 7x - 6}{(x - 1)^2 (x^2 + 1)^2}$$

$$= \frac{A_1}{x - 1} + \frac{A_2}{(x - 1)^2} + \frac{B_1 x + C_1}{x^2 + 1} + \frac{B_2 x + C_2}{(x + 1)^2}$$

$$= \{(A_1 + B_1)x^5 + (-A_1 + A_2 - 2B_1 + C_1)x^4$$

$$+ (2A_1 + 2B_1 - 2C_1 + B_2)x^3$$

$$+ (-2A_1 + 2A_2 - 2B_1 + 2C_1 - 2B_2 + C_2)x^2$$

$$+ (A_1 + B_1 - 2C_1 + B_2 - 2C_2)x$$

$$+ (-A_1 + A_2 + C_1 + C_2)\} \div \{(x - 1)^2 (x^2 + 1)^2\}$$

By equating the coefficients in the numerators of this equation we get the following system of six linear equations in six unknowns.

A_1		$+ B_1$		$= 2$
$- A_1$	$+ A_2$	$- 2B_1 + C_1$		$= -7$
$2A_1$		$+ 2B_1 - 2C_1 + B_2$		$= 7$
$-2A_1$	$+ 2A_2$	$- 2B_1 + 2C_1 - 2B_2 + C_2$		$= -19$
A_1		$+ B_1 - 2C_1 + B_2 - 2C_2$		$= 7$
$- A_1$	$+ A_2$	$+ C_1 + C_2$		$= -6$

We can now make use of any of the standard elementary, but tedious, techniques for solving systems of linear equations to find that $A_1 = 1$, $A_2 = -4$, $B_1 = 1$, $C_1 = 0$, $B_2 = 3$, and $C_2 = -1$.

Consequently,

$$\int \frac{2x^5 - 7x^4 + 7x^3 - 19x^2 + 7x - 6}{(x-1)^2(x^2+1)^2}\, dx$$

$$= \int \frac{1}{x-1}\, dx - 4\int \frac{1}{(x-1)^2}\, dx + \int \frac{x}{x^2+1}\, dx$$

$$+ 3\int \frac{x}{(x^2+1)^2}\, dx - \int \frac{dx}{(x^2+1)^2}$$

$$= \log|x-1| + \frac{4}{x-1} + \frac{1}{2}\log(x^2+1)$$

$$- \frac{3}{2}\frac{1}{x^2+1} - \left[\frac{1}{2}\frac{x}{x^2+1} + \tan^{-1}x\right] + C$$

[we have used Eq. (5-33) with $n = 2$, $a = 1$ on the fifth antiderivative].

EXERCISES

5-9 Find $\int [9x^2 + 7x - 6)/(x^3 - x)]\, dx$.

5-10 Find $\int [(5x^3 - 5x^2 + 2x - 1)/(x^4 + x^2)]\, dx$.

Now that we have some experience with each of the standard techniques of antidifferentiation, it will be instructive to examine them collectively and concentrate on the characteristics they have in common rather than on the differences between them. Each of these methods changes a given antiderivative into one that may or may not be easier to evaluate. In other words, it may happen that any one of these methods can be applied and yet fail to solve the primary problem of computing the antiderivative. Consequently it is not enough to know the methods of antidifferentiation, it is also necessary to know which is most likely to be successful.

With these two requirements in mind, we have arranged the first eight problems in this chapter in the following way: The first seven problems are each concerned exclusively with one method. Problem 5-8 is designed to give the reader skill in the selection of the appropriate method. It includes quite easy antiderivatives, and fairly hard ones (some of which require the application of more than one method), but in each case the reader has to decide first which technique to use.

We conclude by considering the applications of the antiderivative. These applications deal with reversing the interpretation of the derivative as the rate of change of a quantity. If the derivative is so regarded, then its antiderivative must be the quantity in question. For instance, the antiderivative of velocity would be distance, or the antiderivative of acceleration would be velocity.

EXAMPLE 5-11   The following problem is typical of the application of the antiderivative.   A box slides down an inclined chute 200 feet long with an acceleration of 22 feet/sec². If it reaches the bottom in 4 seconds, how fast was it launched down the chute?

To solve this, notice first that $dv/dt = 22$. It follows that $v = 22t + v_0$, where v_0 is the initial velocity we seek. But $ds/dt = v = 22t + v_0$, and so $s = 11t^2 + v_0 t + s_0$. However, we can take the initial position s_0 to be the origin of a coordinate line; in other words, we can put $s_0 = 0$. We then must have $200 = 11(4)^2 + 4v_0$, or $v_0 = 6$ feet/sec.

Remark   From this example it is easy to see that if a particle moves on a coordinate line with constant acceleration a, initial velocity v_0, and initial position s_0, then its velocity v is given by $v = at + v_0$, and its position s by $s = (a/2)t^2 + v_0 t + s_0$.

EXERCISE 5-11   If a current from a battery is flowing in a circuit with resistance R and inductance L, and the battery is cut from the circuit, then the current i subsequently obeys the law $L\, di/dt + Ri = 0$. If i_0 is the current when the battery is cut out, find i.

SOLUTIONS TO EXERCISES

5-1   We have

$$\frac{f(x)}{x^p} = \sum_{k=0}^{n} a_k x^{k-p}$$

If $k - p \neq -1$ for $k = 0, 1, 2, \ldots, n$, then

$$\int \frac{f(x)}{x^p}\, dx = \sum_{k=0}^{n} a_k \int x^{k-p}\, dx = \sum_{k=0}^{n} \frac{a_k}{k-p+1} x^{k-p+1} + C$$

Otherwise, $m - p = -1$, or $p = m + 1$, for some integer m between 0 and n. In this case

$$\int \frac{f(x)}{x^p}\, dx = \sum_{k=0}^{m-1} a_k \int x^{k-(m+1)}\, dx + a_m \int \frac{1}{x}\, dx$$

$$+ \sum_{k=m+1}^{n} a_k \int x^{k-(m+1)}\, dx$$

$$= \sum_{k=0}^{m-1} \frac{a_k}{k-m} x^{k-m} + a_m \log |x|$$

$$+ \sum_{k=m+1}^{n} \frac{a_k}{k-m} x^{k-m} + C$$

5-2 If we let $u = \log x$, then $du = (1/x)\,dx$, and

$$\int (\log x/x)\,dx = \int u\,du = u^2/2 + C = \tfrac{1}{2}(\log x)^2 + C$$

5-3 We obtain Eq. (5-21) as follows: $\int \cot x\,dx = \int (\cos x/\sin x)\,dx$. Now put $u = \sin x$. Then $du = \cos x\,dx$, and $\int (\cos x/\sin x)\,dx = \int (1/u)du = \log |u| + C = \log |\sin x| + C$.
As for Eq. (5-23),

$$\int \operatorname{cosec} x\,dx = \int \frac{\operatorname{cosec}^2 x + \operatorname{cosec} x \cot x}{\operatorname{cosec} x + \cot x}\,dx$$

Now let $u = \operatorname{cosec} x + \cot x$. Then $du = -(\operatorname{cosec}^2 x + \operatorname{cosec} x \cot x)\,dx$, and

$$\int \cot x\,dx = -\int (1/u)\,du = -\log |u| + C = -\log |\operatorname{cosec} x + \cot x| + C$$

$$= \log |(\operatorname{cosec} x - \cot x)/(\operatorname{cosec}^2 x - \cot^2 x)| + C$$

$$= \log |\operatorname{cosec} x - \cot x| + C$$

5-4 We have the following: $\int \sin^3 x\,dx = \int \sin x (1 - \cos^2 x)\,dx = \int \sin x\,dx - \int \sin x \cos^2 x\,dx = -\cos x + \tfrac{1}{3}\cos^3 x + C$.

5-5 Let $u = x$, $dv = \sec^2 x\,dx$. Then $du = dx, v = \tan x$, and $\int x \sec^2 x\,dx = x \tan x - \int \tan x\,dx = x \tan x - \log |\sec x| + C$.

5-6 Here put $u = \sin^{-1} x$ and $dv = x(1 - x^2)^{1/2}\,dx$. Then $du = [1/(1 - x^2)^{1/2}]\,dx$ and $v = -(1/3)(1 - x^2)^{3/2}$. Therefore we have $\int x(1 - x^2)^{1/2} \sin^{-1} x\,dx = -(1/3)\sin^{-1} x(1 - x^2)^{3/2} + (1/3) \int [(1 - x^2)^{3/2}/(1 - x^2)^{1/2}]\,dx = -(1/3) \sin^{-1} x (1 - x^2)^{3/2} + \tfrac{1}{3}x - \tfrac{1}{9}x^3 + C$.

5-7 Make the substitution $x + 1 = u^2$, or $u = (x + 1)^{1/2}$. Then $x = u^2 - 1$, $dx = 2u\,du$, and

$$\int [x^2/(x + 1)^{1/2}]\,dx = \int [(u^2 - 1)^2/u]2u\,du$$

$$= 2 \int (u^4 - 2u^2 + 1)\,du = (2/5)u^5 - (4/3)u^3 + 2u + C$$

$$= (2/5)(x + 1)^{5/2} - (4/3)(x + 1)^{3/2} + 2(x + 1)^{1/2} + C$$

5-8   Here we make the standard substitution $x = \sec \theta$. Then $dx = \sec \theta \tan \theta \, d\theta$, $\theta = \sec^{-1} x$, and

$$\int x^3/(x^2 - 1)^{1/2} \, dx$$

$$= \int (\sec^3 \theta/\tan \theta) \sec \theta \tan \theta \, d\theta = \int \sec^4 \theta \, d\theta$$

$$= \int \sec^2 \theta[\tan^2 \theta + 1] \, d\theta = \int \tan^2 \theta \sec^2 \theta \, d\theta + \int \sec^2 \theta \, d\theta$$

$$= (1/3) \tan^3 \theta + \tan \theta + C = (1/3) \tan^3 (\sec^{-1} x) + \tan (\sec^{-1} x) + C$$

$$= (1/3)(x^2 - 1)^{3/2} + (x^2 - 1)^{1/2} + C$$

(This exercise can also be solved by means of integration by parts.)

5-9   Since $x^3 - x = x(x + 1)(x - 1)$, we must have

$$\frac{9x^2 + 7x - 6}{x^3 - x} = \frac{A}{x} + \frac{B}{x + 1} + \frac{C}{x - 1}$$

$$= \frac{A(x + 1)(x - 1) + Bx(x - 1) + Cx(x + 1)}{x^3 - x}$$

Hence $9x^2 + 7x - 6 = A(x + 1)(x - 1) + Bx(x - 1) + Cx(x + 1)$. In this equation we put $x = 0$ and see that $-6 = -A$, or $A = 6$. Similarly, if $x = 1$, then $10 = 2C$, or $C = 5$, and if $x = -1$, then $-4 = 2B$, or $B = -2$. Therefore,

$$\int \frac{9x^2 + 7x - 6}{x^3 - x} \, dx = 6 \int \frac{1}{x} \, dx - 2 \int \frac{1}{x + 1} \, dx + 5 \int \frac{1}{x - 1} \, dx$$

$$= 6 \log |x| - 2 \log |x + 1| + 5 \log |x - 1| + C$$

Remark   When the roots of the denominator of a rational function are all real and distinct, the technique used in the solution of this exercise is a very rapid one for finding the constants in the partial fraction expansion.

5-10   We first note that $x^4 + x^2 = x^2(x^2 + 1)$. Since $x^2 + 1$ is irreducible, we must have

$$\frac{5x^3 - 5x^2 + 2x - 1}{x^4 + x^2} = \frac{A}{x} + \frac{B}{x^2} + \frac{Cx + D}{x^2 + 1}$$

$$= \frac{(A + C)x^3 + (B + D)x^2 + Ax + B}{x^4 + x^2}$$

From this equation we see that $B = -1$, $A = 2$, $B + D = -5$, and $A + C = 5$. It can now be seen that $C = 3$ and $D = -4$. Thus

$$\int \frac{5x^3 - 5x^2 + 2x - 1}{x^4 + x^2}\, dx = 2\int \frac{1}{x}\, dx - \int \frac{1}{x^2}\, dx$$

$$+ 3\int \frac{x}{x^2 + 1}\, dx - 4\int \frac{1}{x^2 + 1}\, dx$$

$$= 2\log|x| + 1/x + (3/2)\log(x^2 + 1)$$

$$- 4\tan^{-1} x + C$$

5-11 Since $(1/i)\, di/dt = -R/L$, we have $\int (1/i)\, di/dt = \int (-R/L)\, dt$, or $\log i = -(R/L)t + C$. In other words, $i = e^{-(R/L)t + C} = e^C e^{-(R/L)t}$. But $i_0 = e^C e^{-(R/L)0} = e^C$ and so, $i = i_0 e^{-(R/L)t}$.

PROBLEMS

5-1   Use the change of variable method to find:

(a) $\displaystyle\int \frac{x\log(1 + x^2)}{1 + x^2}\, dx$

(b) $\displaystyle\int \frac{1}{x^{1/3}(1 + x^{2/3})}\, dx$

(c) $\displaystyle\int \frac{1}{x^3}\left(\frac{1 - x^2}{x^2}\right)^{10}\, dx$

*(d) $\displaystyle\int \frac{1}{x(x^4 - 1)^{1/2}}\, dx$

*(e) $\displaystyle\int (x - a)^{p-1}(x - b)^{-p-1}\, dx, p \geq 0, a \neq b$

5-2   Use the change of variable method to find:

(a) $\displaystyle\int (\sin x/\cos^{1/2} x)\, dx$

(b) $\displaystyle\int (\sec x + \tan x)^2\, dx$

(c) $\displaystyle\int \sec^2 3x \tan^5 3x\, dx$

*(d) $\displaystyle\int [1/(1 + \cos x)]\, dx$

(e) $\displaystyle\int \sin^{1/2} x \cos^3 x\, dx$

(f) $\displaystyle\int \sin^2 x \cos^2 x\, dx$

(g) $\displaystyle\int \sec^5 x \tan^3 x\, dx$

(h) $\displaystyle\int \tan^2 x \sec^4 x\, dx$

(i) $\displaystyle\int \cos ax \cos bx\, dx, a \neq b$

*(j) $\displaystyle\int [1/(a^2 \sin^2 x + b^2 \cos^2 x)]\, dx,$
$ab \neq 0$

5-3   Use the change of variable method to find:

(a) $\displaystyle\int \sec^2 x e^{\tan x}\, dx$

*(b) $\displaystyle\int [(e^{2x} - 1)/(e^{2x} + 1)]\, dx$

(c) $\displaystyle\int e^{x + e^x}\, dx$

(d) $\displaystyle\int a^x b^{2x} e^{3x}\, dx, a > 0, b > 0$

5-4   Use the change of variable method to find:

(a) $\displaystyle\int \frac{\sec^2 x}{9 + \tan^2 x}\, dx$

(b) $\displaystyle\int \frac{1}{x(4 - \log^2 x)^{1/2}}\, dx$

(c) $\displaystyle\int \frac{1}{(e^{2x} - 16)^{1/2}}\, dx$

*(d) $\displaystyle\int \frac{1}{(8 + 2x - x^2)^{1/2}}\, dx$

5-5   Use integration by parts to find:

(a) $\displaystyle\int \tan^{-1} x\, dx$

(b) $\displaystyle\int x \log x\, dx$

(c) $\displaystyle\int x^2 e^x\, dx$

(d) $\displaystyle\int e^x \sin x\, dx$

*(e) $\displaystyle\int [x/(2x + 1)^{1/2}]\, dx$

(f) $\displaystyle\int [x^3/(1 + x^2)^{1/2}]\, dx$

(g) $\displaystyle\int x^3 e^{-x^2}\, dx$

(h) $\displaystyle\int \sec^3 x\, dx$

5-6   Use the method of substitution to find:

(a) $\displaystyle\int x(x + 1)^{1/3}\, dx$

(b) $\displaystyle\int [1/(1 + x^{1/2})]\, dx$

(c) $\displaystyle\int [1/(x^2 + a^2)^{1/2}]\, dx$

(d) $\displaystyle\int [1/(x^2 - a^2)^{1/2}]\, dx$

(e) $\displaystyle\int [(a^2 - x^2)^{1/2}/x^2]\, dx$

(f) $\displaystyle\int [1/(x^2 - 4x + 5)^2]\, dx$

*(g) $\displaystyle\int \{1/[(x - a)(b - x)]^{1/2}\}\, dx$

5-7   Use the method of partial fractions to find:

(a) $\displaystyle\int \frac{1}{x^2 - a^2}\, dx$

(b) $\displaystyle\int \frac{x^3}{x^2 - 2x - 3}\, dx$

(c) $\displaystyle\int \frac{2x^3 + x^2 + 5x + 4}{x^4 + 8x^2 + 16}\, dx$

(d) $\displaystyle\int \frac{4x^2 - 3x}{(x + 2)(x^2 + 1)}\, dx$

5-8   Find:

*(a) $\displaystyle\int x^2 \sin^{-1} x\, dx$

(b) $\displaystyle\int \frac{x \log x}{(1 - x^2)^{1/2}}\, dx$

(c) $\displaystyle\int x^{1/2}(1 + x^2)^2\, dx$

*(d) $\displaystyle\int x^4 \log^2 x\, dx$

*(e) $\displaystyle\int \frac{x + \sin x}{3x^2 - 2 \cos 3x}\, dx$

(f) $\displaystyle\int \frac{x^{1/2} + 1}{x^{1/2}(x + 1)}\, dx$

(g) $\displaystyle\int \frac{1}{x^3(x^2 - 4)^{1/2}}\, dx$

(h) $\displaystyle\int \frac{1}{(x + 3)^{1/2} - (x + 2)^{1/2}}\, dx$

(i) $\displaystyle\int \sin 2x e^{\sin^2 x}\, dx$

(j) $\displaystyle\int \frac{1}{e^x - 16e^{-x}}\, dx$

*(k) $\displaystyle\int \frac{1}{[1 + (1 + x)^{1/2}]^{1/2}}\, dx$

(l) $\displaystyle\int \sin (\log x)\, dx$

(m) $\int \dfrac{(x^2+4)^{1/2}}{x^4}\,dx$

(n) $\int \dfrac{\tan^{-1} x}{(1+x^2)^{3/2}}\,dx$

(o) $\int \dfrac{x+\sin x}{1+\cos x}\,dx$

(p) $\int \dfrac{x+1}{(x+2)^2(x+3)}\,dx$

(q) $\int \dfrac{x^3+x^2+x-1}{x^2(x^2+1)}\,dx$

(r) $\int (e^x - e^{-2x})\,^2 dx$

(s) $\int \dfrac{1}{x[(1+x)^{1/2}-2]}\,dx$

(t) $\int \dfrac{7x-2}{(7-2x^2)^{1/2}}\,dx$

(u) $\int \sin^6 x\,dx$

(v) $\int \dfrac{x}{(x-4)^{1/2}}\,dx$

(w) $\int \dfrac{x^3-3x^2-5x+5}{x^2+2x-3}\,dx$

(x) $\int (1+x)\cos x^{1/2}\,dx$

(y) $\int \dfrac{\sec^3 x}{\tan x}\,dx$

*(z) $\int \dfrac{(1+x^3)^{1/2}}{x}\,dx$

*5-9   Find $\displaystyle\int \dfrac{1}{2+3\cos x}\,dx$

5-10   Find $\displaystyle\int \left(\dfrac{x-1}{x+1}\right)^{1/2}\dfrac{1}{x}\,dx$

5-11   Find $\displaystyle\int \dfrac{x-1}{x+1}\dfrac{1}{[x(x^2+x+1)]^{1/2}}\,dx$ by means of the substitution $u^2 = x+1+1/x$.

5-12   If p and q are positive rational numbers, find $\int (1+x)^p x^q\,dx$ when

    (a) p is an integer.

    (b) q is an integer.

    If $p+q$ is an integer,

    *(c)   show how $\int (1+x)^p x^q\,dx$ can be found in principle.

5-13   If $I_n = \int \tan^n x\,dx$, show that $(n-1)(I_n + I_{n-2}) = \tan^{n-1} x$

*5-14   If $I_{m,n} = \int \dfrac{x^m}{(1+x^2)^n}\,dx$, show that

$$2(n-1)I_{m,n} = -x^{m-1}(1+x^2)^{-(n-1)} + (m-1)I_{m-2,n-1}$$

5-15   Show that $\int f''(x)F(x)\,dx = f'(x)F(x) - f(x)F'(x) + \int f(x)F''(x)\,dx$

*5-16   Let P and Q be polynomials, with P of lower degree than Q. Show that if the roots $\alpha_1, \ldots, \alpha_n$ of Q are all real and distinct, then

$$\int \dfrac{P(x)}{Q(x)}\,dx = \sum_{k=1}^{n} \dfrac{P(\alpha_k)}{Q'(\alpha_k)} \log |x-\alpha_k| + C$$

5-17 We have $\int \sin 2x \, dx = (-1/2) \cos 2x + C$, and $\int 2 \sin x \cos x \, dx = \sin^2 x$
$+ C$.   But $\sin 2x = 2 \sin x \cos x$, and so $\int \sin 2x \, dx = \int 2 \sin x \cos x \, dx$.
Explain the difference in answers.

5-18 The slope of the tangent line to a curve at (x, y) is $2xy$.   If the curve contains
the point $(2, 1)$, what is its equation?

5-19 What constant acceleration is required to bring a truck to rest in 300 ft if it
is initially running at 88 ft/sec?

5-20 An object is thrown directly downward from the roof of a building with an
initial velocity of v_0 ft/sec.   If it hits the ground in 2.5 sec with a velocity of
110 ft/sec, how high is the building?

5-21 A body projected upward reaches a certain height after 4 sec and returns
to this same height, on its descent, 5 sec later.   Find the height in question
and the initial velocity of upward projection.

5-22 A man runs the 100 yd dash in 10 sec.   If his acceleration was constant for
the first 20 yd and was thereafter zero, what was the acceleration?

5-23 Water is running out of a bathtub according to the law $dy/dt = -0.025 \, y^{1/2}$,
where y is the depth of the water measured in inches and t is measured in
seconds.   When will the tub be empty if the water is 9 in. deep?

5-24 The velocity of a delayed-action bomb is retarded at a rate proportional to
the square root of its velocity when it strikes the ground.   If the bomb will
penetrate to a depth of 3 (3/8) ft when the impact speed is 225 ft/sec, find the
time required for the bomb to come to rest.   Also find the corresponding
time and depth of penetration if the impact speed is 400 ft/sec.

*5-25 A point P moves on the x axis in such a way that $d^2x/dt^2 = -\omega^2 x$, where
$\omega > 0$.   Find x as a function of t.

Chapter 6

THE INTEGRAL AND ITS APPLICATIONS

We know that the derivative is a new function obtained from a given function by means of a limiting process. In this chapter we are going to examine another limiting process that will enable us to associate another new function with a given function. The process is called *integration*, and it results in what is called the *integral*.

To define the integral we need some preliminary terminology. Let f be defined on the closed interval $[a, b]$. Any numbers $x_0, x_1, \ldots, x_n$ such that $a = x_0 < x_1 < x_2 < \cdots < x_n = b$ will be called a *partition* of $[a, b]$. A partition decomposes $[a, b]$ into n subintervals $[x_0, x_1], [x_1, x_2], \cdots,$ $[x_{n-1}, x_n]$ having lengths $x_1 - x_0, x_2 - x_1, \ldots, x_n - x_{n-1}$ respectively; we call the largest of these lengths the *norm* of the partition. Let z_k, for $k = 1, 2, \ldots, n$ satisfy the inequality $x_{k-1} \leq z_k \leq x_k$. The sum

$$\sum_{k=1}^{n} f(z_k)(x_k - x_{k-1})$$

is called a *Riemann sum* of f relative to the partition $x_0, x_1, \ldots, x_n$. For a fixed partition there are clearly an infinite number of Riemann sums.

Definition 6-1 *The function f defined on $[a, b]$ has integral L on $[a, b]$ if given any sequence of partitions of $[a, b]$ such that the sequence of their norms has limit zero, and given any Riemann sum relative to each of these partitions, then this sequence of Riemann sums has limit L.*

Functions that have an integral on $[a, b]$ are said to be *integrable* on $[a, b]$. If f is integrable on $[a, b]$, we denote its integral by

$$\int_a^b f(x) \, dx$$

97

In other words, if L is the number in Definition 6-1, then

$$L = \int_a^b f(x)\,dx \tag{6-1}$$

If f is integrable on $[a, b]$, then its integral can be computed by finding the limit of a conveniently chosen sequence of Riemann sums (provided of course that the sequence of norms of the underlying partitions has limit zero); the reason for this is that L is the same for *all* such sequences of Riemann sums.

The basic properties of the integral are enumerated in the following theorem.

Theorem 6-1 *Let f and g be integrable on $[a, b]$, let k be a constant, and let $a < c < b$. Then*

$$\int_a^b k f(x)\,dx = k \int_a^b f(x)\,dx \tag{6-2}$$

$$\int_a^b f(x)\,dx = \int_a^c f(x)\,dx + \int_c^b f(x)\,dx \tag{6-3}$$

$$\int_a^b [f(x) + g(x)]\,dx = \int_a^b f(x)\,dx + \int_a^b g(x)\,dx \tag{6-4}$$

Since an integral can be viewed as the limit of a sequence of Riemann sums, it is not hard to see that Eq. (6-2) can be obtained from the relation

$$\sum_{m=1}^n k a_m = k \sum_{m=1}^n a_m$$

that Eq. (6-3) can be obtained from the identity

$$\sum_{m=1}^n a_m = \sum_{m=1}^r a_m + \sum_{m=r+1}^n a_m$$

and that Eq. (6-4) will follow from the equation

$$\sum_{m=1}^n (a_m + b_m) = \sum_{m=1}^n a_m + \sum_{m=1}^n b_m$$

Equation (6-3) has a particularly interesting consequence. It implies that if a function f is integrable on $[a, b]$, then it is integrable on $[a, x]$, where $a \le x \le b$. This means that we can associate with an integrable function f a new function g defined by the equation $g(x) = \int_a^x f(y)\,dy$ on the domain $[a, b]$.

In view of the complexity of Definition 6-1, it would be very desirable, and indeed almost necessary, to have a simple condition that would ensure integrability. The next theorem meets this requirement.

Theorem 6-2 *If f is continuous on $[a, b]$, then f is integrable on $[a, b]$.*

Since most functions encountered in practice are continuous, this theorem provides a fairly satisfactory answer to the question of which functions are integrable. At the same time, it should be noticed that the condition of continuity is sufficient but not necessary for integrability. As a matter of fact, there are integrable functions that are not continuous.

EXAMPLE 6-1 We know that the function $f(x) = x$ is continuous on $(-\infty, \infty)$. In particular, then, it is continuous on $[0, 1]$, and so $\int_0^1 x\, dx$ exists. We will compute it by finding the limit of a particular sequence of Riemann sums. We first choose the partitions. For any integer n the numbers $0, 1/n, 2/n, \ldots, n/n$ are a partition of $[0, 1]$ with norm $1/n$. Put $z_k = k/n$ for $k = 1, 2, \ldots, n$, and form the Riemann sum

$$R_n = \sum_{k=1}^{n} \frac{k}{n} \cdot \frac{1}{n}$$

Since $\{1/n\} \to 0$, it follows that $\{R_n\} \to \int_0^1 x\, dx$. Now

$$R_n = \frac{1}{n^2} \sum_{k=1}^{n} k = \frac{1}{n^2} \frac{n(n+1)}{2} = \frac{1}{2}\left(1 + \frac{1}{n}\right)$$

[recall Eq. (1-12)]. Clearly, $\{R_n\} \to \frac{1}{2}$ and so $\int_0^1 x\, dx = \frac{1}{2}$ as well.

EXERCISE 6-1 Find $\int_0^2 x^2\, dx$ by employing the method of Example 6-1.

It is clear that we cannot count on being able to determine the limit of a sequence of Riemann sums of an arbitrary continuous function on an interval (recall that in Chapter 2 we had difficulties with just establishing the existence of the limits of some sequences, and that we were not able to give the value of these limits). In other words, in order for the integral to become computationally practical, we need another method for evaluating it. As far as continuous functions are concerned, the following theorem provides such a method, as well as a great deal more.

Theorem 6-3 *Let f be continuous on $[a, b]$. Then f has an antiderivative F on $[a, b]$, and*

$$F(b) - F(a) = \int_a^b f(x)\, dx \tag{6-5}$$

This theorem is known as the *Fundamental Theorem of Calculus*; it would be difficult to overemphasize its importance.

In the first place, as we have already indicated, one consequence of this theorem is a fairly effective method for computing the integrals of continuous functions.   Namely, Eq. (6-5) reduces the problem of finding integrals to the problem of finding antiderivatives.   This means that while we were learning some techniques of antidifferentiation in Chapter 5 we were, at the same time, learning techniques of integration.   In this connection, we will denote $F(b) - F(a)$ by $F(x)|_a^b$ and rewrite Eq. (6-5) as

$$F(x)\bigg|_a^b = \int_a^b f(x)\,dx$$

But it is the theoretical consequences of Theorem 6-3 that justify its name.   On the one hand, if $a < x < b$, we can conclude that $F(x) - F(a) = \int_a^x f(y)\,dy$.   If we differentiate both sides of this equation with respect to x, we get

$$f(x) = \frac{d}{dx}\left[\int_a^x f(y)\,dy\right]$$

On the other hand, since f is an antiderivative of Df, it follows from Eq. (6-5) that if Df is continuous, then

$$f(x) - f(a) = \int_a^x \frac{d}{dy} f(y)\,dy$$

Thus the Fundamental Theorem means that the integral and the derivative are inverses of one another in the sense that the effect of one is to undo the effect of the other.   In this way the two operations that form the body of calculus are linked together.   It is Theorem 6-3 that makes calculus one subject—without it we would have a "differential" calculus and an "integral" calculus, and the one would have little to do with the other.

EXAMPLE 6-2   We can now check the calculation of Example 6-1.   Since $x^2/2$ is an antiderivative of x, it follows that

$$\int_0^1 x\,dx = x^2/2\bigg|_0^1 = \tfrac{1}{2} - 0 = \tfrac{1}{2}$$

EXERCISE 6-2   Use Theorem 6-3 to check your answer in Exercise 6-1.

Inasmuch as the two very important Theorems 6-2 and 6-3 both impose the hypothesis that the function involved should be continuous on a closed interval, it is natural to ask what can be done about the integrals of functions that do not satisfy this condition.   One investigation along such lines leads to what are called *improper integrals.*

There are two basic improper integrals; the others can be developed from them.   In the first we permit the function to have a discontinuity at one end of an interval. That is, we assume that f is continuous on either $(a, b]$, or $[a, b)$. If f is continuous on $(a, b]$, say, and if $a < t < b$, then $\int_t^b f(x)\, dx$ exists because f is continuous on $[t, b]$. If

$$\lim_{t \to a^+} \int_t^b f(x)\, dx$$

exists and equals I, we say that the improper integral is *convergent*, and we put

$$\int_a^b f(x)\, dx = I$$

If $\lim_{t \to a^+} \int_t^b f(x)\, dx$ does not exist, then we say that $\int_a^b f(x)\, dx$ is *divergent*. Similarly, when f is continuous on $[a, b)$, we examine $\lim_{t \to b^-} \int_a^t f(x)\, dx$. If this limit exists, we set its value equal to $\int_a^b f(x)\, dx$; if the limit fails to exist, the improper integral diverges. The second type of improper integral deals with functions that are continuous on intervals of the form $[a, \infty)$, or $(-\infty, a]$. Suppose for definiteness that f is continuous on $[a, \infty)$. If $a < t$, then f is continuous on $[a, t]$, and so $\int_a^t f(x)\, dx$ exists. If

$$\lim_{t \to \infty} \int_a^t f(x)\, dx$$

exists and equals I, then $\int_a^\infty f(x)\, dx$ is *convergent* and we put

$$\int_a^\infty f(x)\, dx = I$$

If $\lim_{t \to \infty} \int_a^t f(x)\, dx$ fails to exist, then $\int_a^\infty f(x)\, dx$ is *divergent*. In the same way, if f is continuous on $(-\infty, a]$, then the improper integral $\int_{-\infty}^a f(x)\, dx$ is convergent or divergent according to whether $\lim_{t \to -\infty} \int_t^a f(x)\, dx$ exists or not. When this limit exists, its value is the value of the improper integral.

We can now define the other improper integrals. Let us first take f to be continuous on $[a, b]$ except at c, where $a < c < b$. In this case we say that $\int_a^b f(x)\, dx$ converges if *both* the improper integrals $\int_a^c f(x)\, dx$ and $\int_c^b f(x)\, dx$ are convergent, and we put $\int_a^b f(x)\, dx = \int_a^c f(x)\, dx + \int_c^b f(x)\, dx$. If one, or both, of these improper integrals diverges, then so does $\int_a^b f(x)\, dx$. By combining such improper integrals we can examine the convergence or divergence of the improper integral of a function having a finite number of discontinuities in the interior, or at the endpoints, of an interval.   Next, let us suppose that f is continuous on an interval of the form (a, ∞), or $(-\infty, a)$. If the first of these is the case, take $a < b < \infty$.   If *both* the improper integrals $\int_a^b f(x)\, dx$ and $\int_b^\infty f(x)\, dx$ converge, then $\int_a^\infty f(x)\, dx$ is convergent, and its value is the sum of their values.   If one, or both, diverges, then so does $\int_a^\infty f(x)\, dx$.   The convergence or divergence of $\int_{-\infty}^a f(x)\, dx$ is similarly defined in the case where f is continuous on $(-\infty, a)$.   Finally, if f is continuous on $(-\infty, \infty)$, we pick a number c and determine whether $\int_{-\infty}^c f(x)\, dx$ and $\int_c^\infty f(x)\, dx$ are convergent or divergent.   If *both* are convergent, we say that $\int_{-\infty}^\infty f(x)\, dx$ is convergent, and we put $\int_{-\infty}^\infty f(x)\, dx = \int_{-\infty}^c f(x)\, dx + \int_c^\infty f(x)\, dx$; otherwise, $\int_{-\infty}^\infty f(x)\, dx$ is divergent.

EXAMPLE 6-3   Since the function $f(x) = 1/(x-1)^{1/2}$ is discontinuous at 1, the integral $\int_1^2 [1/(x-1)^{1/2}]\, dx$ is improper.   If $1 < t < 2$, then $\int_t^2 [1/(x-1)^{1/2}]\, dx = 2(x-1)^{1/2}|_t^2 = 2[1 - (t-1)^{1/2}]$.   Inasmuch as $\lim_{t \to 1+} 2[1 - (t-1)^{1/2}] = 2$, it follows that $\int_1^2 [1/(x-1)^{1/2}]\, dx = 2$.

EXAMPLE 6-4   On the other hand, consider $\int_2^\infty [1/(x-1)^{1/2}]dx$. If $t > 2$, then $\int_2^t 1/(x-1)^{1/2}\, dx = 2[(t-1)^{1/2} - 1]$ and, since

$$\lim_{t \to \infty} 2[(t-1)^{1/2} - 1] = \infty$$

it follows that $\int_2^\infty [1/(x-.1)^{1/2}]\, dx$ is divergent.   This means that even though $\int_1^2 [1/(x-1)^{1/2}]\, dx$ converges, we must regard $\int_1^\infty [1/(x-1)^{1/2}]\, dx$ as divergent.

EXERCISES

6-3   Determine if

$$\int_0^1 \frac{\log x}{x^{1/2}}\, dx$$

is convergent or divergent.   Find its value if it converges.

6-4   Do the same for $\int_2^\infty [1/(x^2 - x)]\, dx$.

The applications of the integral are more varied than the applications of the derivative.   The reason for this is the large number of quantities that can be approximated by sums. For instance, the area under a curve can be approximated by a sum of areas of rectangles, the length of a curve can be approximated by a sum of lengths of line segments, and so forth. If the approximating sum is properly constructed, it can be viewed as a Riemann sum of some continuous function. Consequently, when the approximation is made better by inserting more terms in the sum, the sum should also approach an integral. This integral can then be defined to be the quantity being approximated.

The following list of formulas will give some indication of the diversity of the applications of the integral.

The area A of the region R under the graph of a positive continuous function f between a and b, $a < b$, is given by

$$A = \int_a^b f(x)\,dx \tag{6-6}$$

When this region R is rotated about the x axis, the volume V_x of the resulting solid is given by

$$V_x = \pi \int_a^b f^2(x)\,dx \tag{6-7}$$

When R is rotated about the y axis, the volume V_y of the resulting solid is given by

$$V_y = 2\pi \int_a^b x f(x)\,dx \tag{6-8}$$

provided that either $a \geq 0$, or $b \leq 0$ (that is, the axis of rotation should not pass through R). Next, denote by C the graph of the function f. If f is differentiable and f' is continuous on $[a, b]$, then the length L of the curve C is given by

$$L = \int_a^b (1 + [f'(x)]^2)^{1/2}\,dx \tag{6-9}$$

(this formula is valid even if f is not a positive function). If the curve C is rotated about the x axis, then the surface area S_x of the resulting surface of revolution is given by

$$S_x = 2\pi \int_a^b f(x)(1 + [f'(x)]^2)^{1/2}\,dx \tag{6-10}$$

If C is rotated about the y axis, then the surface area S_y of the resulting surface of revolution is given by

$$S_y = 2\pi \int_a^b x(1 + [f'(x)]^2)^{1/2}\,dx \tag{6-11}$$

provided that either $a \geq 0$, or $b \leq 0$.

Equations (6-7) and (6-8) are actually instances of a more general formula. If a solid lies between a and b on the x axis, and its cross-sectional area at x is $A(x)$, where A is a continuous function, then its volume V is given by

$$V = \int_a^b A(x)dx \qquad (6\text{-}12)$$

[Notice that when A is a constant function, Eq. (6-12) becomes: volume equals area of base multiplied by height. In other words, Eq. (6-12) generalizes the elementary definition of volume.]

In addition to the above applications, integrals can also be used to find centers of mass, moments of inertia, work, fluid pressure, and so forth. The reader interested in such applications of the integral will find them in most calculus books or physics texts that assume a knowledge of calculus.

Formulas for area and arc length can also be obtained when the region, or curve, is specified in terms of polar coordinates. Suppose that the region R is bounded by the lines $\theta = \alpha$, $\theta = \beta$, and by the curve C which is the graph of $\rho = f(\theta)$. Then the area A of R can be obtained from

$$A = \frac{1}{2} \int_\alpha^\beta f^2(\theta)\, d\theta \qquad (6\text{-}13)$$

while the length L of C is given by

$$L = \int_\alpha^\beta \{f^2(\theta) + [f'(\theta)]^2\}^{1/2}\, d\theta \qquad (6\text{-}14)$$

There is a variant of Eq. (6-14) that is occasionally useful. If C is the graph of $\theta = g(\rho)$ for ρ between ρ_1 and ρ_2, then

$$L = \int_{\rho_1}^{\rho_2} (\rho^2[g'(\rho)]^2 + 1)^{1/2}\, d\rho \qquad (6\text{-}15)$$

More generally, if the coordinates of the curve C are given by the parametric equation $x = g(t)$ and $y = h(t)$, if $a = g(t_1)$ and $b = g(t_2)$, then the area A of the region bounded by C, the lines $x = a$, $x = b$, and the x axis are given by

$$A = \left| \int_{t_1}^{t_2} h(t)g'(t)\, dt \right| \qquad (6\text{-}16)$$

provided that g and h have continuous derivatives, and g is strictly increasing, all on $[t_1, t_2]$. The length L of C between a and b is given by

$$L = \int_{t_1}^{t_2} \{[h'(t)]^2 + [g'(t)]^2\}^{1/2}\, dt \qquad (6\text{-}17)$$

provided that g and h have continuous derivatives on $[t_1, t_2]$. It can be seen that, apart from the absolute value signs, Eq. (6-16) is what would be obtained from Eq. (6-6) by means of the substitution $x = g(t)$. Eq. (6-17) can likewise be obtained formally from Eq. (6-9). By means of such formal substitutions parametric versions of other of the Eqs. (6-6)–(6-11) can be derived; the conditions needed to ensure the validity of the resulting formulas are of the sort that follow Eq. (6-16) or Eq. (6-17).

Lastly, certain of the formulas in the above list can be adapted to figures or curves that are unbounded provided that the resulting improper integral converges. In this way it is possible to speak of the area of an unbounded region, the volume of an unbounded solid, and so forth.

EXAMPLE 6-5 We will show that defining area and length by means of integrals is consistent with the elementary notions of area and length as far as the area and the circumference of a circle are concerned. A semicircle of radius r can be described analytically as the graph of $f(x) = (r^2 - x^2)^{1/2}$ for x in $[-r, r]$. The area A of this semicircle can be found by means of Eq. (6-6). Indeed, $A = \int_{-r}^{r} (r^2 - x^2)^{1/2} \, dx$. By a result of Example 5-8 we have

$$A = (r^2/2) \sin^{-1}(x/r) + (x/2)(r^2 - x^2)^{1/2} |_{-r}^{r}$$

$$= (r^2/2) \sin^{-1} 1 - (r^2/2) \sin^{-1}(-1) = \pi r^2/4 + \pi r^2/4 = \pi r^2/2$$

As for the length of this semicircle, instead of using Eq. (6-9), we will express the semicircle parametrically and use Eq. (6-17). The parametric equations are $x = r \cos \theta$ and $y = r \sin \theta$ for $0 \le \theta \le \pi$. Thus $dx/d\theta = -r \sin \theta$, $dy/d\theta = r \cos \theta$ and $(dx/d\theta)^2 + (dy/d\theta)^2 = r^2 \cos^2 \theta + r^2 \sin^2 \theta = r^2$. Therefore, Eq. (6-17) becomes $L = \int_0^{\pi} r \, d\theta = \pi r$.

Remark This computation of the perimeter of a semicircle illustrates how effective the introduction of parametric equations can be. Had we not done this, the computation of L by means of Eq. (6-9) would have led to the more complicated integral $\int_{-r}^{r} [r/(r^2 - x^2)^{1/2}] \, dx$ (which is even improper). We could also have used Eq. (6-16) in place of Eq. (6-6) to calculate A. Actually, the easiest way to compute the area and circumference of a circle is to observe that in polar coordinates the circle is the graph of $\rho = r$ for $0 \le \theta < 2\pi$, and use Eqs. (6-13) and (6-14).

In case extreme difficulties arise in the application of any of the above formulas, the reader should keep in mind the fact that sometimes great simplifications can result from a change of coordinates, or the introduction of parametric equations.

EXERCISES

6-5 Find the area of the region bounded by the graph of $f(x) = (a/2)(e^{x/a} + e^{-x/a})$, where $a > 0$ (this graph is called a *catenary*), the x axis, and the lines $x = a$, $x = 0$.

6-6 Find the volume of the solid generated when the region of Exercise 6-5 is rotated about the x axis, the y axis.

6-7 Find the length of the catenary $y = (a/2)(e^{x/a} + e^{-x/a})$ from $x = 0$ to $x = a$.

6-8 Find the area of the surface generated by rotating the curve of Exercise 6-7 about the x axis; about the y axis.

Our final application of the integral brings us full-circle. We began this chapter by defining the integral as the limit of sequences. We conclude it by evaluating limits of sequences as integrals. An example will illustrate the details of this procedure.

EXAMPLE 6-6 Consider the interval $[a, b]$, where $a > 0$. Let n be a positive integer, and let $a = x_0, x_1, x_2, \ldots, x_n = b$ be those numbers that partition $[a, b]$ into n subintervals of equal lengths. We form the geometric mean

$$G_n = \left[\prod_{k=0}^{n} x_k \right]^{1/(n+1)}$$

of these numbers, and enquire about the limit of $\{G_n\}$. We attack this problem by observing that

$$\log G_n = \frac{1}{n+1} \sum_{k=0}^{n} \log x_k$$

Now this sum looks like a Riemann sum. If we notice that $x_k - x_{k-1} = (b - a)/n$, we can write

$$\log G_n = \frac{\log a}{n+1} + \frac{n}{n+1} \frac{1}{b-a} \sum_{k=1}^{n} \log x_k (x_k - x_{k-1})$$

and the sum that appears is a Riemann sum of the function $f(x) = \log x$ on the interval $[a, b]$. Since the norm of the underlying partition is $(b - a)/n$, and $\{(b - a)/n\} \to 0$, we can conclude, in view of the obvious relations $\{\log a/(n + 1)\} \to 0$ and $\{n/(n + 1)\} \to 1$, that

$$\{\log G_n\} \to \frac{1}{b-a} \int_a^b \log x \, dx$$

The fact that the exponential function is continuous means that

$$\{G_n\} \to e^{\frac{1}{b-a} \int_a^b \log x \, dx}$$

But (see Example 5-5)

$$\frac{1}{b-a} \int_a^b \log x \, dx = \frac{1}{b-a} (x \log x - x \Big|_a^b$$

$$= \frac{1}{b-a} [b \log b - b - a \log a + a]$$

$$= \log \left[\frac{b^b}{a^a}\right]^{1/(b-a)} - 1$$

Consequently,

$$\{G_n\} \to e^{\log\left[\frac{b^b}{a^a}\right]^{1/(a-b)} - 1} = \frac{1}{e} \left[\frac{b^b}{a^a}\right]^{1/(b-a)}$$

EXERCISE 6-9 Let $[a, b]$ and $x_1, x_1, \ldots, x_n$ be as in Example 6-6. Let

$$A_n = \frac{1}{n+1} \sum_{k=0}^n x_k$$

and

$$H_n = \frac{n+1}{\sum_{k=0}^n \frac{1}{x_k}}$$

be the arithmetic and harmonic means of $x_0, x_1, \ldots, x_n$ respectively. Find the limits of $\{A_n\}$ and $\{H_n\}$.

SOLUTIONS TO EXERCISES

6-1 If we partition $[0, 2]$ into n equal subintervals, then the partition consists of $0, 2/n, 2(2/n), \ldots, n(2/n)$. A Riemann sum R_n for x^2 on $[0, 2]$ relative to this partition is, in the notation of the remark following the solution of Problem 1-5,

$$R_n = \sum_{k=1}^n \left(k\frac{2}{n}\right)^2 \frac{2}{n} = \frac{8}{n^3} \sum_{k=1}^n k^2 = \frac{8}{n^3} S_n^2$$

$$= \frac{8}{n^3} \frac{n(n+1)(2n+1)}{6} = \frac{4}{3}\left(1 + \frac{1}{n}\right)\left(2 + \frac{1}{n}\right)$$

From the last equation it can be seen that $\{R_n\} \to (4/3)1 \cdot 2 = 8/3$. Consequently, $\int_0^2 x^2 \, dx = 8/3$.

6-2 It follows from Theorem 6-3 that $\int_0^2 x^2 \, dx = x^3/3 \Big|_0^2 = 8/3$.

6-3 The function $f(x) = \log x/x^{1/2}$ is discontinuous at zero. Consequently, we have to determine whether or not $\lim\limits_{t \to 0+} \int_t^1 [\log x / x^{1/2}] \, dx$ exists.   Now

$\int_t^1 [\log x / x^{1/2}] \, dx = 2x^{1/2} \log x \Big|_t^1 - 2 \int_t^1 x^{-1/2} \, dx$ (we have integrated by

parts) $= 2x^{1/2} \log x \Big|_t^1 - 4x^{1/2} \Big|_t^1 = -4 + 4t^{1/2} - 2t^{1/2} \log t$.   We know, from

Problem 4-46a, that $\lim\limits_{t \to 0+} [-2t^{1/2} \log t] = 0$,   and clearly $\lim\limits_{t \to 0} 4t^{1/2} = 0$.

Therefore, $\int_0^1 [\log x / x^{1/2}] \, dx = -4$.

6-4   Here we are interested in $\lim\limits_{t \to \infty} \int_2^t [1/(x^2 - x)] \, dx$.   Since $1/(x^2 - x) = -1/x + 1/(x - 1)$, we have

$$\int_2^t [1/(x^2 - x)] \, dx = -\int_2^t (1/x) \, dx + \int_2^t [1/(x - 1)] \, dx$$

$$= -\log x \Big|_2^t + \log (x - 1) \Big|_2^t = \log 2 - \log t + \log (t - 1)$$

$$= \log 2 + \log [(t - 1)/t] = \log 2 + \log (1 - 1/t)$$

Now $\lim\limits_{t \to \infty} \log (1 - 1/t) = \log 1 = 0$.   Thus $\int_2^\infty [1/(x^2 - x)] \, dx = \log 2$.

6-5   The area is

$$\int_0^a (a/2)[e^{x/a} + e^{-x/a}] \, dx = (a^2/2)[e^{x/a} - e^{-x/a}] \Big|_0^a$$

$$= (a^2/2)[e - 1/e - e^0 + e^0] = (a^2/2)[e - 1/e]$$

6-6 $V_x = \pi \int_0^a (a^2/4)[e^{x/a} + e^{-x/a}]^2 \, dx$

$$= (\pi a^2/4) \int_0^a [e^{2x/a} + 2 + e^{-2x/a}] \, dx$$

$$= (\pi a^2/4) \left[(a/2)e^{2x/a} + 2x - (a/2)e^{-2x/a} \right] \Big|_0^a$$

$$= (\pi a^2/4)[(a/2)(e^2 - e^0 - e^{-2} + e^0) + 2a] = (\pi a^3/8)[e^2 + 4 + e^{-2}]$$

$V_y = \pi \int_0^a x(a/2)[e^{x/a} + e^{-x/a}] \, dx$

$$= (\pi a/2) \int_0^a xe^{x/a} \, dx + (\pi a/2) \int_0^a xe^{-x/a} \, dx$$

$$= (\pi a/2) \left[xae^{x/a} \Big|_0^a - a \int_0^a e^{x/a} \, dx \right]$$

$$+ (\pi a/2) \left[-xae^{-x/a} \Big|_0^a + a \int_0^a e^{-x/a} \, dx \right]$$

(we have integrated by parts)

$$= (\pi a^3/2)e - (\pi a^3/2)e^{x/a}\Big|_0^a - (\pi a^3/2)e^{-1} - (\pi a^3/2)e^{-x/a}\Big|_0^a$$

$$= \pi a^3 - \pi a^3 e^{-1} = \pi a^3 [1 - e^{-1}]$$

6-7   The length we want is

$$\int_0^a \left[1 + \left(\frac{e^{x/a} - e^{-x/a}}{2} \right)^2 \right]^{1/2} dx$$

$$= (1/2) \int_0^a [(e^{x/a} + e^{-x/a})^2]^{1/2} \, dx = (1/2) \int_0^a (e^{x/a} + e^{-x/a}) \, dx$$

$$= (a/2)[e - e^{-1}]$$

The last integral was evaluated in the solution of Exercise 6-5.

6-8   About the x axis the surface area is

$$2\pi \int_0^a \frac{a}{2} (e^{x/a} + e^{-x/a}) \left[1 + \left(\frac{e^{x/a} - e^{-x/a}}{2} \right)^2 \right]^{1/2} dx$$

If we use the calculation in the solution of Exercise 6-7, this becomes $(\pi a/2) \int_0^a (e^{x/a} + e^{-x/a})^2 \, dx = (\pi a^2/4)[e^2 + 4 + e^{-2}]$. The last equation is from the solution of Exercise 6-6.   About the y axis the surface area is

$$2\pi \int_0^a x \left[1 + \left(\frac{e^{x/a} - e^{-x/a}}{2} \right)^2 \right]^{1/2} dx$$

$$= \pi \int_0^a x(e^{x/a} + e^{-x/a}) \, dx \text{ [again by the calculation in the solution of Exercise}$$
6-7] $= 2\pi a^2 (1 - e^{-1})$.   The last equation is also from the solution of Exercise 6-6.

6-9   As in Example 6-6, we may write

$$A_n = \frac{a}{n+1} + \frac{n}{n+1} \frac{1}{b-a} \sum_{k=1}^{n} x_k (x_k - x_{k-1})$$

Since the sum is a Riemann sum of the function $f(x) = x$ on $[a, b]$, it follows that

$$\{A_n\} \to [1/(b-a)] \int_a^b x \, dx = [1/(b-a)](x^2/2)\Big|_a^b$$

$$= (1/2)(b^2 - a^2)/(b-a) = (b+a)/2$$

(Note here that it is not necessary to require that $a > 0$.)   In the same way

$$\frac{1}{H_n} = \frac{1}{a} \frac{1}{n+1} + \frac{n}{n+1} \frac{1}{b-a} \sum_{k=1}^{n} \frac{1}{x_k} (x_k - x_{k-1})$$

and the sum is a Riemann sum of the function $f(x) = 1/x$ on $[a, b]$. Therefore,

$$\left\{\frac{1}{H_n}\right\} \to \frac{1}{b-a} \int_a^b \frac{1}{x} \, dx = \frac{\log b - \log a}{b - a}$$

By the quotient limit theorem for sequences, we can conclude that

$$\{H_n\} \to \frac{b - a}{\log b - \log a}$$

PROBLEMS

6-1 Find $\displaystyle\int_{-1}^{3} (x^3 + 2x) \, dx$ by following the method of Example 6-1.

*6-2 Find $\displaystyle\int_a^b x^p \, dx$ where p is a positive integer and $0 < a < b$ by putting $(b/a)^{1/n} = r$, partitioning $[a, b]$ by the numbers $a, ar, ar^2, \ldots, ar^{n-1}, ar^n$, forming the Riemann sum associated with this partition wherein x^p is evaluated at the left-hand endpoint of each subinterval, and then taking the limit of these sums as $n \to \infty$.

*6-3 Find $\displaystyle\int_a^b \sin x \, dx$, where $a < b$, by finding the limit of the Riemann sums

$$R_h = h \sum_{k=1}^{n} \sin (a + kh)$$

where $h = (b - a)/n$, as $n \to \infty$.

6-4 (a) If m and n are positive integers, show that

$$\int_0^{2\pi} \cos (mx) \sin (nx) \, dx = 0$$

$$\int_0^{2\pi} \cos (mx) \cos (nx) \, dx = \begin{cases} 0 \text{ if } m \neq n \\ \pi \text{ if } m = n \end{cases}$$

and

$$\int_0^{2\pi} \sin (mx) \cos (nx) \, dx = \begin{cases} 0 \text{ if } m \neq n \\ \pi \text{ if } m = n \end{cases}$$

(b) Consequently, show that if

$$f(x) = \frac{a_0}{2} + \sum_{m=1}^{n} a_m \cos (mx) + \sum_{m=1}^{n} b_m \sin (mx)$$

then $\displaystyle\int_0^{2\pi} f(x) \, dx = \pi a_0$, $\displaystyle\int_0^{2\pi} \cos (kx) f(x) \, dx = \pi a_k$, and

$$\int_0^{2\pi} \sin (kx) f(x) \, dx = \pi b_k$$ when k is a positive integer $\leq n$. If $k > n$, the value of each of the last two integrals is zero.

6-5 Show that

(a) $\displaystyle\int_{-a}^{a} f(x^2)\, dx = 2\int_{0}^{a} f(x^2)\, dx$

(b) $\displaystyle\int_{-a}^{a} xf(x^2)\, dx = 0$

(The integrals here, and in the next three problems, are assumed to exist.)

6-6 Show that $\displaystyle\int_{a}^{b} F(x)\, dx = \int_{a}^{b} F(a+b-x)\, dx$

*6-7 (a) If $f_1(x) = \displaystyle\int_{0}^{x} f(t)\, dt$ and $f_2(t) = \displaystyle\int_{0}^{x} f_1(t)\, dt$, show that $f_2(x) = \displaystyle\int_{0}^{x} (x-t)f(t)\, dt$. More generally, if $f_k(x) = \displaystyle\int_{0}^{x} f_{k-1}(t)\, dt$ for $k = 2$, 3, ..., then

$$f_k(x) = \frac{1}{(k-1)!}\int_{0}^{x} (x-t)^{k-1} f(t)\, dt$$

(b) As an application, if m and n are positive integers, show that

$$\int_{0}^{1} (1-x)^n x^m\, dx = \frac{m!\,n!}{(m+n)!}$$

*6-8 (a) If $f(x) \le g(x)$ for x in $[a, b]$, show that

$$\int_{a}^{b} f(x)\, dx \le \int_{a}^{b} g(x)\, dx$$

(b) Show that

$$\left|\int_{a}^{b} f(x)\, dx\right| \le \int_{a}^{b} |f(x)|\, dx$$

6-9 Show that

$$\int_{-1}^{1} \frac{1}{(1-2\alpha x + \alpha^2)^{1/2}}\, dx = \begin{cases} 2 \text{ if } |\alpha| \le 1 \\ 2/|\alpha| \text{ if } |\alpha| > 1 \end{cases}$$

6-10 If $U_n = \displaystyle\int_{0}^{\pi/2} \sin^n x\, dx$, where n is a positive integer, show that

$$U_n = \frac{n-1}{n} U_{n-2}$$

Deduce that

$$U_n = \begin{cases} \dfrac{2\cdot 4\cdot 6\cdots (n-1)}{3\cdot 5\cdot 7\cdots\ \ n} & \text{if } n \text{ is odd} \\[2ex] \dfrac{\pi}{2}\dfrac{1\cdot 3\cdot 5\cdots (n-1)}{2\cdot 4\cdot 6\cdots\ \ n} & \text{if } n \text{ is even} \end{cases}$$

*6-11 If $n > 1$, show that

$$\frac{1}{2} < \int_0^{1/2} \frac{dx}{(1 - x^{2n})^{1/2}} < .524$$

*6-12 Show that if $0 < \alpha < \pi/2$, and $0 < \phi < \pi/2$, then

$$\phi < \int_0^\phi \frac{dx}{(1 - \sin^2 \alpha \sin^2 x)^{1/2}} < \frac{\phi}{(1 - \sin^2 \alpha \sin^2 \phi)^{1/2}}$$

*6-13 Show that $\lim\limits_{x \to \infty} e^{-x^2} \int_0^x e^{t^2} \, dt = 0$

6-14 Show that

$$\lim\limits_{T \to \infty} \frac{1}{T} \int_0^T \sin \alpha x \sin \beta x \, dx = 0$$

if $|\alpha| \neq |\beta|$. What is the limit if $|\alpha| = |\beta|$?

*6-15 (a) If a is a fixed, positive number, show that

$$\lim\limits_{h \to 0} \int_{-a}^a \frac{h}{h^2 + x^2} \, dx = \pi$$

(b) If f is continuous on $[-a, a]$, show further that

$$\lim\limits_{h \to 0} \int_{-a}^a \frac{h}{h^2 + x^2} f(x) \, dx = \pi f(0)$$

*6-16 (a) Show that $\int_0^\infty f(x) \, dx$ converges if $\int_0^\infty |f(x)| \, dx$ does.

(b) If $0 \leq f(x) \leq g(x)$ for all $x \geq 0$, and $\int_0^\infty g(x) \, dx$ converges, prove that $\int_0^\infty f(x) \, dx$ converges.

6-17 Determine which of the following improper integrals are convergent and which are divergent. Evaluate each convergent integral.

(a) $\displaystyle\int_0^{\pi/2} \sec x \, dx$ (b) $\displaystyle\int_0^1 x \log x \, dx$

(c) $\displaystyle\int_0^a \frac{1}{(a^2 - x^2)^{1/2}} \, dx$ *(d) $\displaystyle\int_0^1 \frac{1}{e^x + e^{-x}} \, dx$

6-18 Follow the instructions of Problem 6-17 for the following improper integrals:

(a) $\displaystyle\int_0^\infty e^{-x} \sin x \, dx$ (b) $\displaystyle\int_0^\infty \frac{x}{1 + x^2} \, dx$

(c) $\displaystyle\int_0^\infty \frac{1}{a^2 + b^2 x^2} \, dx$ (d) $\displaystyle\int_e^\infty \frac{1}{x \log x} \, dx$

6-19 Follow the instructions of Problem 6-17 for the improper integrals:

(a) $\displaystyle\int_0^{3a} \frac{2x}{(x^2 - a^2)^{2/3}} \, dx$ (b) $\displaystyle\int_0^{2a} \frac{dx}{(a - x^2)}$

6-20 Follow the instructions of Problem 6-17 for the improper integrals:

(a) $\displaystyle\int_0^\infty \frac{1}{(x+1)x^{1/2}}\,dx$ (b) $\displaystyle\int_a^\infty \frac{x}{(x^2-a^2)^{1/2}}\,dx$

6-21 Follow the instructions of Problem 6-17 for the improper integrals:

(a) $\displaystyle\int_{-\infty}^\infty \sin x\,dx$ *(b) $\displaystyle\int_{-\infty}^\infty \frac{1}{e^x+e^{-x}}\,dx$

6-22 Criticize:

(a) $\displaystyle\int_{-a}^a \frac{dx}{x} = \log|x|\ \Big|_{-a}^a = \log|a| - \log|-a| = 0$

(b) $\displaystyle\int_{-\infty}^\infty \frac{4x^3}{1+x^4}\,dx = \lim_{t\to\infty}\int_{-t}^t \frac{4x^3}{1+x^4}\,dx$

$$= \lim_{t\to\infty}\log(1+x^4)\ \Big|_{-t}^t$$

$$= \lim_{t\to\infty}[\log(1+t^4) - \log(1+(-t)^4] = 0$$

6-23 If $a>0$, then for a certain value of α the integral

$$\int_0^\infty \left(\frac{1}{(1+ax^2)^{1/2}} - \frac{\alpha}{(x+1)}\right)dx$$

is convergent. Find this α and evaluate the resulting integral.

6-24 For which values of $\alpha > 0$ are

(a) $\displaystyle\int_1^\infty \frac{dx}{x^\alpha}$ (b) $\displaystyle\int_0^1 \frac{dx}{x^\alpha}$

convergent?

*6-25 For which positive integers n is $\displaystyle\int_0^\infty x^n e^{-x}\,dx$ convergent?

6-26 For which values of $\alpha > 0$ is

$$\int_0^\infty \frac{x^{\alpha-1}}{1+x}\,dx$$

convergent?

*6-27 Show that

$$\lim_{\alpha\to\infty}\int_0^\infty \frac{dx}{1+\alpha x^{1+\delta}} = 0$$

for $\delta > 0$.

*6-28 Show that

$$\int_0^\infty \frac{\sin x}{x^{1/2}}\,dx$$

is convergent.

6-29 Find the area included between the two parabolas $y^2 = ax$ and $x^2 = by$.

6-30 Find the area of the region bounded by the rectangular hyperbola $x^2 - y^2 = a^2$, the x axis, and the line drawn from the origin to the point (x_0, y_0) on the hyperbola.

6-31 Find the area enclosed by the loop of the curve whose equation is $y^2 = x(x - 2)^2$.

6-32 Find the area enclosed by the loop of the curve whose equation is $4y^2 = x^4(4 - x)$.

6-33 Find the area bounded by one arch of the cycloid $x = a(\theta - \sin \theta)$, $y = a(1 - \cos \theta)$ and the x axis.

*6-34 Find the area bounded by one arch of the "companion to the cycloid" $x = a\theta$, $y = a(1 - \cos \theta)$ and the x axis.

*6-35 Find the area of the ellipse whose semi-axes are a and b.

*6-36 Find the area of the hypocycloid $x^{2/3} + y^{2/3} = a^{2/3}$.

6-37 Find the area enclosed by the curve $\rho = a \cos 3\theta$.

6-38 Find the area enclosed by the cardioid $\rho = a(1 + \cos \theta)$.

6-39 Find the area bounded by the curves $\rho^2 = \cos 2\theta$ and $\rho^2 = \sin 2\theta$.

6-40 For the spiral of Archimedes, $\rho = a\theta$, what area is swept out, starting from $\theta = 0$, in
(a) 1 revolution, (b) 2 revolutions, (c) 3 revolutions

6-41 Find the volume of the solid generated by revolving the area bounded by the ellipse $b^2x^2 + a^2y^2 = a^2b^2$ about the x axis.

6-42 Find the volume of the solid generated by revolving the area within the hypocycloid $x^{2/3} + y^{2/3} = a^{2/3}$ about the x axis.

6-43 Find the volume generated by revolving the area bounded by one arch of the cycloid $x = a(\theta - \sin \theta)$, $y = a(1 - \cos \theta)$ and the x axis about the x axis.

6-44 Revolve the area bounded by the two parabolas $y^2 = 4x$ and $y^2 = 5 - x$ about each axis and find the respective volumes.

6-45 Find the volume of the solid generated when the area bounded by $y = e^{-x^2}$, $x = 0$, $x = 1/2^{1/2}$ and the x axis is revolved about the y axis.

6-46 A hole of radius r is drilled through the center of a sphere of radius R. Find the volume of the solid remaining.

6-47 Find the volume of the torus generated by revolving a circle of radius a about a line in its plane whose distance from the center is b, $b > a$.

6-48 Find the volume of the solid generated when the area bounded by the cycloid $x = a(\theta - \sin \theta)$, $y = a(1 - \cos \theta)$, the x axis and the lines $x = 0$, $x = 2\pi a$ is revolved about the y axis.

6-49 Two right circular cylinders of radius r have their axes meeting at right angles. Find the volume of the common part.

6-50 In felling a tree a lumberjack first saws halfway through at right angles to the trunk. He then makes a second cut in a plane inclined at an angle θ to the first cut, the two planes meeting in a line which intersects the axis of the tree. Find the volume of the wedge removed if the tree is assumed to be a cylinder of radius r.

6-51 Find the length of arc of $y = \log (\cos x)$ from $x = 0$ to $x = \pi/3$.

6-52 Find the length of the arc of the parabola $x^2 = 2py$ from the vertex to one end of the latus rectum.

*6-53 Find the length in one quadrant of the curve $(x/a)^{2/3} + (y/b)^{2/3} = 1$.

*6-54 Find the length between $x = a$ and $x = b$ of the curve $e^y = (e^x + 1)/(e^x - 1)$ where $b > a > 0$.

6-55 The equations of the involute of a circle are $x = a(\cos \theta + \theta \sin \theta)$, $y = a(\sin \theta - \theta \cos \theta)$. Find the length of the arc from $\theta = 0$ to $\theta = \alpha > 0$.

6-56 Find the length of one arch of the cycloid $x = a(\theta - \sin \theta)$, $y = a(1 - \cos \theta)$.

6-57 Find the perimeter of the cardiod $\rho = a(1 + \cos \theta)$.

6-58 Find the length of the spiral of Archimedes $\rho = a\theta$ from the origin to the end of the first revolution.

6-59 Find the length of the spiral $\rho = e^{a\theta}$ from the origin to (ρ_0, θ_0).

6-60 Find the area of the surface generated by revolving about the x axis the arc of the parabola $y^2 = 2px$ from $x = 0$ to $x = 4p$.

6-61 Find the area of the surface generated by revolving the ellipse $x^2/a^2 + y^2/b = 1$, $a > b$, about the x axis; about the y axis.

6-62 Find the area of the surface generated by revolving the loop of $9ay^2 = x(3a - x)^2$ about the x axis.

6-63 Find the area of the surface generated by revolving one arch of the cycloid $x = a(\theta - \sin \theta)$, $y = a(1 - \cos \theta)$ about the x axis.

6-64 Find the area of the region bounded by the graph of $f(x) = x^{-3/4}$, the x axis and the lines $x = 0$, $x = 1$. What is the volume of the solid generated by revolving this region about the x axis?

6-65 Find the volume of the solid generated when the region bounded by the graph of $f(x) = 1/x$, the x axis and lying to the right of the line $x = 1$ is revolved about the x axis. What is the surface area of this solid?

6-66   Find the volume of the solid generated when the region bounded by the graph of $f(x) = e^{-x^2}$, the x axis and the line $x = 0$ is revolved about the y axis.

*6-67   Find the area of the surface generated when the graph of $f(x) = e^{-x}$ from $x = 0$ to $x = \infty$ is revolved about the x axis.

6-68   Find the area of the surface generated by revolving the hypocycloid $x^{2/3} + y^{2/3} = a^{2/3}$ about the x axis.

*6-69   Find the area of the loop of the folium of Descartes, $x^3 + y^3 = 3axy$.

6-70   (a)   Show that

$$\left\{ \frac{1}{n} \sum_{k=0}^{n} \cos\left(\frac{k}{n} x\right) \right\} \to \frac{\sin x}{x}$$

(b)   Find the limit of

$$\left\{ \frac{1}{n^{\alpha+1}} \sum_{k=1}^{n} k^\alpha \right\}$$

when $\alpha > -1$.

(c)   Show that

$$\left\{ \sum_{k=0}^{n-1} \frac{n}{n^2 + k^2} \right\} \to \frac{\pi}{4}$$

(d)   Show that

$$\left\{ \left(\frac{n!}{n^n}\right)^{1/n} \right\} \to \frac{1}{e}$$

In all parts $n \to \infty$.

Chapter 7

INFINITE SERIES

The subject of infinite series is of great interest, extent, and importance in its own right, but one's first exposure to it is usually as an adjunct of calculus. We will follow convention with this chapter, but the reader will see that calculus does not make an essential appearance until power series are encountered. Nearly everything before depends only on the notion of the limit of a sequence; a notion that is a more elementary concept than those of "integral" and "derivative."

Let us begin with an arbitrary sequence $\{a_k\}$; this sequence will be called the sequence of *terms*. We can associate with $\{a_k\}$ the sequence $\{S_n\}$ defined by

$$S_n = \sum_{k=1}^{n} a_k$$

$\{S_n\}$ is called the sequence of *partial sums* generated by $\{a_k\}$.

Definition 7-1 *The infinite series*

$$\sum_{k=1}^{\infty} a_k$$

associated with the sequence $\{a_k\}$ *is identical with the sequence of partial sums* $\{S_n\}$.

In other words, an infinite series is simply a sequence that is formed from another sequence in a particular way. The notation

$$\sum_{k=1}^{\infty} a_k$$

does not mean that we have defined the sum of infinitely many numbers; it is only an alternate notation for the sequence of partial sums.

We use the symbol

$$\sum_{k=m}^{\infty} a_k$$

as an alternate notation for the sequence $\{T_n\}$, where

$$T_n = \sum_{k=m}^{m+n-1} a_k$$

Since

$$S_n = \sum_{k=1}^{m-1} a_k + \sum_{k=m}^{n} a_k$$

if $n > m$, we have

$$S_n = \sum_{k=1}^{m-1} a_k + T_{n-m+1}$$

In the summation notation for series, this last equation can be written as

$$\sum_{k=1}^{\infty} a_k = \sum_{k=1}^{m-1} a_k + \sum_{k=m}^{\infty} a_k$$

We now make the *notational convention* that, from now on, $\sum a_k$ *will stand for*

$$\sum_{k=1}^{\infty} a_k$$

In the event that it is necessary to start an infinite series with a_m instead of a_1, it will be specifically indicated.

EXAMPLE 7-1 Let us consider several sequences of terms $\{a_k\}$ and the infinite series associated with them. Take first $a_k = 1$ for all k; then $S_n = n$, and $\Sigma 1$ is the sequence $\{n\}$. If we put $a_k = (-1)^{k-1}$, then $S_1 = 1$, $S_2 = 1 - 1 = 0$, $S_3 = 1 - 1 + 1 = 1$, and it can be seen that

$$S_n = \begin{cases} 1 & \text{if } n \text{ is odd} \\ 0 & \text{if } n \text{ is even} \end{cases}$$

In other words, $\Sigma(-1)^{k-1}$ is the sequence $\{[1 + (-1)^{n-1}]/2\}$. If $a_k = k^2$, then

$$S_n = \sum_{k=1}^{n} k^2$$

and, as we know from the remark following the solution of Problem 1-5,

$$\sum_{k=1}^{n} k^2 = \frac{n(n+1)(2n+1)}{6}$$

Consequently, Σk^2 is the sequence $\{n(n+1)(2n+1)/6\}$. The examples up to here are somewhat misleading in that it is usually not possible to determine $\{S_n\}$ explicitly. For instance, let us modify the last sequence of terms somewhat and put $a_k = 1/k^2$. Then we have no formula for determining

$$\sum_{k=1}^{n} \frac{1}{k^2}$$

in general. Hence, even though we can calculate a given S_n, at least in principle, it does not seem that we can write down all the terms of the sequence $\{S_n\}$.

Remark The fact that it is not always possible, at least from a practical point of view, to calculate $\{S_n\}$ given $\{a_k\}$ means that, of necessity, the questions that arise in connection with infinite series must be dealt with by examining the sequence of terms rather than the sequence of partial sums. It is for this reason that we use the notation Σa_k as an alternate notation for $\{S_n\}$. On the other hand, if $\{S_n\}$ is known, then $\{a_k\}$ can be calculated from the relation

$$a_k = S_k - S_{k-1}$$

Inasmuch as an infinite series is a special sequence, the first question we should examine is whether or not it converges. The following definition will establish the requisite terminology.

Definition 7-2 *The series $\sum a_k$ converges if $\{S_n\}$ converges; if $\{S_n\}$ diverges, then so does Σa_k. Further, if $\{S_n\} \to S$, then Σa_k is said to have sum S. We denote this fact by putting*

$$\sum a_k = S \tag{7-1}$$

If follows from this definition and the relation

$$S_n = \sum_{k=1}^{m-1} a_k + T_{n-m+1}$$

if $n > m$, that $\Sigma a_k \begin{cases} \text{converges} \\ \text{diverges} \end{cases}$ if and only if

$$\sum_{k=m}^{\infty} a_k$$

$\begin{cases} \text{converges} \\ \text{diverges} \end{cases}$.

EXAMPLE 7-2 If we examine the series of Example 7-1, we see that $\Sigma 1$ diverges because $\{n\}$ diverges; $\Sigma(-1)^{k-1}$ diverges because $\{[1+(-1)^{n-1}]/2\}$

diverges; and Σk^2 diverges because $\{n(n+1)(2n+1)/6\}$ diverges.   We are not yet able to determine whether $\Sigma 1/k^2$ converges or diverges.

The problem then is how to determine if a given series converges or not. Now in Chapter 2 we considered the corresponding problem for arbitrary sequences, and we were able to develop techniques for attacking this more general question.   Most of the following results are based on adaptations of these general principles that take advantage of the fact that an infinite series is a rather special sequence.   At the same time, our experience with sequences should lead us to expect that we will not be able to determine the sum of most convergent series.

We will begin by stating some general properties of convergent series.

Theorem 7-1   *Let $\sum a_k$ and $\sum b_k$ converge and have sums S and T respectively*
(a)   *If c is a constant, then Σca_k converges, and its sum is cS.*
(b)   *$\Sigma(a_k \pm b_k)$ converges, and its sum is $S \pm T$.*

Theorem 7-1 is the beginning and end of the purely theoretical aspect of convergent series of constant terms at least at this first encounter.   Until we take up power series the rest of our account will be concerned with various tests for the convergence (or divergence) of series.   These tests bear the same relation to series as the techniques of Chapter 5 did to antiderivatives in that they too are only guidelines.   For this reason their effective use requires the same sort of persistence, flexibility, and foresight that the effective use of the techniques of antidifferentiation requires.

Theorem 7-2   *If $\sum a_k$ converges, then $\{a_k\} \to 0$.*

This theorem is the subject of a great deal of misapprehension.   Beginners tend to confuse it with its converse "If $\{a_k\} \to 0$, then Σa_k converges," which as we shall presently see happens to be a *false* statement.   In other words, the condition $\{a_k\} \to 0$ is necessary, but not sufficient, for the convergence of Σa_k.   To emphasize this, we reformulate Theorem 7-2 as follows: *If $\{a_k\}$ does not have limit zero, then Σa_k diverges.*   That is, Theorem 7-2 is a test for divergence, not convergence.

From the standpoint of convergence, we know that monotone sequences are particularly easy to treat.   For this reason the investigation of series begins with sequences $\{a_k\}$ such that $\{S_n\}$ is monotone.   If $\{S_n\}$ is increasing, we must have $0 \le S_n - S_{n-1} = a_n$, and, similarly, if $\{S_n\}$ is decreasing, then $a_n \le 0$.   This means that we can concentrate on series whose terms are either all nonnegative, or all nonpositive.   For definiteness we restrict ourselves to series Σa_k such that $a_k \ge 0$ for all k; the case $a_k \le 0$ for all k can be reduced to the case $a_k \ge 0$ by means of the series $\Sigma(-1)a_k$ and an application of Theorem 7-1a.

If $a_k \geq 0$ for all k, then to show that Σa_k converges it is sufficient to show that $\{S_n\}$ is bounded, and to show that Σa_k diverges it is sufficient to show that $\{S_n\}$ is unbounded. One way of doing this is given by the next theorem, which is called the *comparison test*.

Theorem 7-3 *If* $\sum c_k$ *is a convergent series of nonnegative terms, and if* $0 \leq a_k \leq c_k$ *for all* k, *then* Σa_k *converges. If* Σd_k *is a divergent series of nonnegative terms, and if* $a_k \geq d_k$ *for all* k, *then* Σa_k *diverges.*

It follows from the observation made after Definition 7-2 that the conclusions of the comparison test continue to hold even if we only have $0 \leq a_k \leq c_k$ for all $k \geq N$, or $a_k \geq d_k$ for all $k \geq N$, where N is a fixed positive integer.

Another test based on the same general property of monotone sequences is the *integral test*.

Theorem 7-4 *If* f *is continuous, positive, and decreasing on* $[1, \infty)$, *and if* $a_k = f(k)$, *then* Σa_k *converges if* $\int_1^\infty f(x)\,dx$ *converges, and* Σa_k *diverges if* $\int_1^\infty f(x)\,dx$ *diverges.*

In order to be able to use the comparison test, it is obvious that we must have at our disposal some series whose convergence or divergence is known. It so happens that we have already developed enough machinery to settle the convergence question for two classes of series of nonnegative terms.

The first class consists of all series of the form $\Sigma 1/n^\alpha$, where α is any positive number. Now the function f defined on $[1, \infty)$ by $f(x) = 1/x^\alpha$ satisfies the conditions of the integral test. Consequently, $\Sigma 1/n^\alpha$ converges or diverges with $\int_1^\infty 1/x^\alpha\,dx$. But we know from Problem 6-24a that this integral converges if $\alpha > 1$, diverges if $\alpha \leq 1$. In other words, $\Sigma 1/n^\alpha$ converges if $\alpha > 1$, diverges if $\alpha \leq 1$.

We can use this fact to shed more light on Theorem 7-2. The series $\Sigma 1/n$ is known as the *harmonic series* (because its terms form a harmonic progression). Clearly $\{1/n\} \to 0$, but $\Sigma 1/n$ diverges. The harmonic series is thus an example of a divergent series whose sequence of terms has limit zero. (When $\alpha \neq 1$, the series $\Sigma 1/n^\alpha$ is called a *generalized harmonic* series.)

The second class of series are all those of the form

$$\sum_{k=0}^{\infty} r^k$$

where r is any nonnegative number. Such a series is called a *geometric series* (because its terms form a geometric progression). If $r = 1$, then, as we know from Example 7-2, the resulting series $\Sigma 1$ diverges. If $r > 1$, then, as we know from Problem 2-5a, $\{r^k\} \to \infty$ and so Theorem 7-2 implies that the geometric series diverges if $r > 1$. Finally, if $r < 1$, then we know from

Eq. (1-9) that the sequence of partial sums is $\{1 - r^{n+1})/(1 - r)\}$.   But, from Problem 2-5a again, $\{r^{n+1}\} \to 0$ if $0 \leq r < 1$.   Hence $\{(1 - r^{n+1})/(1 - r)\} \to 1/(1 - r)$, or the geometric series converges if $0 \leq r < 1$.   Moreover, we have

$$\sum_{k=0}^{\infty} r^k = \frac{1}{1 - r} \tag{7-2}$$

if $0 \leq r < 1$.   [Even though we have agreed for now to consider only series of nonnegative terms, this seems like the appropriate time to point out that Eq. (7-2) also holds for $-1 < r < 0$.   The reason for this is that if $-1 < r < 0$, then $-|r|^n \leq r^n \leq |r|^n$, and this inequality implies that $\{r^n\} \to 0$. Inasmuch as Eq. (1-9) is true for any $r \neq 1$, the sequence of partial sums of our geometric series is likewise $\{(1 - r^{n+1})/(1 - r)\}$, and it likewise has limit $1/(1 - r)$ when $-1 < r < 0$.]   It can be seen that a geometric series is one of the rare series whose partial sums can explicitly be evaluated.

Remark   Another class of series whose partial sums can explicitly be evaluated are the so-called *telescoping series*.   A series Σa_k is telescoping if there is a sequence $\{b_k\}$ such that $a_k = b_k - b_{k-1}$ for all k.   Then

$$\sum_{k=1}^{n} a_k = b_n - b_0$$

and it can be seen that a telescoping series converges if and only if $\{b_n\} \to L$, and if this is the case, then $\Sigma a_k = L - b_0$.

EXAMPLE 7-3   We can now illustrate the use of the comparison test. Consider the series $\Sigma \log n/n(n + 1)$.   At first sight it seems as though we cannot compare it with a generalized harmonic series.   But recall from Problem 4-44a that $\lim_{x \to \infty} \log x/x^{\alpha} = 0$ for $\alpha > 0$.   In terms of sequences this result can be stated as $\{\log n/n^{\alpha}\} \to 0$ for $\alpha > 0$.   Now put

$$\log n/n(n + 1) = [\log n/n^{1/2}][1/n^{1/2}(n + 1)]$$

Since $\{\log n/n^{1/2}\} \to 0$, we can find N such that if $n > N$, then $\log n/n^{1/2} < 1$. On the other hand, $1/n^{1/2}(n + 1) < 1/n^{3/2}$ for all n.   Therefore, if $n > N$, then $\log n/n(n + 1) < 1/n^{3/2}$.   Inasmuch as $3/2 > 1$, $\Sigma 1/n^{3/2}$ converges, and so $\Sigma \log n/n(n + 1)$ also converges by Theorem 7-3.

EXAMPLE 7-4   Equation (7-2) is more flexible than it appears. For instance, we have

$$\sum_{n=0}^{\infty} \frac{u^2}{(1 + u^2)^n} = u^2 \sum_{n=0}^{\infty} \left(\frac{1}{1 + u^2}\right)^n$$

and if $u \neq 0$, then $1/(1 + u^2) < 1$.   Consequently,

$$u^2 \sum_{n=0}^{\infty} \left(\frac{1}{1 + u^2} \right)^n = \frac{u^2}{1 - \dfrac{1}{1 + u^2}} = 1 + u^2$$

Thus the series converges for all $u \neq 0$ and has sum $1 + u^2$.   If $u = 0$, the series clearly has sum 0.

Remark   If we define a function f on $(-\infty, \infty)$ by the equation

$$f(u) = \sum_{n=0}^{\infty} \frac{u^2}{(1 + u^2)^n}$$

then $\lim_{u \to 0} f(u) = 1 \neq 0 = f(0)$.   This shows that a function having a removable discontinuity can arise in a natural way.

EXAMPLE 7-5   Geometric series can arise out of decimal expansions. For simplicity, let us consider only numbers in the interval $(0, 1)$.   Every such number can be expressed as a decimal $.a_1 a_2 a_3 \ldots$, where a_k is an integer, or digit, satisfying the inequality $0 \leq a_k \leq 9$.   Now such a decimal expansion is only another way of writing the series

$$\sum_{k=1}^{\infty} \frac{a_k}{10^k}$$

(observe that this series is always convergent because $a_k/10^k \leq 9/10^k$, and $\Sigma 1/10^k$ is a convergent geometric series).   A decimal is periodic if the digits $a_1 a_2 \cdots a_n$ repeat in order.   For example, $. 12589\ 12589\ 12589 \cdots$ is a periodic decimal.   If $x = . a_1 a_2 \cdots a_n a_1 a_2 \cdots a_n \cdots$ is a periodic decimal, then

$$x = \frac{a_1}{10} + \frac{a_2}{10^2} + \cdots + \frac{a_n}{10^n} + \frac{a_1}{10^{n+1}} + \frac{a_2}{10^{n+2}} + \cdots + \frac{a_n}{10^{n+n}} + \cdots$$

$$+ \frac{a_1}{10^{kn+1}} + \frac{a_2}{10^{kn+2}} + \cdots + \frac{a_n}{10^{kn+n}} + \cdots$$

$$= \sum_{k=0}^{\infty} \left[\sum_{m=1}^{n} \frac{a_m}{10^{kn+m}} \right] = \left(\sum_{m=1}^{n} a_m 10^{-m} \right) \sum_{k=0}^{\infty} \left[\frac{1}{10^n} \right]^k$$

Now

$$\sum_{k=0}^{\infty} \left[\frac{1}{10^n} \right]^k = \frac{1}{1 - 10^{-n}}$$

because $10^{-n} < 1$. Hence

$$x = \sum_{m=1}^{n} a_m 10^{-m} \frac{1}{1 - 10^{-n}} = \frac{\sum_{m=1}^{n} a_m 10^{n-m}}{10^n - 1}$$

Inasmuch as $10^n - 1$ is an integer, and $a_m 10^{n-m}$ is an integer for $m = 1, 2, \cdots,$ n, we see that x is the quotient of two integers. In other words, any periodic decimal represents a rational number. It can be shown in a similar way that decimals possessing other regularities also represent rational numbers.

EXERCISES

7-1   Determine whether $\Sigma \tan^{-1} n/(n+1)^{1/2}$ converges or diverges.

7-2   Determine whether $\Sigma n!/n^n$ converges or diverges.

7-3   For $u \geq 0$ and arbitrary p and q, investigate the convergence of

$$\sum_{n=0}^{\infty} \frac{u^p}{(1 + u^q)^n}$$

Find its sum whenever it converges.

There are two other tests for the convergence or divergence of series of nonnegative terms. The first is known as the nth *root test*.

Theorem 7-5   *Let* $\sum a_n$ *be a series of positive terms. If* $a_n^{1/n} \leq r < 1$ *for all sufficiently large* n, *then* Σa_n *converges. If* $a_n^{1/n} \geq 1$ *for infinitely many* n, *then* a_n *diverges.*

The following special case of this theorem is what is more commonly called "the nth root test" in calculus texts.

Corollary 7-5   *Let* $\sum a_n$ *be a series of positive terms. If* $\{a_n^{1/n}\} \to L$, *then* $\sum a_n$ *converges if* $L < 1$, *diverges if* $L > 1$. *If* $L = 1$, *no conclusion is possible.*

EXERCISE 7-4   Deduce Corollary 7-5 from Theorem 7-5 when $L \neq 1$.

Our second test is called the *ratio test*.

Theorem 7-6   *Let* $\sum a_n$ *be a series of positive terms. If* $a_{n+1}/a_n \leq r < 1$ *for all sufficiently large* n, *then* Σa_n *converges. If* $a_{n+1}/a_n \geq 1$ *for all sufficiently large* n, *then* Σa_n *diverges.*

As with the nth root test, this theorem also has a special case that is more commonly called "the ratio test."

Corollary 7-6   *Let* $\sum a_n$ *be a series of positive terms. If* $\{a_{n+1}/a_n\} \to L$, *then* $\sum a_n$ *converges if* $L < 1$ *and diverges if* $L > 1$. *If* $L = 1$, *no conclusion is possible.*

EXERCISE 7-5 Deduce Corollary 7-6 from Theorem 7-6 when $L \neq 1$.

EXAMPLE 7-6 We will show why Corollaries 7-5 and 7-6 are inconclusive when $L = 1$. We know that the harmonic series $\Sigma 1/n$ diverges. Now $\{1/(n+1)/1/n\} \to 1$, and, from Problem 2-5e, $\{n^{1/n}\} \to 1$. On the other hand, $\Sigma(1/n^2)$ converges. Again, $\{1/(n+1)^2/1/n^2\} \to 1$, and $\{(n^2)^{1/n}\} \to 1$. Here then are two series, one convergent, the other divergent, for which $\{a_{n+1}/a_n\} \to 1$, and $\{a_n^{\ 1/n}\} \to 1$.

Remark In Problem 2-14 we saw that if $\{a_{n+1}/a_n\} \to L$, *then* $\{a_n^{1/n}\} \to L$. This means that if Corollary 7-6 is inconclusive, then Corollary 7-5 will also be inconclusive.

EXAMPLE 7-7 Let us apply the nth root test to $\sum [n/(n+1)]^{n^2}$. We have

$$\left[\left(\frac{n}{n+1}\right)^{n^2}\right]^{1/n} = \left(\frac{n}{n+1}\right)^n = \frac{1}{(1+1/n)^n}$$

Inasmuch as $\{(1 + 1/n)^n\} \to e$, we see that the sequence of nth roots of the terms of this series has limit $1/e < 1/2$. The series is thus convergent.

EXAMPLE 7-8 Let k be a positive integer, and $\alpha > 0$. We will use the ratio test on the series

$$\sum \binom{n+k}{n}\alpha^n$$

If we denote the general term of this series by a_n, then

$$\frac{a_{n+1}}{a_n} = \frac{\binom{n+1+k}{n+1}\alpha^{n+1}}{\binom{n+k}{n}\alpha^n} = \frac{n+1+k}{n+1}\alpha$$

Now $\{(n+1+k)\alpha/(n+1)\} \to \alpha$ and so this series converges if $\alpha < 1$, diverges if $\alpha > 1$. If $\alpha = 1$, Corollary 7-6 tells us nothing; however, we can conclude from Theorem 7-6 that the series then diverges because $a_{n+1}/a_n = (n+1+k)/(n+1) > 1$ for all n.

EXERCISES

7-6 Use the nth root test to determine the values of $\alpha > 0$ for which the series $\Sigma \alpha^n/n^n$ converges.

7-7   Use the ratio test to determine the convergence or divergence of the series

$$\sum \frac{1 \cdot 3 \cdot 5 \cdots (2n + 1)}{3 \cdot 6 \cdot 9 \cdots (3n)}$$

Let us now consider series Σa_n such that the sequence of terms $\{a_n\}$ contains infinitely many positive members and infinitely many negative members. Except for one special circumstance, which we will presently point out, we will not test such series directly for convergence.   Rather, we will try to establish the convergence of Σa_n by examining the series $\Sigma |a_n|$.   That is, we try to reduce the general situation to the convergence of series of nonnegative terms.   The next definition and theorem will indicate precisely how this is done.

Definition 7-3   *The series $\sum a_n$ is absolutely convergent if $\sum |a_n|$ converges.*

Theorem 7-7   *If a series is absolutely convergent, then it is convergent.*

Otherwise stated, this theorem means that the convergence of $\Sigma |a_n|$ implies the convergence of Σa_n.   Compare it with Problem 6-16a, its analog for improper integrals.

The special circumstances whereby we can directly test the convergence of a series of arbitrary terms is set forth in the next theorem.   It is known as the *alternating series test* because it is concerned with series of the form $\Sigma (-1)^{n-1}a_n$, where $a_n > 0$, and the terms of such a series alternate in sign.

Theorem 7-8   *If $\{a_n\}$ is such that $a_n > 0$ for all n, then the series $\sum (-1)^{n-1}a_n$ [or $\Sigma(-1)^n a_n$] is convergent if $\{a_n\}$ is decreasing (that is, $a_{n+1} \leq a_n$ for all n), and $\{a_n\} \to 0$.*

There is an interesting corollary to Theorem 7-8 that has many applications.

Corollary 7-8   *If $\sum (-1)^{n-1}a_n$ satisfies the conditions of Theorem 7-8, then*

$$\left| \sum_{n=1}^{\infty} (-1)^{n-1}a_n - \sum_{n=1}^{m} (-1)^{n-1}a_n \right| \leq a_{m+1} \tag{7-3}$$

EXAMPLE 7-9   The alternating series $\sum (-1)^{n-1}1/n$ is sometimes called the alternating harmonic series.   Since it clearly satisfies the conditions of Theorem 7-8, it is a convergent series.   However, it is not absolutely convergent because $\Sigma |(-1)^{n-1}1/n|$ is the divergent harmonic series.   This shows that absolute convergence is sufficient but not necessary for convergence.   In other words, the converse of Theorem 7-7 is false.

In view of Example 7-10, we introduce the following terminology: A series which is convergent but not absolutely convergent is said to be *conditionally convergent*.

Up to now the integral test has been the only result about series that involves any of the notions of calculus. To connect these two fields it is necessary to study series whose terms are functions. In this book we shall only consider the simplest class of such series; namely, the so-called power series. Power series arise very naturally out of the following theorem, which is known as *Taylor's theorem.*

Theorem 7-9 *If $D^n f$ is continuous on* $[a, b]$, *and $D^{n+1} f$ exists on* (a, b), *then there exists a number z such that $a < z < b$, and*

$$f(b) = \sum_{k=0}^{n} D^k f(a) \frac{(b-a)^k}{k!} + \frac{(b-a)^{n+1}}{(n+1)!} D^{n+1} f(z) \qquad (7\text{-}4)$$

[Here $Df^0(a)$ is interpreted as $f(a)$; also, recall that $0! = 1$.] One thing that should be emphasized about Eq. (7-4) is that the number z depends on n. That is, if we were to write down Eq. (7-4) with n replaced by m, the number z would, in general, be a different number. Also, observe that when $n = 0$ this theorem becomes the Mean Value Theorem.

The expression

$$\sum_{k=0}^{n} D^k f(a) \frac{(b-a)^k}{k!}$$

that appears in Eq. (7-4) is a polynomial of degree n in b. It is called the nth *degree Taylor's polynomial of f at a,* and we will denote it by $T_a^n(b)$ The other expression that appears on the right-hand side of Eq. (7-4) is called the nth *order remainder of f at a.* It will be denoted by $R_a^n(b)$. Thus,

$$R_a^n(b) = \frac{(b-a)^{n+1}}{(n+1)!} D^{n+1} f(z)$$

If we replace b by x in Theorem 7-9, and $a < x \leq b$, then Eq. (7-4) can be rewritten in this notation as

$$f(x) = T_a^n(x) + R_a^n(x) \qquad (7\text{-}5)$$

EXAMPLE 7-10 To show that the Taylor's polynomials depend on where they are formed, we will calculate T_1^3 and T_2^3 for the function $f(x) = \log x$. Now $f'(x) = 1/x$, $f''(x) = -1/x^2$, and $f^{(3)}(x) = 2/x^3$. For $a = 1$, we have $f(1) = 0$, $f'(1) = 1$, $f''(1) = -1$, and $f^{(3)}(1) = 2$. Consequently,

$$T_1^3(x) = 1(x-1) - 1\frac{(x-1)^2}{2!} + 2\frac{(x-1)^3}{3!}$$

For $a = 2$, we have $f(2) = \log 2$, $f'(2) = 1/2$, $f''(2) = -1/4$, and $f^{(3)}(2) = 1/4$.

Hence

$$T_2{}^3(x) = \log 2 + \frac{1}{2}(x - 2) - \frac{1}{4}\frac{(x - 2)^2}{2!} + \frac{1}{4}\frac{(x - 2)^3}{3!}$$

It can be seen that $T_2{}^3(x) \neq T_2{}^3(x)$.   Needless to say, the remainder term also depends on where it is formed.

EXERCISE 7-8   If f is a polynomial of degree n, prove that $f(x) = T_a^n(x)$ for all a.

Let us suppose that the function f has all its derivatives on a fixed interval. We can then consider the possibility of extending the Taylor's polynomial of f to an infinite series.   Such a series is known as a *power series*.   More particularly, the expression

$$\sum_{n=0}^{\infty} a_n(x - a)^n \tag{7-6}$$

is called a power series with center a.   Since a power series with center a can always be transformed into a power series with center 0 by means of the simple substitution $y = x - a$, we shall mostly consider power series with center 0.   When dealing with power series having center 0, let us agree to drop the phrase "with center 0."   That is, we make the convention that "*power series*" *always stands for* "*power series with center* 0."   When it is necessary to deal with power series having center other than 0, it will be specifically indicated.   Further, we make the *notational convention that* $\Sigma a_n(x - a)^n$ *always stands for*

$$\sum_{n=0}^{\infty} a_n(x - a)^n$$

As far as the convergence of power series is concerned, the following theorem is basic.

Theorem 7-10   *Given a power series* $\sum a_n x^n$, *either it converges absolutely for all* x, *or it converges only for* $x = 0$, *or there exists a number* $\rho > 0$ *such that it converges absolutely for* $|x| < \rho$ *and diverges for* $|x| > \rho$.

The number ρ is called the *radius of convergence* of $\Sigma a_n x^n$; if $\Sigma a_n x^n$ converges for all x, we say that it has an infinite radius of convergence; if it converges only for $x = 0$, we say that its radius *of convergence* is zero.   The interval $(-\rho, \rho)$ is called the *interval of convergence* of $\Sigma a_n x^n$.   Observe that Theorem 7-10 makes no mention about the convergence or divergence of $\Sigma a_n x^n$ at $x = \pm\rho$.   The reason for this is that at an endpoint of the interval of convergence the power series may converge absolutely, converge conditionally, or diverge.

EXERCISE 7-9   If the interval of convergence of $\sum a_n x^n$ is $(-\rho, \rho)$, what is the interval of convergence of $\sum a_n (x - a)^n$?

The ratio test and the nth root test are particularly suited to power series. If we test the absolute convergence of the power series $\sum a_n x^n$ by means of the ratio test, we are led to $|a_{n+1}/a_n||x|$.   Suppose that $\{|a_{n+1}/a_n|\} \to L$.   Then $\{|a_{n+1}/a_n||x|\} \to |x|L$, and $\sum a_n x^n$ will converge absolutely if $|x|L < 1$, and diverge if $|x|L > 1$.   Hence $\sum a_n x^n$ converges for all x if $L = 0$, and if $L \neq 0$, then its radius of convergence ρ is given by $\rho = 1/L$.   The nth root test applied to $\sum a_n x^n$ leads to $\{|a_n|^{1/n}|x|\} \to |x|L$, if $\{|a_n|^{1/n}\} \to L$ and we can draw the same conclusions about the radius of convergence of $\sum a_n x^n$ as we just did with the ratio test.   We can thus find the radius of convergence of any power series $\sum a_n x^n$ for which $\{|a_{n+1}/a_n|\}$ has a limit, or $\{|a_n|^{1/n}\}$ has a limit.   Now most of the power series that are ordinarily encountered satisfy one or both of these conditions.   Consequently, we have a fairly effective method for determining the radius of convergence.   In more advanced books it is shown how the radius of convergence of any power series can be determined.

EXAMPLE 7-11   Let us find the radius of convergence of the power series $\sum \alpha^{n^2} x^n$, where $\alpha > 0$.   The nth root test leads to the sequence $\{\alpha^n |x|\}$. We know from Problem 2-5a that $\{\alpha^n\}$ has limit 0 if $0 < \alpha < 1$; has limit 1 if $\alpha = 1$; and has limit infinity if $\alpha > 1$.   Therefore, $\sum \alpha^{n^2} x^n$ has an infinite radius of convergence if $\alpha < 1$; has radius of convergence 1 if $\alpha = 1$; and has zero radius of convergence if $\alpha > 1$.   We thus see that all three possibilities for the radius of convergence can be realized.

EXERCISE 7-10   Find the radius of convergence of the power series

$$\sum \left(\frac{1 \cdot 3 \cdot 5 \cdots (2n + 1)}{3 \cdot 6 \cdot 9 \cdots (3n + 3)} \right)^\alpha x^n$$

where $\alpha > 0$.

Let us return to Theorem 7-9 and examine it in the light of what we have learned about power series.   If a function f has all its derivatives in some interval including a, then the power series

$$\sum D^n f(a) \frac{(x - a)^n}{n!}$$

is called the *Taylor's series of f at a*; let us denote it by T_a^∞.   The Taylor's series of f at 0 is called the *Maclaurin's series of f*.   Since T_a^∞ can be obtained by formally replacing n by ∞ in T_a^n, it is natural to ask what happens to Eq. (7-5) when the Taylor's polynomial is replaced by the Taylor's series. More precisely, we are interested in the behavior of the right-hand side of Eq. (7-5) when $n \to \infty$.

A number of outcomes are conceivable: (1) T_a^∞ converges only for the number a; (2) T_a^∞ has a positive or infinite radius of convergence, but there is no number $\delta > 0$ such that $f(x) = T^a(x)$ for all x in $(a - \delta, a + \delta)$; (3) T_a^∞ has a positive or infinite radius of convergence, and there is a number $\delta > 0$ such that $f(x) = T_a^\infty(x)$ for all x in $(a - \delta, a + \delta)$. Of these the one we are most interested in is, of course, (3), and it turns out that it is by far the most frequent outcome for functions that are generally encountered. However, there are functions for which either (1) or (2) holds. In Exercise 7-12 we give a fairly simple example of a function for which (2) holds, and in Example 7-17 we give an example of a function for which (1) holds.

Once we realize that (1) and (2) are possible as well as conceivable, the problem is to determine when (3) will occur. Now it is clear from Eq. (7-5) that $f(x) = T_a^\infty(x)$ if and only if $\{R_a^n(x)\} \to 0$. This means that if $\{R_a^n(x)\} \to 0$ for all x in $(a - \delta, a + \delta)$, where $\delta > 0$, then $f(x) = T_a^\infty(x)$ for all x in $(a - \delta, a + \delta)$, and conversely. We state this result as the next theorem.

Theorem 7-11 *If f has all its derivatives in some open interval including a, then*

$$f(x) = \sum D^n f(a) \frac{(x - a)^n}{n!}$$

for all x, and only those x, such that $\{R_a^n(x)\} \to 0$.

EXAMPLE 7-12 We will illustrate the use of this theorem, together with one of its drawbacks, with the function $f(x) = \log x$ at the number $a = 1$. Our first task is to calculate $D^n \log x$ for $n = 1, 2, \ldots$. If we compute the first few of them, we find that $D \log x = 1/x$, $D^2 \log x = -1/x^2$, $D^3 \log x = 2/x^3$, $D^4 \log x = -3!/x^4$, $D^5 \log x = 4!/x^5$, and so on. In this way we can perceive the formula $D^n \log x = (-1)^{n-1}(n-1)!/x^n$; once it has been found it can be proved by an easy induction. Thus, $D^n \log 1 = (-1)^{n-1}(n-1)!$ for all $n \geq 1$. Since $\log 1 = 0$, we have

$$T_1^\infty(x) = \sum_{n=1}^\infty (-1)^{n-1}(n-1)! \frac{(x-1)^n}{n!} = \sum_{n=1}^\infty (-1)^{n-1} \frac{(x-1)^n}{n}$$

It is not hard to see, by employing the ratio test or the nth root test, that the interval of convergence of T_1^∞ is $(0, 2)$. For x in $(0, 2)$ $R_1^n(x) = (-1)^n(x-1)^{n+1}/z^{n+1}(n+1)$, where z is between 1 and x. Now if $1 \leq x \leq 2$, we have $1 < z < 2$, or $1/z < 1$. Since $x - 1 \leq 1$, this means that $|R_1^n(x)| < 1/(n+1)$ for $1 \leq x \leq 2$. Consequently, $\{R_1^n(x)\} \to 0$ for $1 \leq x \leq 2$, or

$$\log x = \sum_{n=1}^\infty (-1)^{n-1} \frac{(x-1)^n}{n}$$

for $1 \le x \le 2$. Suppose now that $0 < x < 1$. Then $x < z < 1$, or $x^{n+1} < z^{n+1} < 1$, or $1/z^{n+1} < 1/x^{n+1}$. Hence

$$|R_1{}^n(x)| < \frac{(x-1)^{n+1}}{(n+1)x^{n+1}} = \frac{1}{n+1}\left|1 - \frac{1}{x}\right|^{n+1}$$

When $1/2 \le x < 1$, we can still conclude that $|R_1{}^n(x)| < 1/(n+1)$. As above, this inequality implies that $\log x = T_1{}^\infty(x)$ for $1/2 \le x < 1$. It is when we examine the case $0 < x < 1/2$ that the drawback manifests itself because the estimate $|R_1{}^n(x)| < [1/(n+1)]|1 - 1/x|^{n+1}$ is, while still true, of no use because now $|1 - 1/x| > 1$ and, as is easy to see, this means that $\{|1 - 1/x|^{n+1}/(n+1)\} \to \infty$. Clearly we can conclude nothing about the limit of $\{|R_1{}^n(x)|\}$ from this relation. The alternative would be to examine $|x - 1|^{n+1}/(n+1)z^{n+1}$ directly. But since z is both indeterminate and depends on n, such an approach would be difficult.

We can thus see that one defect of Theorem 7-11 is that it sometimes is very difficult to determine the limiting behavior of $\{R_a{}^n(x)\}$. At the same time, it should be pointed out that there are alternate forms of $R_a{}^n$ which are sometimes easier to treat, but this is a matter for a more advanced text.

Remark Note we have shown that for $f(x) = \log x$ at $a = 1$ we have $\{|R_1{}^n(2)|\} \to 0$. This means that $\log 2 = T_1{}^\infty(2)$, or

$$\log 2 = \sum_{n=1}^{\infty} \frac{(-1)^{n-1}}{n} \tag{7-7}$$

We have thus determined the sum of the alternating harmonic series of Example 7-10.

We can use Eq. (7-7) to determine the limit of the sequence

$$\left\{\sum_{k=1}^{n} \frac{1}{n+k}\right\}$$

of Problem 2-8c. Indeed, if we denote the general term of this sequence by a_n, then $a_n = S_{2n}$, where

$$S_{2n} = \sum_{k=1}^{2n} \frac{(-1)^{k-1}}{k}$$

We prove this claim by induction. Since $S_2 = 1 - 1/2 = 1/(1+1)$ the identity is true for $n = 1$. Suppose that $S_{2n} = a_n$. Notice that

$$a_{n+1} - a_n = \sum_{k=1}^{n+1} \frac{1}{(n+1)+k} - \sum_{k=1}^{n} \frac{1}{n+k} = \frac{1}{2n+2} + \frac{1}{2n+1} - \frac{1}{n+1}$$

$$= \frac{1}{2n+1} - \frac{1}{2n+2}$$

This equation, together with our induction hypothesis, gives us

$$S_{2n+2} = S_{2n} + \frac{1}{2n+1} - \frac{1}{2n+2} = a_n + \frac{1}{2n-1} - \frac{1}{2n+2} = a_{n+1}$$

Our identity is thus proved. Inasmuch as $\{S_{2n}\} \to \log 2$, $\{a_n\} \to \log 2$ as well.

EXERCISES

7-11   Use Theorem 7-11 to prove that

$$e^x = \sum \frac{x^n}{n!} \tag{7-8}$$

for all x.

7-12   Let $f(x) = \begin{cases} 0 & \text{if } x = 0 \\ e^{-1/x^2} & \text{if } x \neq 0 \end{cases}$

(a)   Show that $D^n f(0) = 0$ for every positive integer n.
(b)   Find the Maclaurin's series of f and show that it converges for all x.
(c)   Show that the Maclaurin's series of f is not equal to f on any interval.
[Hint for part (a): Prove by induction that if $x \neq 0$, then $D^n f(x) = P(x)e^{-1/x^2}/x^{3n}$, where P is a polynomial; now use L'Hospital's rule.]

Another drawback of Theorem 7-11 is that even if a function f is known to have all its derivatives in some interval including a, it may not be practically possible to compute $D^n f(a)$ for all integers n. For instance, consider how tedious it would be to compute something as straightforward as $D^8 \tan x$, to say nothing of $D^{80} \tan x$, or $D^{800} \tan x$, and so forth. Thus, Theorem 7-11 leaves something to be desired from a computational point of view. What is needed are some indirect methods of computing power series expansions. The following theorem provides us with two such methods.

Theorem 7-12   *Let* $\sum a_n (x - a)^n$ *have a positive or infinite radius of convergence, and put* $f(x) = \sum a_n(x - a)^n$. *If x is in the interior of the interval of convergence, then*

$$Df(x) = \sum a_n D[(x - a)^n] = \sum_{n=1}^{\infty} n a_n (x - a)^{n-1} \tag{7-9}$$

and

$$\int_a^x f(t)\, dt = \sum a_n \int_a^x (t - a)^n\, dt = \sum_{n=0}^{\infty} \frac{a_n}{n+1} (x - a)^{n+1} \tag{7-10}$$

In other words, the power series representing f can be differentiated or integrated term-by-term within its interval of convergence. Moreover, the resulting power series will have the same radius of convergence, and they will represent respectively the derivative of f or the integral of f.

We will digress from computational considerations to point out an important theoretical application of this theorem.

Corollary 7-12 *If $\sum a_n(x-a)^n$ has a positive or infinite radius of convergence, and if*

$$f(x) = \Sigma a_n(x-a)^n$$

then

$$a_n = \frac{D^n f(a)}{n!}$$

This corollary means that if a function can be represented by some power series having center a in some open interval including a, then that power series is necessarily T_a^∞. Consequently, if $\Sigma a_n(x-a)^n = f(x) = \Sigma b_n(x-a)^n$ in some interval including a, then $\{a_n\} = \{b_n\}$; that is, power series representations are unique.

EXERCISE 7-13 Deduce Corollary 7-12 from Theorem 7-12.

Theorem 7-12 provides two ways of analytically obtaining new power series expansions from known ones. In the following theorem we will see a way of doing this algebraically.

Theorem 7-13 *Suppose that $f(x) = \sum a_n(x-a)^n$ and $g(x) = \sum b_n(x-a)^n$, and that both power series have a positive or infinite radius of convergence. Let ρ be the smaller of the two radii of convergence (if both power series have an infinite radius of convergence, take ρ to be ∞). Then, for all x in $(a-\rho, a+\rho)$,*

$$f(x)g(x) = \sum c_n(x-a)^n \tag{7-11}$$

where

$$c_n = \sum_{k=0}^{n} a_k b_{n-k}$$

for $n = 1, 2, \ldots$.

Recall from Chapter 2 that $\{c_n\}$ is the convolution of the sequences $\{a_n\}$ and $\{b_n\}$. That is, $\{c_n\} = \{a_n\} * \{b_n\}$.

Remark   We have already seen another way of obtaining a new power series from known power series by algebraic means. Namely, if f and g are as in Theorem 7-13, and ρ has the same meaning as it did there, then

$$f(x) \pm g(x) = \sum (a_n \pm b_n)(x - a)^n$$

for all x in $(a - \rho, a + \rho)$.

In order to be able to use these results we must have a reasonable supply of power series expansions at our disposal. Power series expansions for the exponential, sine, and cosine functions have been pointed out. In addition, we can now view the geometric series as a power series and interpret Eq. (7-2) as being the power series expansion of the function $f(x) = 1/(1 - x)$ on the interval $(-1, 1)$. Later, in Exercise 7-17, we shall see how to determine the power series expansion of the function $f(x) = (1 + x)^\rho$, where ρ is any real number. All of these expansions can be considerably augmented. If $f(x) = \Sigma a_n(x - a)^n$ has $(a - \rho, a + \rho)$ for its interval of convergence, and if g is a function such that $a - \rho < g(x) < a + \rho$ for x in some interval I, then $f(g)$ has the expansion $f(g(x)) = \Sigma a_n(g(x) - a)^n$ for all x in I. Now this expansion is not a power series in general, but if g is a polynomial then $\Sigma a_n(g(x) - a)^n$ will be a power series. [A rearrangement of terms is necessary to change $\Sigma a_n(g(x) - a)^n$ into a power series in this case, unless g is a monomial, but it can be justified because a power series is absolutely convergent within its interval of convergence. See the remark following the solution of Problem 7-17.]

EXAMPLE 7-13   Since $1/(1 - x^3) = \sum x^{3n}$ for $|x| < 1$, it follows that

$$\frac{3x^2}{(1 - x^3)^2} = D\frac{1}{1 - x^3} = \sum_{n=1}^{\infty} 3nx^{3n-1}$$

for $|x| < 1$.

EXAMPLE 7-14   We know that $D \log x = 1/x = 1/(1-(1 - x)) = \Sigma(1 - x)^n = \Sigma(-1)^n(x - 1)^n$, and that the expansion is valid for $-1 < x - 1 < 1$, or $0 < x < 2$. Therefore,

$$\log x - \log 1 = \log x = \int_1^x \frac{1}{t}\, dt = \sum (-1)^n \int_1^x (t - 1)^n\, dt$$

$$= \sum \frac{(-1)^n(x - 1)^{n+1}}{(n + 1)}$$

for all x in $(0, 2)$. In other words, $\log x = \Sigma(-1)^n(x - 1)^{n+1}/(n + 1)$ not only for x in $[1/2, 2]$, as we know from Example 7-13, but for all x in $(0, 2]$.

This result is usually written in the form

$$\log (1 + x) = \sum_{n=1}^{\infty} \frac{(-1)^{n-1}}{n} x^n \qquad (7\text{-}12)$$

for $|x| < 1$.

EXAMPLE 7-15 As an illustration of Theorem 7-13 we will compute the power series expansion of $e^x/(1-x)$. In the first place, notice that even though the series representing e^x converges for all x, the series representing the product of e^x and $1/(1-x)$ will converge only for $|x| < 1$ because 1 is the radius of convergence of the geometric series that represents $1/(1-x)$. It is sufficient to compute $\{1/n\} * \{1\}$. Now the nth term of this convolution is clearly

$$\sum_{k=0}^{n} \frac{1}{k!}$$

Consequently,

$$\frac{e^x}{1-x} = \sum \left(\sum_{k=0}^{n} \frac{1}{k!} \right) x^n$$

for $|x| < 1$.

Remark It is easy to see that the last equation generalizes. That is, if $f(x) = \Sigma a_n x^n$, and the radius of convergence of this power series is ρ (possibly $\rho = \infty$), and if

$$S_n = \sum_{k=0}^{n} a_k$$

then

$$\frac{f(x)}{1-x} = \sum S_n x^n$$

for $|x|$ less than the smaller of ρ and 1.

EXAMPLE 7-16 Now that we have encountered Theorem 7-12, we can better appreciate why the following function f is an example of a function whose Maclaurin's series converges only for the value zero. The function is

$$f(x) = \sum \frac{\cos (n^2 x)}{e^n}$$

Since $|\cos (n^2 x)| \le 1$, it can be seen that the series converges for all x. In view of Theorem 7-12 and the fact that

$$\sum \frac{-n^2 \sin (n^2 x)}{e^n}$$

is convergent for all x (because, as can be seen from the nth root test, $\Sigma n^2/e^n$ is convergent), it is plausible that the derivative of f can be computed term-by-term. It can be proved that this is indeed the case. Further, since $\Sigma n^{2k}/e^n$ converges for all integers k (this can also be proved by the nth root test), it can be shown that all the derivatives of f can be computed by differentiating term-by-term. In this way we find that

$$D^k f(x) = \begin{cases} \left| \sum \dfrac{(-1)^{k/2} n^{2k} \cos(n^2 x)}{e^n} \right. & \text{if } k \text{ is even} \\[4mm] \left| \sum \dfrac{(-1)^{(k+1)/2} n^{2k} \sin(n^2 x)}{e^n} \right. & \text{if } k \text{ is odd} \end{cases}$$

Hence $D^k f(0) = 0$ if k is odd, and $D^k f(0) = (-1)^{k/2} \Sigma n^{2k}/e^n$ if k is even. Therefore, in the Maclaurin's series of f only even powers of x appear. Let x be any fixed nonzero number, and let m be a positive integer such that $m > |e/2x|$. The absolute value of the term associated with x^{2m} is

$$\frac{x^{2m}}{(2m)!} \sum \frac{n^{4m}}{e^n}$$

Now the series is larger than any one of its terms, and in particular it is larger than the term for which $n = 2m$. In other words,

$$\frac{x^{2m}}{(2m)!} \sum \frac{n^{4m}}{e^n} > \frac{x^{2m}}{(2m)!} \frac{(2m)^{4m}}{e^{2m}} > \left(\frac{(2m)^2 x}{2me} \right)^{2m}$$

[the last inequality is true because $1/(2m)! > 1/(2m)^{2m}$]. Inasmuch as $2mx/e > 1$, it follows that

$$\frac{x^{2m}}{(2m)!} \sum \frac{n^{4m}}{e^n} > 1^{2m} = 1$$

Hence if $x \neq 0$ the sequence of terms of the Maclaurin's series of f cannot have limit zero. Theorem 7-2 thus implies that the Maclaurin's series of f diverges for all $x \neq 0$.

EXERCISES

7-14   Find the power series expansion of $f(x) = \tan^{-1} x$ by integrating the appropriate power series term-by-term. State the interval of convergence of this expansion.

7-15   If $f(x) = \Sigma x^{4n}/(4n)!$, prove that $D^4 f(x) = f(x)$ for all x.

7-16   Find the power series expansion of $f(x) = (\tan^{-1} x)^2$ and state its interval of convergence.

7-17 If n is a positive integer, generalize the binomial coefficient $\binom{p}{n}$ to arbitrary p by setting

$$\binom{p}{n} = \frac{p(p-1)\cdots(p-n+1)}{n!}$$

Put $\binom{p}{0} = 1$

(a) Show that $\Sigma \binom{p}{n} x^n$ converges for $|x| < 1$.

(b) If $f(x) = \Sigma \binom{p}{n} x^n$ for $|x| < 1$, show that $(1+x)f'(x) - pf(x) = 0$.

(c) Deduce from (b) that $f(x) = (1+x)^p$ for $|x| < 1$. This will prove that

$$(1+x)^p = \Sigma \binom{p}{n} x^n \tag{17-3}$$

for $|x| < 1$ and any p. This series is known as the *binomial series* because it generalizes the binomial theorem from positive integers to arbitrary numbers p.

We will conclude this chapter by stating a theorem that is useful for determining the sum of certain convergent series. It is known as *Abel's Theorem.*

Theorem 7-14 *Let $f(x) = \sum a_n x^n$ have radius of convergence 1. If $\sum a_n$ converges, then*

$$\lim_{x \to 1-} f(x) = \sum a_n$$

This theorem also holds for the right-hand limit at -1 if $\Sigma(-1)^n a_n$ converges.

EXAMPLE 7-17 We have already seen one instance of the validity of Abel's Theorem. Namely, we know from Example 7-13 that $\log(1+x) = \Sigma(-1)^{n-1} x^n/n$, and that $\log 2 = \Sigma(-1)^{n-1}/n$. Inasmuch as

$$\lim_{x \to 1-} \log(1+x) = \log 2$$

this result is in agreement with Theorem 7-14.

EXAMPLE 7-18 It turns out that the converse of Theorem 7-14 is false. That is, it is not true that if $f(x) = \Sigma a_n x^n$ for $|x| < 1$, and $\lim_{x \to 1-} f(x) = L$, then Σa_n converges and has sum L. The standard counterexample to this statement is $f(x) = 1/(1+x) = \Sigma(-1)^n x^n$ for $|x| < 1$. Now $\lim_{x \to 1-} f(x) = 1/2$, but $\Sigma(-1)^n$ is a divergent series (as we saw in Example 7-2).

EXERCISE 7-18   Use Theorem 7-14 to prove that

$$\frac{\pi}{4} = 1 - \frac{1}{3} + \frac{1}{5} - \frac{1}{7} + \cdots \tag{7-14}$$

This remarkable series, which establishes a relation between π and the odd integers, is known as *Leibniz' series.*

SOLUTIONS TO EXERCISES

7-1   We know that $\lim\limits_{x \to \infty} \tan^{-1} x = \pi/2 > 1$.   Therefore, we can find $X > 0$ such that if $x > X$, then $\tan^{-1} x > 1$.   Hence if $n > X$, then $\tan^{-1} n/(n + 1)^{1/2} > 1/(n + 1)^{1/2}$.   But the series $\Sigma 1/(n + 1)^{1/2}$ is a generalized harmonic series with exponent $1/2$, and so it diverges.   Our series must also diverge by the comparison test.

7-2   We have

$$\frac{n!}{n^n} = \frac{n-1}{n}\frac{n-2}{n} \cdots \frac{2}{n} \cdot \frac{1}{n} < \frac{2}{n^2}$$

if $n > 2$.   Since $\Sigma 1/n^2$ is convergent, so is the series in question.

7-3   If $u = 0$, the series clearly has sum 0.   Otherwise we may write it as $u^p \Sigma [1/(1 + u^q)]^n$.   Now, apart from u^p, this is a geometric series.   Accordingly, it will converge whenever $1/(1 + u^q) < 1$, or $1 + u^q > 1$.   But if $u > 0$, then $1 + u^q$ always exceeds 1.   Consequently, our series converges for all $u > 0$.   Further, its sum is

$$u^p \cfrac{1}{1 - \cfrac{1}{1 + u^q}} = u^p \frac{1 + u^q}{u^q} = u^p + u^{p-q}$$

for $u > 0$.

7-4   If $\{a_n^{1/n}\} \to L < 1$, we can pick a number r such that $L < r < 1$, and find an integer N such that if $n > N$, then $a_n^{1/n} < r$.   Similarly, if $L > 1$, we can find an integer M such that if $n > N$, then $a_n^{1/n} \geq 1$.

7-5   This proof is word for word the same as the proof of the previous exercise if $\{a_n^{1/n}\}$ is replaced by $\{a_{n+1}/a_n\}$.

7-6   We have

$$\left\{\left(\frac{\alpha^n}{n^n}\right)^{1/n}\right\} = \left\{\frac{\alpha}{n}\right\} \to 0 < 1$$

for all $\alpha > 0$.   Therefore, $\Sigma \alpha^n/n^n$ converges for all $\alpha > 0$.

7-7   If we denote the general term of this series by a_n, then

$$\left\{\frac{a_{n+1}}{a_n}\right\} = \left\{\frac{1\cdot 3\cdot 5\cdots(2n+1)(2n+3)}{3\cdot 6\cdot 9\cdots(3n)(3n+3)} \div \frac{1\cdot 3\cdot 5\cdots(2n+1)}{3\cdot 6\cdot 9\cdots(3n)}\right\}$$

$$= \left\{\frac{2n+3}{3n+3}\right\} \to \frac{2}{3} < 1$$

Therefore, the series converges.

7-8   If f is a polynomial of degree n, then $D^n f$ is continuous on $(-\infty, \infty)$, and $D^{n+1}f(x) = 0$ for all x. Therefore, f satisfies the conditions of Taylor's theorem, and $R_a^n(x) = 0$ regardless of the value of a. It now follows from Eq. (7-5) that $f(x) = T_a^n(x)$ for all x.

7-9 $\Sigma a_n(x-a)^n$ will converge if $|x-a| < \rho$, diverge if $|x-a| > \rho$. That is, if x is in the interval $(a-\rho, a+\rho)$, this power series will converge; if x is not in this interval and $x \neq a \pm \rho$, this power series will diverge. The interval of convergence is thus seen to be $(a-\rho, a+\rho)$.

7-10   If we denote the coefficient of x^n in this power series by a_n, then $|a_{n+1}/a_n|\,|x| = |x|[(2n+3)/(3n+6)]^\alpha$. Now $\{[(2n+3)/(3n+6)]^\alpha\} \to (2/3)^\alpha$. Therefore, the radius of convergence of our power series is $\rho = (3/2)^\alpha$.

7-11   It is clear that $D^n e^x = e^x$ for all n. In particular, $D^n e^0 = 1$ for all n. This means that

$$T_0^\infty(x) = \Sigma x^n/n! \quad\text{and}\quad R_0^n(x) = e^z x^{n+1}/(n+1)!$$

If $x > 0$, then $0 < z < x$, and $e^z < e^x$; if $x < 0$, then $e^z < e^0 = 1$. In either case there is a number $K > 0$ such that $|R_0^n(x)| < K|x|^{n+1}/(n+1)!$. We know from Problem 2-9a that $\{a^n/n!\} \to 0$ for all $a > 0$. Therefore, $\{|R_0^n(x)|\} \to 0$ for all x. Hence $e^x = T_0^\infty(x)$ for all x.

Remark   It can be proved in a similar way that

$$\sin x = \sum(-1)^n \frac{x^{2n+1}}{(2n+1)!}$$

for all x, and

$$\cos x = \sum(-1)^n \frac{x^{2n}}{(2n)!}$$

for all x.

7-12   (a)   For any positive integer m it follows from L'Hospital's rule that

$$0 = \lim_{y\to\infty}\frac{y^{m/2}}{e^y} = \lim_{x\to 0+}\frac{1}{|x|^m e^{1/x^2}}$$

In other words, we have

$$\lim_{x \to 0} \frac{e^{-1/x^2}}{x^m} = 0$$

for all positive integers m.   Inasmuch as the nth derivative of f at $x \neq 0$ is of the form

$$\frac{P(x)}{x^{3n}} e^{-1/x^2}$$

for some polynomial P, the above limit implies that

$$\lim_{x \to 0} \frac{P(x)}{x^{3n}} e^{-1/x^2} = 0$$

Consequently, we can regard f as having all its derivatives at zero with $D^n f(0) = 0$ for $n = 0, 1, 2, \ldots$.

(b)   In view of (a), the Maclaurin's series of f is identically zero for all x. Hence, it converges for all x.

(c)   Since $f(x) \neq 0$ for $x \neq 0$, f is not equal to its (convergent) Maclaurin's series for any nonzero x.

Remark   We can modify this example to obtain another function whose Maclaurin's series exhibits peculiar behavior.   This function is defined by the equation

$$g(x) = \begin{cases} 0 & \text{if } x = 0 \\ e^{-1/x^2} \sin 1/x & \text{if } x \neq 0 \end{cases}$$

It is not hard to see that g also has all its derivatives equal to zero at zero. Consequently, its Maclaurin's series is also identically zero.   But for $x \neq 0$, $g(x) = 0$ if and only if $x = \pm 1/n\pi$ for $n = 1, 2, \ldots$.   Therefore, since $\{1/n\pi\} \to 0$, there are always numbers in any interval $(-b, b)$, $b > 0$, at which the (convergent) Maclaurin's series of f is equal to f, and numbers at which it is not equal to f.

7-13   Let n be a positive integer.   In the view of Theorem 7-12 we have

$$D^n f(x) = \sum_{k=n}^{\infty} k(k-1) \cdots (k-n+1) a_k (x-a)^{k-n}$$

for all x in some interval $(a - r, a + r)$, where $r > 0$.   Hence

$$D^n f(a) = \sum_{k=n}^{\infty} k(k-1) \cdots (k-n+1) a_k (a-a)^{k-n} = n! a_n$$

7-14 Since $D \tan^{-1} x = 1/(1 + x^2) = \Sigma(-1)^n x^{2n}$, and this power series converges for all $|x| < 1$, we have

$$\tan^{-1}x = \tan^{-1}x - \tan^{-1}0 = \int_0^x \Sigma(-1)^n t^{2n}\,dt = \Sigma \frac{(-1)^n}{2n+1} x^{2n+1}$$

for $|x| < 1$.

7-15 It is easy to see (from the ratio test) that this power series converges for all x. Consequently, for all x we have

$$Df(x) = \sum_{n=1}^{\infty} \frac{4nx^{4n-1}}{(4n)!} = \sum_{n=1}^{\infty} \frac{x^{4n-1}}{(4n-1)!}$$

$$D^2f(x) = \sum_{n=2}^{\infty} \frac{x^{4n-2}}{(4n-2)!}$$

$$D^3f(x) = \sum_{n=3}^{\infty} \frac{x^{4n-3}}{(4n-3)!}$$

and

$$D^4f(x) = \sum_{n=4}^{\infty} \frac{x^{4n-4}}{(4n-4)!} = \sum_{m=0}^{\infty} \frac{x^{4m}}{(4m)!} = f(x)$$

7-16 We know from Exercise 7-14 that

$$\tan^{-1}x = \Sigma \frac{(-1)^n}{2n+1} x^{2n+1}$$

for $|x| < 1$. We can compute the power series expansion of $f(x) = (\tan^{-1} x)^2$ by computing the product

$$\left(\Sigma \frac{(-1)^n}{2n+1} x^{2n+1} \right)\left(\Sigma \frac{(-1)^n}{2n+1} x^{2n+1} \right)$$

If we denote the coefficient of x^n in the power series expansion of $\tan^{-1} x$ by a_n, then

$$a_n = \begin{cases} 0 & \text{if } n \text{ is even} \\ \dfrac{(-1)^k}{2k+1} & \text{if } n = 2k+1 \end{cases}$$

where k is a nonnegative integer. Thus, to find the power series expansion of $(\tan^{-1} x)^2$, we must compute $\{a_n\} * \{a_n\}$. Denote the nth term of this convolution by c_n; that is, put

$$c_n = \sum_{k=0}^{n} a_k a_{n-k}$$

Notice that if n and k are odd, then $n - k$ is even.   Thus if n is odd and k is odd, then $a_k a_{n-k} = 0$ because $a_{n-k} = 0$.   On the other hand, if n is odd and k is even, then $a_k a_{n-k} = 0$ because $a_k = 0$.   In other words, $c_n = 0$ if n is odd. Since $c_0 = a_0 d_0 = 0$, it can be seen that the series expansion of $(\tan^{-1} x)^2$ contains only even powers of x.   Suppose then that n is even.   We can write $n = 2m$ for some integer m, and examine

$$c_{2m} = \sum_{k=0}^{2m} a_k a_{2m-k}$$

Again, if j is even, then $a_j = 0$.   Consequently, only terms of the form $a_{2k+1} a_{2m-(2k+1)}$ appear in c_{2m}.   Thus

$$c_{2m} = \sum_{k=0}^{m-1} \frac{(-1)^k}{2k+1} \frac{(-1)^{m-k-1}}{2m-(2k+1)} = (-1)^{m-1} \sum_{k=0}^{m-1} \frac{1}{2k+1} \frac{1}{2m-(2k+1)}$$

While there is nothing wrong with this last expression, it can be somewhat simplified by making use of the identity

$$\frac{1}{2k+1} \frac{1}{2m-(2k+1)} = \frac{1}{2m} \left[\frac{1}{2k+1} + \frac{1}{2m-(2k+1)} \right]$$

[This identity is the partial fraction expansion of $1/x(a - x)$ when $x = 2k + 1$ and $a = 2m$.]   Since

$$\sum_{k=0}^{m-1} \frac{1}{2k+1} = \sum_{k=0}^{m-1} \frac{1}{2m-(2k+1)}$$

we see that

$$c_{2m} = \frac{(-1)^{m-1}}{m} \sum_{k=0}^{m-1} \frac{1}{2k+1}$$

In other words,

$$(\tan^{-1}x)^2 = \sum_{n=1}^{\infty} \frac{(-1)^{n-1}}{n} \left(\sum_{k=0}^{n-1} \frac{1}{2k+1} \right) x^{2n}$$

for $|x| < 1$.

7-17   (a)   We apply the ratio test.   Since

$$\left\{ \left| \frac{\binom{p}{n+1} x^{n+1}}{\binom{p}{n} x^n} \right| \right\} = \left\{ \left| \frac{p-n}{n+1} \right| \right\} |x| \to |x|$$

we see that the series has radius of convergence 1.

(b) Since

$$f'(x) = \sum_{n=1}^{\infty} n\binom{p}{n} x^{n-1}$$

we have, for $|x| < 1$,

$$(1 + x)f'(x) = \sum_{n=1}^{\infty} n\binom{p}{n} x^{n-1} + \sum_{n=1}^{\infty} n\binom{p}{n} x^n$$

$$= p + \sum_{n=1}^{\infty} \left[n\binom{p}{n} + (n+1)\binom{p}{n+1} \right] x^n$$

Now it is easy to see that

$$n\binom{p}{n} = p\binom{p-1}{n-1} \qquad \text{and} \qquad (n+1)\binom{p}{n+1} = p\binom{p-1}{n}$$

Thus

$$n\binom{p}{n} + (n+1)\binom{p}{n+1} = p\left[\binom{p-1}{n-1} + \binom{p-1}{n} \right]$$

$$= p\left[\frac{(p-1)\cdots(p-n+1)}{(n-1)!} + \frac{(p-1)\cdots(p-n)}{n!} \right]$$

$$= p\left[(p-1)\cdots(p-n+1)\frac{n+(p-n)}{n!} \right] = p\left[\frac{p(p-1)\cdots(p-n+1)}{n!} \right]$$

$$= p\binom{p}{n}$$

and so

$$\sum_{n=1}^{\infty} \left[n\binom{p}{n} + (n+1)\binom{p}{n+1} \right] x^n = p\sum_{n=1}^{\infty} \binom{p}{n} x^n$$

Therefore, $(1 + x)f'(x) = pf(x)$ for $|x| < 1$.

(c) Since $(1 + x)f'(x) = pf(x)$, we have

$$\frac{f'(x)}{(1+x)^p} - \frac{p}{(1+x)^{p+1}} f(x) = 0$$

or

$$D\left[\frac{f(x)}{(1+x)^p} \right] = 0$$

Thus $f(x)/(1 + x)^p = C$ for some constant C. But $f(0) = 1$ and this means that $C = 1$. Hence $f(x) = (1 + x)^p$ for $|x| < 1$.

7-18 By means of the alternating series test, it is easy to see that $\Sigma(-1)^n/(2n + 1)$ converges. On the other hand, we know from Exercise 7-14 that $\tan^{-1} x = \Sigma(-1)^n x^{2n+1}/(2n+1)$ for $|x| < 1$. Consequently, we can apply Abel's Theorem and conclude that

$$\frac{\pi}{4} = \lim_{x \to 1-} \tan^{-1} x = \Sigma \frac{(-1)^n}{2n+1}$$

PROBLEMS

7-1   Each of the following sequences is a sequence of partial sums. Determine the terms of each series and state whether it is convergent or divergent. Find the sum of each convergent series.

(a) $\left\{\dfrac{n}{2n+1}\right\}$

(b) $\{\log(n+1)\}$

(c) $\left\{\dfrac{(-1)^n}{2^n}\right\}$

7-2   Show that each of the following series diverges.

(a) $\sum \cos n$

(b) $\sum (\tfrac{1}{2})^{1/n}$

(c) $\sum 1/n \log[1 + 1/n]$

(d) $\displaystyle\sum_{n=2}^{\infty} \dfrac{n^{n-1/n}}{(n - 1/n)^n}$

7-3   Use the comparison test to determine the convergence or divergence of each of the following series.

(a) $\displaystyle\sum \dfrac{1}{n^2 + a^2}, a \neq 0$

(b) $\displaystyle\sum \dfrac{n}{n^2 + 1}$

(c) $\displaystyle\sum \dfrac{2 + \cos n}{n^2 + 2}$

*(d) $\displaystyle\sum \sin \dfrac{\pi}{n}$

(e) $\sum \sin^2[\pi(n + 1/n)]$

(f) $\sum [(1 + n^2)^{1/2} - n]$

(g) $\displaystyle\sum \dfrac{1}{n^{1+1/n}}$

*(h) $\displaystyle\sum_{n=2}^{\infty} \dfrac{1}{(\log n)^\alpha}, \alpha > 0$

*(i) $\displaystyle\sum_{n=2}^{\infty} \dfrac{1}{(\log n)^{1/n}}$

*(j) $\displaystyle\sum_{n=2}^{\infty} \dfrac{1}{(\log n)^{\log n}}$

7-4   Use the integral test to determine the convergence or divergence of each of the following series.

(a) $\displaystyle\sum_{n=2}^{\infty} \dfrac{1}{n(\log n)^\alpha}, \alpha > 0$

(b) $\displaystyle\sum_{n=3}^{\infty} \dfrac{1}{n \log n[\log(\log n)]^2}, \alpha > 0$

7-5   Use the nth root test to determine the convergence or divergence of each of the following series

(a) $\sum (n^{1/n} - 1)^n$

(b) $\sum n/2^n$

(c) $\sum 2^n n!/n^n$

(d) $\sum 3^n n!/n^n$

(e) $\sum \alpha^n n^\alpha, \alpha > 0$

7-6 Use the ratio test to determine the convergence or divergence of each of the following series.

(a) $\sum \dfrac{n!}{3 \cdot 5 \cdot 7 \cdots (2n+1)}$

(b) $\sum n!/\alpha^n,\ \alpha > 0$

(c) $\sum (n!)^2/(2n)!$

(d) $\sum n^n/n!\,n^x,\ \alpha > 0$

(e) $\sum 1/a_n$, where a_n is the nth Fibonacci number (see Example 2-2).

7-7 Determine the convergence or divergence of each of the following series.

(a) $\sum \sin^2 \dfrac{\pi}{n}$

(b) $\sum \left[\dfrac{1}{bn} - \dfrac{1}{an}\right],\ a > b > 0$

(c) $\sum \dfrac{2n}{2^n \alpha^n},\ \alpha > 0$

(d) $\sum n^{n+1/n}/(n+1/n)^n$

(e) $\sum \dfrac{n+1}{n^2}$

(f) $\sum \dfrac{n+1}{n^3}$

(g) $\sum \dfrac{n+1}{2^n}$

(h) $\sum n^p/(n^q + a),\ a > 0$

*(i) $\sum \left[\dfrac{1}{n} - \dfrac{1}{e^{n^2}}\right]$

(j) $\sum \dfrac{1}{(a+bn)^x},\ a > 0,\ b > 0,\ \alpha > 0$

(k) $\sum 2^n(1+n^2)/n!$

(l) $\sum \left(\dfrac{1}{2}\right)^{n/2}$

(m) $\sum n^x/n!$

(n) $\sum \dfrac{n+1}{n+2}\dfrac{1}{2^n}$

*(o) $\sum\limits_{n=2}^{\infty} \dfrac{\log(n+1) - \log n}{(\log n)^2}$

(p) $\sum [(n^x + 1)^{1/2} - (n^x)^{1/2}],\ \alpha > 0$

*(q) $\sum \dfrac{\log n}{n^x},\ \alpha > 0$

7-8 Determine the convergence or divergence of each of the following alternating series. Test each convergent series for absolute convergence.

(a) $\sum (-1)^{n-1} \dfrac{2n-1}{9n+1}$

(b) $\sum (-1)^{n-1} \log \dfrac{n}{n+1}$

(c) $\sum\limits_{n=2}^{\infty} (-1)^{n-1} \dfrac{\log n}{n}$

(d) $\sum (-1)^{n-1} \dfrac{1}{n^x},\ \alpha > 0$

7-9 If $a_n > 0$ for all n and $\sum a_n$ converges, show that $\sum 1/a_n$ diverges.

*7-10 Show that if $a_n > 0$ for all n and $\sum a_n$ converges, then so does $\sum (a_n a_{n+1})^{1/2}$.

*7-11 Show that if $a_n > 0$ for all n and $\sum a_n$ converges, then so does $\sum [a_n/n^{1+\delta}]^{1/2}$ for $\delta > 0$.

*7-12 If $\{a_n\}$ is a strictly decreasing sequence of positive terms, and

$$b_n = \prod_{k=1}^{n} a_k$$

when does $\sum b_n$ converge?

7-13   If $\{a_n\}$ is a decreasing sequence of positive terms and $\sum a_n$ converges, show that $\{na_n\} \to 0$.

7-14   If $\sum a_n^2$ and $\sum b_n^2$ both converge, show that $\sum a_n b_n$ is absolutely convergent.

7-15   Show that if $\sum a_n$ is absolutely convergent, then so are

(a) $\sum \dfrac{a_n}{1 + a_n}$, where no $a_n = -1$ (b) $\sum a_n^2$

(c) $\sum \dfrac{a_n^2}{1 + a_n^2}$

7-16   If $\sum a_n$ has sum S and $S_n = \sum\limits_{k=1}^{n} a_k$, show that

$$\left\{ \frac{1}{m} \sum_{n=1}^{m} S_n \right\} \to S$$

7-17   Let f be continuous, positive and decreasing on $[1, \infty)$ and put $a_n = f(n)$,

$$S_n = \sum_{k=1}^{n} a_k$$

and $I_n = \int_1^n f(x)\, dx.$ Show that $\{S_n - I_n\}$ is decreasing and has a limit between 0 and a_1.

7-18   If $a_n > 0$ and $\{a_{n+1}/a_n\} \to L < 1$, prove that if $L < r < 1$, then there exists an integer m such that

$$\sum a_n - \sum_{n=1}^{m} a_n < a_m \frac{r}{1-r}$$

7-19   Each of the following series is a geometric series, a telescoping series, or one related to these.   Show that

(a) $\sum \dfrac{1}{n(n+1)} = 1$ (b) $\sum (-1)^{n+1} \dfrac{2n+1}{n(n+1)} = 1$

(c) $\sum \dfrac{2^n + 3^n}{6^n} = \dfrac{3}{2}$ (d) $\sum\limits_{n=2}^{\infty} \dfrac{1}{n^2 - 1} = \dfrac{3}{4}$

(e) $\sum\limits_{n=0}^{\infty} \dfrac{1}{(\alpha + n)(\alpha + n + 1)} = \dfrac{1}{\alpha}$, where $\alpha \neq 0, -1, -2, \ldots$

7-20   Find the interval of convergence of each of the following power series and test for convergence at the endpoints (if any).

(a) $\sum\limits_{n=1}^{\infty} \dfrac{(-1)^n x^n}{n^\alpha}$, $\alpha > 0$ (b) $\sum (-1)^n n!\, x^n$

(c) $\sum n^\alpha x^n$, $\alpha > 0$ (d) $\sum (-1)^n \dfrac{(x-a)^n}{(n+1)2^n}$

*(e) $\sum\limits_{n=2}^{\infty} \dfrac{(x-a)^n}{(\log n)^\alpha}$ *(f) $\sum \dfrac{n^n}{n!} x^n$

7-21 Prove that if an $\begin{cases} \text{even} \\ \text{odd} \end{cases}$ function can be represented by its Maclaurin's series

in some interval including zero, then this power series contains only $\begin{cases} \text{even} \\ \text{odd} \end{cases}$

powers.

7-22 By using geometric series, find the power series expansion of each of the following functions and give its interval of convergence.

(a) $f(x) = \dfrac{1}{a-x}, a > 0$ (b) $f(x) = \dfrac{bx^m}{a-x^k}, a > 0, m \text{ and } k$

positive integers

*(c) $f(x) = \dfrac{x}{1+x-2x^2}$ *(d) $f(x) = \dfrac{26x^2 - 18x + 3}{(1-2x)(1-3x)(1-4x)}$

7-23 By using the binomial series, find the power series expansion of each of the following functions and give its interval of convergence.

(a) $f(x) = (a+x)^p, a > 0$ (b) $f(x) = bx^m(a+x^k)^p, a > 0,$

$m \text{ and } k \text{ positive integers}$

(c) $f(x) = (1-x)^p(1+x)^p$ *(d) $f(x) = \dfrac{(a^2 + b^2)x^2 - 2(a+b)x + 2}{(1-ax)^2(1-bx)^2}$

$a > 0, b > 0$

7-24 Find the power series expansion of each of the following functions and give its interval of convergence.

*(a) $f(x) = \dfrac{x^{m-1}}{(1+x^m)^2}, m$ a positive integer   (b) $f(x) = \sin^{-1} x$

(c) $f(x) = \log [x + (1+x^2)^{1/2}]$ (d) $f(x) = \log \left[\dfrac{1 + (1+x)^{1/2}}{2} \right]$

7-25 If $0 < p < 1$, and $p + q = 1$, prove that
$\sum np^n = p/q^2$, and $\sum n^2 p^n = (p^3 + p)/q^3$

*7-26 Show that $\sum n/n! = e$, $\sum n^2/n! = 2e$, and $\sum n^3/n! = 5e$. More generally, if k is a positive integer, prove that $\sum n^k/n!$ is an integral multiple of e.

7-27 From power series expansions obtained earlier, find the power series expansion of each of the following functions and give its interval of convergence.

(a) $f(x) = \tan^{-1}x^2$ (b) $f(x) = e^{-x^2/2}$

(c) $f(x) = \cos x^{1/2}$ (d) $f(x) = \log \dfrac{1}{1-x}$

7-28 By adding or multiplying power series expansions obtained earlier, find the power series expansion of each of the following functions and give its interval of convergence.

(a) $f(x) = \log \dfrac{1+x}{1-x}$ (b) $f(x) = \left(\dfrac{x+a}{x+b} \right)^p, a > 0, b > 0, a \neq b$

(c) $f(x) = \dfrac{x}{(1-x)(1-x^2)}$ (d) $f(x) = \dfrac{\log{(1+x)}}{1+x}$

(e) $f(x) = \tan^{-1}x \, \log{(1+x^2)}$

*7-29 By integrating the appropriate power series, find the power series expansion of each of the following functions and give its interval of convergence.

(a) $f(x) = [\log{(1+x)}]^2$

(b) $f(x) = (x+a)^{1/2}(x+b)^{1/2} + (a-b)\log{[(x+a)^{1/2}+(x+b)^{1/2}]}$,
$\quad a>0, b>0, a \neq b$

(c) $f(x) = (a-x)^{1/2}(b+x)^{1/2} + (a+b)\sin^{-1}\left(\dfrac{x+b}{a+b}\right)^{1/2}$, $a>0, b>0$

7-30 Using the power series expansions of the exponential, sine, and cosine functions, show that

(a) $e^x e^y = e^{x+y}$ (b) $2 \sin x \cos x = \sin 2x$

7-31 Evaluate each of the following integrals as a convergent series; justify all your steps

(a) $\displaystyle\int_0^1 \dfrac{\tan^{-1}x}{x}\,dx$ (b) $\displaystyle\int_0^1 \dfrac{\log{(1+x)}}{x}\,dx$

(c) $\displaystyle\int_0^1 \dfrac{x^{m-1}}{1+x^k}\,dx$, k and m positive integers

*7-32 If p and q are real numbers, and n is a positive integer, prove that

$$\sum_{k=0}^{n} \binom{p}{k}\binom{q}{n-k} = \binom{p+q}{n}$$

Deduce from this identity the relation

$$\sum_{k=0}^{n} (-1)^k \binom{p}{k}\binom{n-k+q-1}{n-k} = (-1)^n \binom{p-q}{n}$$

7-33 Use Abel's theorem to show that each of the following series has the indicated sum

*(a) $\log\left(\dfrac{1+2^{1/2}}{2}\right) = \displaystyle\sum_{n=1}^{\infty} (-1)^{n-1} \dfrac{1 \cdot 3 \cdot 5 \cdots (2n-1)}{2 \cdot 4 \cdot 6 \cdots (2n)} \dfrac{1}{2n}$

*(b) $\dfrac{\pi}{2} = 1 + \displaystyle\sum_{n=1}^{\infty} \dfrac{1 \cdot 3 \cdot 5 \cdots (2n-1)}{2 \cdot 4 \cdot 6 \cdots (2n)} \dfrac{1}{2n+1}$

*(c) $\dfrac{1}{2}(\log 2)^2 = \displaystyle\sum_{n=1}^{\infty} (-1)^{n-1}\left[1 + \dfrac{1}{2} + \cdots + \dfrac{1}{n}\right] \dfrac{1}{n+1}$

*(d) $\dfrac{\pi}{8}\log 2 = \displaystyle\sum_{n=1}^{\infty} (-1)^{n-1}\left[1 + \dfrac{1}{2} + \cdots + \dfrac{1}{2n}\right] \dfrac{1}{2n+1}$

*(e) $\dfrac{\pi^2}{32} = \displaystyle\sum_{n=1}^{\infty} (-1)^{n-1}\left[1 + \dfrac{1}{3} + \cdots + \dfrac{1}{2n-1}\right] \dfrac{1}{2n}$

*7-34   If $p + 1 > 0$, prove that

$$\sum \binom{p}{n} = 2^p$$

If $p > 0$, prove that

$$\sum (-1)^n \binom{p}{n} = 0$$

What is the behavior of $\sum \binom{p}{n}$ when $p + 1 \leq 0$, and of $\sum (-1)^n \binom{p}{n}$ when $p \leq 0$?

HINTS

1-2   Use Problem 1-1.

1-3   Equation (1-18) is needed.

1-4   Recall the trigonometric identity $\sin 2\psi = 2 \sin \psi \cos \psi$

1-5   Equation (1-12) may be needed.

1-6   Equation (1-18) and the binomial theorem may be useful.

1-11   (e)   Use Problem 1-11d.

1-11   (f)   Use Problem 1-11e.

1-14   Prove the first statement by induction; also, Problem 1-13b may be helpful.

1-17   Use Problem 1-16c.

1-20   Make use of the identity $(b + c)(a^2 - bc) + (a + c)(b^2 - ac) + (a + b)(c^2 - ab) = 0$; also, recall Problem 1-17b.

1-22   Prove first that $(a_k - a_m)(b_k - b_m) \geq 0$ for $1 \leq k \leq n$ and $1 \leq m \leq n$.

1-24, 25, 26   Use the Cauchy–Schwartz inequality (see the remark following the solution of Problem 1-23).

1-27   Use the Cauchy–Schwartz inequality and Problem 1-16c.

1-28, 29   Try a proof by induction.

1-30   Use the inequality of the arithmetic-geometric means (see the remark following the solution of Problem 1-16).

2-5c   To evaluate the limit, consider the sequence $\{b_n\}$ defined by $b_n = n^3/2^n$.

2-5d   Put $a = 1/(1 + b)$

2-5e Let $a_n = (n^{1/n})^{1/2}$, put $a_n = 1 + b_n$, and apply Bernoulli's inequality to $(1 + b_n)^n$ [for Bernoulli's inequality see the remark following the solution of Problem 1-31].

2-7a Recall Problem 2-6c.

2-7b Use part (a)

2-8c Put $a_n = \sum_{k=1}^{n} \dfrac{1}{n+k}$, and consider $a_{n+1} - a_n$.

2-8d If $a_n = (1 + 1/n)^n$, show that $a_{n+1} \leq a_n$; Bernoulli's inequality may be useful.

2-10 Let α be the positive root of $x^2 - x - 1 = 0$. Consider the three cases $0 < a_1 < \alpha$, $a_1 = \alpha$, and $a_1 > \alpha$ separately. If $0 < a_1 < \alpha$, or $a_1 > \alpha$, the sequence is doubly monotone.

2-11 Recall the inequality of the arithmetic-geometric means.

2-12 Put $a_{n+2} = (\alpha_n/2^n)a_1 + (\beta_n/2^n)a_2$, then prove that $\{\alpha_n\}$ and $\{\beta_n\}$ satisfy the same recurrence relation as $\{a_n\}$.

2-13d Recall the inequality $H_n \leq G_n \leq A_n$ given in the remark following the solution of Problem 1-16.

2-14 Use Problem 2-13d.

3-1c Break up the domain of f.

3-5b Use the identity $f(x) = [f(x) + f(-x)]/2 + [f(x) - f(-x)]/2$.

3-13c $|x^2 - a^2| = |x + a|\,|x - a|$. Consider only those x such that $a - 1 < x < a + 1$; then there is a constant M such that $|x + a| \leq M$.

3-13d $|1/x - 1/a| = |x - a|/|ax|$. Restrict your attention to x such that $1/|ax| \leq M$, where M is some constant.

3-13e $|x^{1/2} - a^{1/2}| = |x - a|/(x^{1/2} + a^{1/2})$. Proceed as in parts (c) and (d).

3-17 Put $a_n = 1/2\pi n$, $b_n = 2/\pi(4n + 1)$, and examine the limits of $\{f(a_n)\}$ and $\{f(b_n)\}$.

3-18 If $\{a_n\}$ is monotone, then $\{f(a_n)\}$ is also monotone.

3-19 Prove that $f(nx) = nf(x)$ for any integer n; put $c = f(1)$; then note that $c = f(n/n) = nf(1/n)$. Deduce that if r is a rational number, then $f(r) = cr$.

4-10a Try a proof by induction; also, recall Eq. (1-18).

4-15, 16 $f'(0) = \lim_{h \to 0} [f(0 + h) - f(0)]/h$.

4-19a Consider $D\,f/g$.

4-22a Problem 4-19a may be useful.

5-1d Let $u = x^2$

5-1e   Put $u = (x - a)/(x - b)$

5-2d $1/(1 + \cos x) = (1 - \cos x)/(1 - \cos^2 x)$

5-2j $1/(a^2 \sin^2 x + b^2 \cos^2 x) = \sec^2 x/(a^2 \tan^2 x + b^2)$

5-3b   Let $u = e^x + e^{-x}$

5-4d   Complete the square.

5-5e   Put $u = x$, $dv = dx/(2x + 1)^{1/2}$

5-6g   Let $x - a = (b - a) \sin^2 \theta$

5-8(a)   Let $x = \sin \theta$
   (d)   Put $u = \log^2 x$, $dv = x^4 \, dx$
   (e)   Let $u = 3x^2 - 2 \cos 3x$
   (k)   Let $[1 + (1 + x)^{1/2}]^{1/2} = u$
   (z)   Put $x = \tan^{2/3} \theta$

5-9   Let $u = \tan x/2$

5-12c   Let $(1 + x)/x = u^s$

5-14   Integrate by parts.

5-16 $\dfrac{Q(x)}{x - \alpha_k} = \dfrac{Q(x) - Q(\alpha_k)}{x - \alpha_k} \to Q'(\alpha_k)$ as $x \to \alpha_k$; now consider $(x - \alpha_k)P(x)/Q(x)$.

5-25   Note that $dv/dt = (dv/dx)(dx/dt) = v(dv/dx)$

6-2   Equation (1-9) will be needed.

6-3   Recall that $2 \sin A \sin B = \cos (A - B) - \cos (A + B)$.

6-7   Integrate by parts.

6-8   Consider the limits of sequences of Riemann sums.

6-11 $1/2 = \displaystyle\int_0^{1/2} dx$

6-12   Same trick as in problem 6-11

6-13   Use L'Hospital's rule

6-15   For part (b), put

$$\int_{-a}^a \frac{hf(x)}{(h^2 + x^2)} \, dx = \int_{-a}^a \frac{h[f(x) - f(0)]}{(h^2 + x^2)} \, dx + f(0) \int_{-a}^a \frac{h}{(h^2 + x^2)} \, dx$$

now show that

$$\lim_{h \to 0} \int_{-a}^a \{h[f(x) - f(0)]/(h^2 + x^2)\} \, dx = 0$$

6-16 Suppose that $\int_0^\infty f(x)\,dx$ does not converge. It follows from Problem 6-8b that $\left|\int_0^t f(x)\,dx\right| \le \int_0^t |f(x)|\,dx \le \int_0^\infty |f(x)|\,dx$. Therefore $\int_0^t f(x)\,dx$ must oscillate, and so sequences $\{a_n\}$ and $\{b_n\}$ can be found such that $\{a_n\} \to \infty$, $a_n < b_n < a_{n+1}$ for $n = 1, 2, \ldots$, and $\left\{\int_0^{a_n} f(x)\,dx\right\} \to a$, $\left\{\int_0^{b_n} f(x)\,dx\right\} \to b$ with $a \ne b$. Show that this is not possible.

6-21b Let $x = \log u$

6-25 Use L'Hospital's rule as many times as is necessary.

6-27 $\int_0^\infty f(x)\,dx = \int_0^{\varepsilon/2} f(x)\,dx + \int_{\varepsilon/2}^\infty f(x)\,dx$; now treat both integrals separately.

6-28 $\int_0^\infty \sin x/x\,dx = \int_0^{\pi/2} \sin x/x\,dx + \int_{\pi/2}^\infty \sin x/x\,dx$

6-35 Parametric equations of the ellipse are $x = a \sin \theta$, $y = b \cos \theta$.

6-36 This curve has parametric equations $x = a \sin^3 \theta$, $y = b \cos^3 \theta$.

6-53, 54 Find dy/dx by implicit differentiation.

6-67 When the integral is set up, put $e^x = u$

6-69 Let $y = tx$ and express x, y, and dx in terms of t.

7-3(d) $\{\sin \pi/n / \pi/n\} \to 1$
  (h) $\{\log n/n^\delta\} \to 0$ for all $\delta > 0$ [see Problem 4-44a].
  (i) See part (h)
  (j) $\log (\log n) > 2$ for all sufficiently large n.

7-7(i) Show that $\{n/e^{n^2}\} \to 0$
  (o) $\log (1 + 1/n) < 1/n$ (see Problem 4-56).
  (q) See the hint for Problem 7-3h.

7-10, 11 Recall the inequality of the arithmetic-geometric mean.

7-12 $\{a_n\}$ converges.

7-20(e) Recall Problem 7-3h.
  (f) Recall Stirling's formula (see the remark following the solution of Problem 6-70).

7-22, c, d Use a partial fraction expansion

7-23d Ditto

7-24a Integrate f

7-26 Prove by induction that

$$\sum \frac{n^k}{n!}\, x^{n-1} = e^x p_k(x)$$

where p_k is a polynomial of degree k.

7-29   Find the power series expansion of the derivative.

7-32 $(1 + x)^p(1 + x)^q = (1 + x)^{p+q}$

7-33(a)   See Problem 7-24d
  (b)   See Problem 7-24b
  (c)   See Problem 7-29a; also, Euler's constant may be needed (see the remark following the solution of Problem 7-17).
  (d)   See Problem 7-28e
  (e)   See Exercise 7-16; also, recall Euler's constant.

7-34   Consider

$$\left| \frac{\binom{p}{n+1}}{\binom{p}{n}} \right| = \left| \frac{p-n}{n+1} \right|$$

SOLUTIONS TO PROBLEMS

CHAPTER 1

1-1   (a) $\sum_{k=1}^{n}(a_k - a_{k-1}) = \sum_{k=1}^{n} a_k - \sum_{k=1}^{n} a_{k-1} = \sum_{k=1}^{n-1} a_k + a_n - \sum_{k=1}^{n-1} a_k - a_0 = a_n - a_0$

     (b) $\prod_{k=1}^{n} \dfrac{a_k}{a_{k-1}} = \prod_{k=1}^{n} a_k \bigg/ \prod_{k=1}^{n} a_{k-1} = a_n \prod_{k=1}^{n-1} a_k \bigg/ a_0 \prod_{k=1}^{n-1} a_k = a_n/a_0$

1-2   (a)   Since $1/k(k+1) = 1/k - 1/(k+1)$, we have, from Problem 1-1a,

$$\sum_{k=1}^{n} \frac{1}{k(k+1)} = \sum_{k=1}^{n}\left[\frac{1}{k} - \frac{1}{k+1}\right] = 1 - \frac{1}{n+1}$$

     (b)   Here we have $(-1)^{k+1}(2k+1)/k(k+1) = (-1)^{k-1}/k - (-1)^k/(k+1)$, and so

$$\sum_{k=1}^{n} (-1)^{k+1} \frac{2k+1}{k(k+1)} = 1 - \frac{(-1)^n}{n+1} = 1 + \frac{(-1)^{n+1}}{n+1}$$

     (c) $\prod_{k=2}^{n}(1 - 1/n^2) = \prod_{k=2}^{n}(1 - 1/n)(1 + 1/n) = \prod_{k=2}^{n}(1 - 1/n)\prod_{k=2}^{n}(1 + 1/n)$

$$= \prod_{k=2}^{n}\frac{n-1}{n} \prod_{k=2}^{n}\frac{n+1}{n} = \frac{1}{n}\frac{n+1}{2} = \frac{n+1}{2n}$$

(we have used Problem 1-1b twice).

     (d)   From the identity $(1 + a^{2^{k-1}})(1 - a^{2^{k-1}}) = 1 - (a^{2^{k-1}})^2 = 1 - a^{2^k}$, it follows that

$$\prod_{k=1}^{n}(1 + a^{2^{k-1}}) = \prod_{k=1}^{n}\frac{1 - a^{2^k}}{1 - a^{2^{k-1}}} = \frac{1 - a^{2^n}}{1 - a}$$

1-3   Let P_n denote the proposition in question. Obviously, P_1 is true. Assume that P_n is true. Then

$$(a+b)^{n+1} = (a+b)(a+b)^n = (a+b) \sum_{k=0}^{n} \binom{n}{k} a^k b^{n-k}$$

$$= \sum_{k=0}^{n} \binom{n}{k} a^{k+1} b^{n-k} + \sum_{k=0}^{n} \binom{n}{k} a^k b^{n+1-k}$$

$$= b^{n+1} + \sum_{k=1}^{n} \left[\binom{n}{k} + \binom{n}{k-1} \right] a^k b^{n+1-k} + a^{n+1}$$

$$= b^{n+1} + \sum_{k=1}^{n} \binom{n+1}{k} a^k b^{n+1-k} + a^{n+1} \text{ [by Eq. (1-18)]}$$

$$= \sum_{k=0}^{n+1} \binom{n+1}{k} a^k b^{n+1-k}$$

Thus P_{n+1} is true if P_n is true and the proof is completed.

Remark   Two interesting consequences of the binomial theorem result from putting $a = b = 1$, and $a = -1, b = 1$.   We then have

$$\sum_{k=0}^{n} \binom{n}{k} = 2^n$$

and

$$\sum_{k=0}^{n} (-1)^k \binom{n}{k} = 0$$

respectively.

1-4   Once again, denote the given proposition by P_n. P_1 then is $\sin \theta = 2 \sin \theta/2 \cos \theta/2$. This equation is a familiar double-angle identity from trigonometry, and so P_1 is true. If P_n is true, we have

$$(2^{n+1} \sin \theta/2^{n+1}) \prod_{k=1}^{n+1} \cos \theta/2^k = (2 \sin \theta/2^{n+1} \cos \theta/2^{n+1}) 2^n \prod_{k=1}^{n} \cos \theta/2^k$$

$$= \sin \theta/2^n \frac{\sin \theta}{\sin \theta/2^n} = \sin \theta$$

The truth of P_{n+1} thus follows from that of P_n.

1-5   If P_n denotes the stated equation, then P_1 is clearly true. Assume that P_n is true. Then

$$\left(\sum_{k=1}^{n+1} k \right)^2 = \left[\sum_{k=1}^{n} k + (n+1) \right]^2 = \left(\sum_{k=1}^{n} k \right)^2 + 2(n+1) \sum_{k=1}^{n} k + (n+1)^2$$

$$= \sum_{k=1}^{n} k^3 + 2(n+1) \frac{n(n+1)}{2} + (n+1)^2 \text{ [by Eq. (1-12)]}$$

$$= \sum_{k=1}^{n} k^3 + (n+1)^3 = \sum_{k=1}^{n+1} k^3$$

Hence if P_n is true, so is P_{n+1} and the formula is established by induction.

Remark We may write this result as

$$\sum_{k=1}^{n} k^3 = \left[\frac{n(n+1)}{2}\right]^2$$

by making use of Eq. (1-12). If we introduce the notation

$$S_n^m = \sum_{k=1}^{n} k^m$$

we have $S_n^1 = n(n+1)/2$ and $S_n^3 = [n(n+1)/2]^2$. These formulas naturally raise the question of whether or not we can evaluate S_n^m for all m. The following provides an affirmative answer. More precisely, we show how to find S_n^m if $S_n^1, \cdots, S_n^{m-1}$ are known, and since S_n^1 is known this means that we can find all S_n^m, at least in principle. To this end, observe that the binomial theorem gives us

$$(k+1)^{m+1} - k^{m+1} = \binom{m+1}{1} k^m + \binom{m+1}{2} k^{m-1} + \cdots + \binom{m+1}{m} k + 1$$

If we sum both sides of this identity from $k=0$ to $k=n$, we get

$$(n+1)^{m+1} = \sum_{k=0}^{n} [(k+1)^{m+1} - k^{m+1}]$$

$$= \binom{m+1}{1} \sum_{k=0}^{n} k^m + \binom{m+1}{2} \sum_{k=0}^{n} k^{m-1} + \cdots + \binom{m+1}{m} \sum_{k=0}^{n} k + (n+1)$$

$$= \binom{m+1}{1} S_n^m + \binom{m+1}{2} S_n^{m-1} + \cdots + \binom{m+1}{m} S_n^1 + (n+1)$$

Therefore,

$$\binom{m+1}{1} S_n^m = (n+1)^{m+1} - (n+1) - \sum_{k=1}^{m-1} \binom{m+1}{m+1-k} S_n^k$$

and from this equation we see that S_n^m can be calculated if we know $S_n^1, \ldots, S_n^{m-1}$. As an illustration, we find S_n^2 from S_n^1 in this way:

$$\binom{3}{1} S_n^2 = (n+1)^3 - (n+1) - \binom{3}{2} \frac{n(n+1)}{2} = \frac{n(n+1)(2n+1)}{6}$$

The reader is invited to re-evaluate S_n^3 by this method, and to check the correctness of the formula

$$S_n^4 = \frac{n(n+1)}{60} [12n^3 + 18n^2 + 2n - 2]$$

1-6   Denote this equation by P_n.   Then P_1 is obviously true, and if we assume that P_n is true, we have

$$\sum_{k=1}^{n+1}(-1)^{k-1}\frac{1}{k}\binom{n+1}{k} - \sum_{k=1}^{n+1}\frac{1}{k}$$

$$= \sum_{k=1}^{n+1}(-1)^{k-1}\frac{1}{k}\binom{n+1}{k} - \sum_{k=1}^{n}(-1)^{k-1}\frac{1}{k}\binom{n}{k} - \frac{1}{n+1}$$

$$= \sum_{k=1}^{n}(-1)^{k-1}\frac{1}{k}\left[\binom{n+1}{k} - \binom{n}{k}\right] + (-1)^n\frac{1}{n+1} - \frac{1}{n+1}$$

$$= \sum_{k=1}^{n}(-1)^{k-1}\frac{1}{k}\binom{n}{k-1} + \frac{(-1)^n}{n+1} - \frac{1}{n+1}\ \ \text{[because of Eq. (1-18)]}$$

Now

$$\frac{1}{k}\binom{n}{k-1} = \frac{1}{k}\frac{n!}{(k-1)![n-(k-1)]!} = \frac{1}{n+1}\frac{(n+1)!}{k!(n+1-k)!} = \frac{1}{n+1}\binom{n+1}{k}$$

Therefore,

$$\sum_{k=1}^{n+1}(-1)^{k-1}\frac{1}{k}\binom{n+1}{k} - \sum_{k=1}^{n+1}\frac{1}{k}$$

$$= \sum_{k=1}^{n}(-1)^{k-1}\frac{1}{n+1}\binom{n+1}{k} + \frac{(-1)^n}{n+1} - \frac{1}{n+1}$$

$$= \frac{1}{n+1}\sum_{k=1}^{n+1}(-1)^{k-1}\binom{n+1}{k} - \frac{1}{n+1} = \frac{1}{n+1}\left[1 - \sum_{k=0}^{n+1}(-1)^k\binom{n+1}{k}\right] - \frac{1}{n+1}$$

$$= \frac{1}{n+1}[1 - (1-1)^{n+1}] - \frac{1}{n+1}\ \ \text{(by the binomial theorem)}$$

$$= \frac{1}{n+1} - \frac{1}{n+1} = 0$$

In other words, if P_n is true, then

$$\sum_{k=1}^{n+1}(-1)^{k-1}\frac{1}{k}\binom{n+1}{k} = \sum_{k=1}^{n+1}\frac{1}{k}$$

or P_{n+1} is true.   The proof is completed.

1-7   (a)   We have, using rule (1) twice, $x - 2 < y - 2 < y + 3$.   On the other hand, $2 - 2 < 0 + 3$ and $0 < 2$.
(b)   Since $-x \le -y$, we have $-x + 4 \le -y + 4 < -y + 5$.   For the second part, let $x = 0$, $y = 1/2$.
(c)   If $x > y$, then $2x > 2y$ and $2x + 5 > 2y + 5 > 2y$.   An example for the second part is $x = 0$, $y = 1$.

Remark The argument used in the solution of this problem does not depend on the numbers involved. We can use it to show that if $a < b$ and $c < d$, then $a + c < b + d$. The proof is $a + c < b + c < b + d$. Needless to say, we can replace $<$ by $\leq$, $>$, or $\geq$. The result may be extended, by induction, as follows: If $a_k < b_k$ for $k = 1, 2, \ldots, n$, then

$$\sum_{k=1}^{n} a_k < \sum_{k=1}^{n} b_k$$

Again, the same is true for $\leq$, $>$, or $\geq$.

1-8   (a)   This inequality is the same as $3x < x + 4$, or $2x < 4$, or $x < 2$.

      (b)   The inequality here becomes $-3x \geq 3$, or $x \leq -1$.

      (c)   Since $2 > 0$, $(3x - 5)/2 \leq 0$ reduces to $3x - 5 \leq 0$, or $3x \leq 5$, or $x \leq 5/3$.

      (d)   We can multiply this inequality by $(x + 4)^2$ without changing its sense and obtain $x + 4 > 0$, or $x > -4$.

      (e)   Since the product of two numbers is negative if and only if one is positive and the other negative, we must have either $x - 1 < 0$ and $x + 3 > 0$, or $x - 1 > 0$ and $x + 3 < 0$. In other words, either $x < 1$ and $x > -3$, or $x > 1$ and $x < -3$. Since the second alternative is impossible, the solution is $-3 < x < 1$.

      (f)   If $x < 0$, both $4x$ and $25/x$ are negative, and so their sum is negative. Hence $4x + 25/x > 20$ is impossible for $x < 0$. If $x > 0$, we can write this inequality as $4x^2 + 25 > 20x$, or $4x^2 - 20x + 25 > 0$, or $(2x - 5)^2 > 0$. The last relation is true for all $x > 0$ except $x = 5/2$.

      (g)   It is easy to see that $1 + x$ and $3x - 1$ are both positive if $x > 1/3$, both negative if $x < -1$; while if $-1 < x < 1/3$, $x + 1 > 0$ and $3x - 1 < 0$. When both are positive or both are negative the inequality becomes $3x - 1 < 2 + 2x$, or $x < 3$. This means that we must have either $x < -1$, or $1/3 < x < 3$. When $x + 1$ and $3x - 1$ have opposite signs, the inequality becomes $3x - 1 > 2 + 2x$, or $x > 3$ which is incompatible with $-1 < x < 1/3$. The solution is thus either $x < -1$, or $1/3 < x < 3$.

      (h)   The product of four numbers is negative if and only if either exactly one is negative, or exactly three are negative. Now $x - 6 < 0$ when $x < 6$, $x - 2 < 0$ when $x < 2$, $x + 3 < 0$ when $x < -3$, and $x + 8 < 0$ when $x < -8$. Hence one factor, namely $x - 6$, is negative if and only if $2 < x < 6$, and all except $x + 8$ are negative if and only if $-8 < x < -3$. The solution then is either $2 < x < 6$, or $-8 < x < -3$.

1-9   (a)   The inequality is the same as $-4 < x - 1 < 4$, or $-3 < x < 5$.

      (b)   Here we have $-1 \leq 1 - 2x \leq 1$, or $-2 \leq -2x \leq 0$, or $0 \leq x \leq 1$.

      (c)   We must have either $x + 2 > 3$, or $x + 2 < -3$. That is, either $x > 1$, or $x < -5$.

      (d)   As in (c), either $(3x - 2)/4 \geq 5$, or $(3x - 2)/4 \leq -5$. That is, either $3x - 2 \geq 20$, or $3x - 2 \leq 20$. In other words, either $x \geq 22/3$, or $x \leq -6$.

1-10   (a)   Since $1/a^2 > 0$, we have $a(1/a^2) > 0(1/a^2)$, or $1/a > 0$.

      (b)   As in (a), $a(1/a^2) < 0(1/a^2)$, or $1/a < 0$

      (c)   Since $1/ab > 0$, we have $0(1/ab) < a/ab < b/ab$, or $0 < 1/b < 1/a$.

1-11   (a) $|-x| = [(-x)^2]^{1/2} = (x^2)^{1/2} = |x|$
      (b) $|xy| = [(xy)^2]^{1/2} = [x^2y^2]^{1/2} = (x^2)^{1/2}(y^2)^{1/2} = |x|\,|y|$
      (c) $|x|^2 = [(x^2)^{1/2}]^2 = x^2$
      (d)   If $x \geq 0$, $x = |x|$, and so $x = |x| \geq 0 \geq -|x|$. If $x < 0$, $-|x| = x < 0 < |x|$. Thus in either case $-|x| \leq x \leq |x|$.
      (e)   Using part (d), we have $-|y| \leq y \leq |y|$ and $-|x| \leq x \leq |x|$. If we add these inequalities, we get $-(|x| + |y|) \leq x + y \leq |x| + |y|$, and this means, by Theorem 1-1, that $|x + y| \leq |x| + |y|$.
      (f)   We want to show that $-|x + y| \leq |x| - |y| \leq |x - y|$. The left-hand inequality can be written as $|y| \leq |x| + |y - x|$, or $|x + (y - x)| \leq |x| + |y - x|$, which is part (e). The right-hand inequality is the same as $|x| \leq |x - y| + |y|$, or $|y + (x - y)| \leq |x - y| + |y|$, which is part (e) again.

Remark   The inequality in part (e) is called the *triangle inequality* because in the proper context, it can be interpreted as expressing the fact that the length of one side of a triangle is less than or equal to the sum of the lengths of the other two sides. This inequality is extremely useful, and it is easy to see that it can be extended by induction to become

$$\left| \sum_{k=1}^{n} x_k \right| \leq \sum_{k=1}^{n} |x_k|$$

1-12   Since $x^2 > 0$, $x \neq 0$, and so $|x| > 0$. As for the other inequality, suppose that $|x| \geq a^{1/2}$. Then $|x|^2 \geq |x|\,a^{1/2} \geq (a^{1/2})^2$, or $x^2 \geq a$ which is contrary to hypothesis. This contradiction means that $|x| < a^{1/2}$.

1-13   (a)   We want to show that $0 < (b - a) + (d - c)$. Now $b - a$ is positive, and $d - c$ is nonnegative, and so their sum must be positive.
      (b)   We must have either $0 < a < b$, or $0 < c < d$. If the first alternative is the case, then $0 < d$, and so $ac < ad$ and $ad < bd$. In other words, $ac < bd$. If the second alternative is the case, the proof is similar.

1-14   Let P_n denote the statement $a^n > b^n$ if $a > b > 0$. Then P_1 is clearly true. The truth of P_{n+1} follows from the truth of P_n by an application of Problem 1-13b. The inequality is thus established by induction. As for the second assertion, let $r = n/m$, where n and m are positive integers, and assume that $a^{n/m} \leq b^{n/m}$. But then the inequality we have just proved gives us $(a^{n/m})^m \leq (b^{n/m})^m$, or $a^n \leq b^n$. This contradiction establishes the inequality $a^r > b^r$ if $a > b > 0$, and r is a positive rational number.

1-15   (a)   We establish the given inequality by reducing it to an equivalent known inequality. Now the inequality holds if and only if $(a^2 + b)^{1/2} + (a^2 - b)^{1/2} < 2a$, and this inequality is true if and only if $a^2 + b + 2(a^2 + b)^{1/2}(a^2 - b)^{1/2} + a^2 - b < 0$, or $(a^2 + b)^{1/2}(a^2 - b)^{1/2} < a^2$. Finally, the last inequality holds if and only if $(a^2 + b)(a^2 - b) < a^4$, or $a^4 - b^2 < a^4$. Inasmuch as the inequality $a^4 - b^2 < a^4$ is true, and all our steps are reversible, the initial inequality must also be true.
      (b)   We follow the same procedure as in part (a). The given inequality is

equivalent to $a^2b/(1 + ab) < a^2b/(1 - ab)$, which is equivalent to $1 - ab < 1 + ab$, or $-ab < ab$ and this last inequality is obviously true.

(c) The inequality here is equivalent to $a^2(1 + bc) < b^2(1 + ac)$, or to $a^2 + a^2bc < b^2 + ab^2c$. Now the last inequality is the same as $0 < (b^2 - a^2) + abc(b - a)$, which is clearly true.

(d) This inequality is equivalent to $(a^2 + c)^{1/2} + a < (b^2 + c)^{1/2} + b$, or to $0 < (b - a) + [(b^2 + c)^{1/2} - (a^2 + c)^{1/2}]$ and the truth of the last inequality follows from the fact that $b^2 + c > a^2 + c$ and an application of Problem 1-14.

1-16 (a) Suppose for definiteness that $a < b$. Then $a + a < a + b < b + b$, or $a < (a + b)/2 < b$.

(b) Again take $a < b$. Then $a \cdot a < a \cdot b < b \cdot b$, or $a^2 < ab < b^2$, or $a < (ab)^{1/2} < b$. As for the other assertion, we have from Part (a) $1/b < (1/a + 1/b)/2 < 1/a$, or $a < 2/(1/a + 1/b) < b$.

(c) This inequality is equivalent to $2(ab)^{1/2} \le a + b$, which is equivalent to $4ab \le (a + b)^2 = a^2 + 2ab + b^2$. Now the last inequality can be written as $0 \le a^2 - 2ab + b^2 = (a - b)^2$, and the truth of this relation not only establishes our inequality, but shows further that there is equality only if $a = b$.

(d) This inequality is the same as $2ab/(a + b) \le (ab)^{1/2}$, which is equivalent to $(ab)^{1/2} \le (a + b)/2$, and this last is part (c).

Remark The numbers $(a + b)/2$ and $(ab)^{1/2}$ are called the *arithmetic mean* of a and b and the *geometric mean* of a and b respectively. Notice that a, $(a + b)/2$, and b form an arithmetic progression while a, $(ab)^{1/2}$, and b form a geometric progression. More generally, if $a_1, a_2, \ldots, a_n$ are n positive numbers, their geometric mean G_n and arithmetic mean A_n are defined by

$$G_n = \left[\prod_{k=1}^{n} a_k \right]^{1/n}$$

and

$$A_n = \frac{1}{n} \sum_{k=1}^{n} a_k$$

It is easy to see that, extending parts (a) and (b), both A_n and G_n are between the greatest and least of the a_k's. More surprisingly, it so happens that the inequality of part (c) extends to $G_n \le A_n$, or

$$\left[\prod_{k=1}^{n} a_k \right]^{1/n} \le \frac{1}{n} \sum_{k=1}^{n} a_k$$

This is the inequality of the arithmetic and geometric means; the most celebrated of all inequalities. We give here an elementary proof of this remarkable relation.

Let a_s be the smallest and a_l the largest of $a_1, a_2, \ldots, a_n$ (if there are several smallest or several largest, pick any one of them). Replace a_s by $a_s' = G_n$ and a_l by $a_l' = a_s a_l / G_n$, and denote by G_n' and A_n' the geometric mean and arithmetic mean of the resulting set of numbers. We then have $a_s' a_l' = a_s a_l$, and so $G_n' = G_n$.

On the other hand,

$$a'_s + a'_l - a_s - a_l = G_n + a_s a_l / G_n - a_s - a_l = (G_n^2 - a_s G_n - a_l G_n + a_s a_l) / G_n$$

$$= (G_n - a_s)(G_n - a_l) / G_n$$

but $G_n - a_s \geq 0$ and $G_n - a_l \leq 0$, and so $a'_s + a'_l - a_s - a_l \leq 0$, or $a'_s + a'_l \leq a_s + a_l$. The last inequality implies that $A'_n \leq A_n$. After at most n repetitions of this procedure we end up with n numbers all equal to G_n. Their arithmetic mean is also G_n, and it cannot be larger than A_n because the arithmetic means do not increase at each step.

The number $2/(1/a + 1/b)$ is called the *harmonic mean* of a and b. It too can be extended to n positive numbers $a_1, a_2, \ldots, a_n$ by setting

$$H_n = \frac{n}{\sum\limits_{k=1}^{n} 1/a_k}$$

We leave it to the reader to establish the inequality $H_n \leq G_n$ by considering the arithmetic and geometric means of the numbers

$$\alpha_k = \frac{1}{a_k}$$

For a more systematic treatment of inequalities involving more general means, see the first chapter of *Inequalities* by Hardy, Littlewood, and Polya.

1-17 (a) From Problem 1-16c, $a + b \geq 2(ab)^{1/2}$, $b + c \geq 2(bc)^{1/2}$, and $a + c \geq 2(ac)^{1/2}$. In view of Problem 1-13b, we may multiply these inequalities and obtain the given inequality.

(b) Again from Problem 1-16c, we have $(a^2 + b^2)/2 \geq ab$, $(b^2 + c^2)/2 \geq bc$, and $(a^2 + c^2)/2 \geq ac$. The addition of these inequalities yields the stated inequality.

(c) Using Problem 1-16c once more, we have $(a^2b^2 + b^2c^2)/2 \geq ab^2c$, $(b^2c^2 + a^2c^2)/2 \geq abc^2$, and $(a^2b^2 + a^2c^2)/2 \geq a^2bc$. The addition of these results in the desired inequality.

1-18 This inequality is equivalent to $|x + 1/x|^2 \geq 4$, or to $x^2 + 2 + 1/x^2 \geq 4$. The last inequality is the same as $x^2 - 2 + 1/x^2 \geq 0$, or $(x - 1/x)^2 \geq 0$.

1-19 Since $|a|^2 = a^2$, $|b|^2 = b^2$, and $|a + b|^2 = (a + b)^2 = a^2 + 2ab + b^2$, this inequality is the same as $a^2 + 2ab + b^2 \leq (1 + c)a^2 + (1 + 1/c)b^2$, or $0 \leq ca^2 - 2ab + b^2/c = (c^{1/2}a - b/c^{1/2})^2$.

1-20 We want to show that $0 \leq a^3 + b^3 + c^3 - 3abc$. Now

$$\begin{aligned} a^3 + b^3 + c^3 - 3abc &= a(a^2 - bc) + b(b^2 - ac) + c(c^2 - ab) \\ &= (a + b + c)(a^2 - bc) + (a + b + c)(b^2 - ac) \\ &\quad + (a + b + c)(c^2 - ab) \\ &= (a + b + c)(a^2 + b^2 + c^2 - ab - bc - ac). \end{aligned}$$

Since $a + b + c$ is positive, and $a^2 + b^2 + c^2 - ab - bc - ac \geq 0$ by Problem 1-17b, the inequality is proved.

1-21 If $x = y$, we have a sum of squares which cannot be negative. If $x \neq y$,

$$\sum_{k=0}^{2n} x^{2n-k} y^k = x^{2n} \sum_{k=0}^{2n} (y/x)^k = x^{2n}[1 - (y/x)^{2n+1}]/(1 - y/x)$$
$$= (x^{2n+1} - y^{2n+1})/(x - y)$$

Suppose for the moment that $x > y$. We then want to show that $x^{2n+1} > y^{2n+1}$ as well. If $y > 0$, this follows from Problem 1-14. If $x > 0 > y$, it is clear that $x^{2n+1} > 0 > y^{2n+1}$. Finally, if $0 > x > y$, then $-y > -x$, or $|y| > |x|$ and we have, from Problem 1-14 again, that $|y|^{2n} > |x|^{2n}$, or $y^{2n} > x^{2n}$. But then $x^{2n+1} > xy^{2n}$, and $xy^{2n} > y^{2n+1}$. Thus $x^{2n+1} > y^{2n+1}$ in all cases, and the quotient $(x^{2n+1} - y^{2n+1})/(x - y)$ is positive. If $y > x$, we can apply the same reasoning to $(y^{2n+1} - x^{2n+1})/(y - x)$. The inequality is established.

1-22 Let k and m be integers such that $1 \leq k \leq n$ and $1 \leq m \leq n$. Then $a_k - a_m$ and $b_k - b_m$ are both positive if $k < m$, both negative if $k > m$. Hence, in any event, $(a_k - a_m)(b_k - b_m) \geq 0$. Thus $a_k b_k + a_m b_m \geq a_m b_k + a_k b_m$ for any two integers k and m between 1 and n inclusive. If we add all the n^2 versions of the last inequality that result as k and m range between 1 and n, we get

$$2n \sum_{k=1}^{n} a_k b_k \geq 2 \sum_{k,m=1}^{n} a_k b_m$$

and since, as is easy to see,

$$\left(\sum_{k=1}^{n} a_k \right) \left(\sum_{k=1}^{n} b_k \right) = \sum_{k,m=1}^{n} a_k b_m$$

our inequality is established.

1-23 If we expand both sides of this inequality, it becomes $a_1^2 b_1^2 + 2a_1 a_2 b_1 b_2 + a_2^2 b_2^2 \leq a_1^2 b_1^2 + a_1^2 b_2^2 + a_2^2 b_1^2 + a_2^2 b_2^2$, and we see that it is equivalent to the inequality $2a_1 a_2 b_1 b_2 \leq a_1^2 b_2^2 + a_2^2 b_1^2$. But Problem 1-16c gives us $2|a_1 a_2 b_1 b_2| \leq a_1^2 b_2^2 + a_2^2 b_1^2$, and since $a_1 a_2 b_1 b_2 \leq |a_1 a_2 b_1 b_2|$, our inequality holds.

Remark We can find φ and ψ such that $a_1/(a_1^2 + a_2^2)^{1/2} = \cos \varphi$, $a_2/(a_1^2 + a_2^2)^{1/2} = \sin \varphi$, $b_1/(b_1^2 + b_2^2)^{1/2} = \cos \psi$, and $b_2/(b_1^2 + b_2^2)^{1/2} = \sin \psi$. Since our inequality can be written as

$$\frac{|a_1 b_1 + a_2 b_2|}{(a_1^2 + a_2^2)^{1/2}(b_1^2 + b_2^2)^{1/2}} \leq 1$$

it follows that $|\cos \varphi \cos \psi + \sin \varphi \sin \psi| \leq 1$, or, by a familiar trigonometric identity, $|\cos (\varphi - \psi)| \leq 1$. Our inequality is thus another way of saying that the cosine of an angle is between -1 and 1. Further, this inequality extends to

$$\left(\sum_{k=1}^{n} a_k b_k \right)^2 \leq \sum_{k=1}^{n} a_k^2 \sum_{k=1}^{n} b_k^2$$

and it is known as the *Cauchy-Schwartz* inequality. Interestingly enough, the general inequality can also be obtained from Problem 1-16c.   Indeed, in $a^{1/2}b^{1/2} \leq a/2 + b/2$, we let successively $a = a_k^2/A$, where

$$A = \sum_{k=1}^{n} a_k^2$$

and $b = b_k^2/B$, where

$$B = \sum_{k=1}^{n} b_k^2$$

for $k = 1, 2, \ldots, n$, and add the resulting inequalities.   This gives us

$$\frac{1}{A^{1/2}B^{1/2}} \sum_{k=1}^{n} |a_k| \, |b_k| \leq \frac{1}{2}\frac{1}{A} \sum_{k=1}^{n} a_k^2 + \frac{1}{2}\frac{1}{B} \sum_{k=1}^{n} b_k^2 = 1$$

and so

$$\sum_{k=1}^{n} |a_k b_k| \leq (AB)^{1/2}$$

and since

$$\left| \sum_{k=1}^{n} a_k b_k \right| \leq \sum_{k=1}^{n} |a_k b_k|$$

we have

$$\left(\sum_{k=1}^{n} a_k b_k \right)^2 \leq \sum_{k=1}^{n} a_k^2 \sum_{k=1}^{n} b_k^2$$

1-24   If we square both sides, this inequality is seen to be equivalent to

$$\sum_{k=1}^{n} [a_k^2 + 2a_k b_k + b_k^2] \leq \sum_{k=1}^{n} a_k^2 + 2\left[\sum_{k=1}^{n} a_k^2 \sum_{k=1}^{n} b_k^2 \right] + \sum_{k=1}^{n} b_k^2$$

or to

$$\sum_{k=1}^{n} a_k b_k \leq \left[\sum_{k=1}^{n} a_k^2 \sum_{k=1}^{n} b_k^2 \right]^{1/2}$$

If

$$\sum_{k=1}^{n} a_k b_k \leq 0$$

the last inequality is obviously true; otherwise we can square both sides and it becomes the Cauchy–Schwartz inequality.

Remark   When $n = 2$ Minkowski's inequality states, in geometric terms, that the distance between the points $(a_1, -b_1)$ and $(a_2, -b_2)$ is not greater than the sum of distance between $(0, 0)$ and $(a_1, -b_1)$ and the distance between $(0, 0)$ and $(a_2, -b_2)$. In other words, it states that the length of one side of a triangle does not exceed the sum of the lengths of the other two sides.

Both the Cauchy–Schwartz inequality and Minkowski's inequality have widespread applications in many branches of higher mathematics. For extensions of them, consult the first chapter of the book of Hardy, Littlewood, and Polya referred to in the remark following Problem 1-16.

1-25(a) We have, using the Cauchy-Schwartz inequality,

$$n^2 = \left(\sum_{k=1}^n 1\right)^2 = \left(\sum_{k=1}^n a_k^{1/2}/a_k^{1/2}\right)^2 \leq \sum_{k=1}^n (a_k^{1/2})^2 \sum_{k=1}^n (1/a_k^{1/2})^2$$

$$= \sum_{k=1}^n a_k \sum_{k=1}^n 1/a_k$$

(b) Again, an application of the Cauchy–Schwartz inequality gives us

$$\left(\sum_{k=1}^n a_k b_k\right)^2 = \left[\sum_{k=1}^n a_k^{1/2}(a_k^{1/2}b_k)\right]^2 \leq \sum_{k=1}^n a_k \sum_{k=1}^n a_k b_k^2$$

1-26 $$\left(\sum_{k=1}^n a_k b_k c_k\right)^4 = \left[\left(\sum_{k=1}^n a_k b_k c_k\right)^2\right]^2 \leq \left[\sum_{k=1}^n a_k^2 c_k^2 \sum_{k=1}^n b_k^2\right]^2$$

$$= \left(\sum_{k=1}^n b_k^2\right)^2 \left(\sum_{k=1}^n a_k^2 c_k^2\right)^2 \leq \left(\sum_{k=1}^n b_k^2\right)^2 \sum_{k=1}^n a_k^4 \sum_{k=1}^n c_k^4$$

We have made two applications of the Cauchy–Schwartz inequality in this calculation.

1-27 We have

$$2[(a+1/a)^2 + (b+1/b)^2] = [1^2 + 1^2][(a+1/a)^2 + (b+1/b)^2] \geq [a+1/a+b+1/b]^2$$

by the Cauchy–Schwartz inequality. Now $a + b + 1/a + 1/b = 1 + (a+b)/ab = 1 + 1/ab = 1 + 1/a(1-a) \geq 1 + 4 = 5$ (the last inequality holds because, by the inequality of the arithmetic-geometric mean, $a(1-a) \leq ([a + (1-a)]/2)^2 = 1/4$).

Remark This inequality can be generalized. If $a_k > 0$ for $k = 1, 2, \ldots, n$ and if $a_1 + a_2 + \cdots + a_n = 1$, then

$$\sum_{k=1}^n (a_k + 1/a_k)^2 \geq \frac{(1+n^2)^2}{n}$$

To prove it, write

$$n = \sum_{k=1}^n 1^2$$

use the Cauchy–Schwartz inequality, and then the fact that the harmonic mean of $a_1, a_2, \ldots, a_n$ is not greater than their arithmetic mean (see the remark following the solution of Problem 1-16).

1-28 When $n = 2$, the inequality if $(1/2)(3/4) < 1/7^{1/2}$. or $63 < 64$. Assume that

$$\prod_{k=1}^n (2k-1)/2k < 1/(3n+1)^{1/2}$$

then

$$\prod_{k=1}^{n+1}(2k-1)/2k < (2n+1)/(2n+2)(3n+1)^{1/2}$$

and we would like to show that $(2n+1)/(2n+2)(3n+1)^{1/2} < 1/[3(n+1)+1]^{1/2}$, or $(2n+1)^2(3n+4) < (2n+2)^2(3n+1)$, or $(4n^2+4n+1)(3n+4) < (4n^2+8n+4)(3n+1)$, or $12n^3+28n^2+19n+4 < 12n^3+28n^2+20n+4$, or $19n < 20n$, or $19 < 20$. The inequality is established by induction.

1-29 We prove this inequality by induction. It is clearly true when $n=1$. If we assume that

$$\prod_{k=1}^{n}(1+a_k) \geq 1+\sum_{k=1}^{n}a_k$$

then

$$\prod_{k=1}^{n+1}(1+a_k) \geq (1+a_{n+1})(1+\sum_{k=1}^{n}a_k) = 1+\sum_{k=1}^{n+1}a_k+\sum_{k=1}^{n}a_k a_{n+1}$$

Now $a_k a_{n+1} \geq 0$ for $k=1, 2, \ldots, n$ since a_k and a_{n+1} have the same sign. Hence

$$\prod_{k=1}^{n+1}(1+a_k) \geq 1+\sum_{k=1}^{n+1}a_k$$

and the proof by induction is complete.

Remark   When $a_1 = a_2 = \ldots = a_n = a > -1$, this inequality becomes

$$(1+a)^n \geq 1+na$$

and it is known as *Bernoulli's inequality*.

1-30 By the inequality of the arithmetic-geometric means, we have

$$\left[\prod_{k=1}^{n}(1+a_k)\right]^{1/n} \leq \frac{1}{n}\sum_{k=1}^{n}(1+a_k) = 1+\frac{1}{n}\sum_{k=1}^{n}a_k$$

and the given inequality follows by Problem 1-14. Note that if $a_k = k-1$ for $k=1, 2, \ldots, n$, then this inequality becomes

$$n! \leq \left[1+\frac{1}{n}S_{n-1}^1\right]^n = \left[1+\frac{1}{n}\frac{(n-1)n}{2}\right]^n = \left[\frac{n+1}{2}\right]^n$$

CHAPTER 2

2-1   (a)   The first five terms of $\{n^2-n\}$ are 0, 2, 6, 12, and 20.
   (b)   The first five terms of $\{n/(n+1)\}$ are 1/2, 2/3, 3/4, 4/5, and 5/6.
   (c)   The first five terms of their sum are $1/2, 2+2/3, 6+3/4, 12+4/5$, and $20+5/6$, or 1/2, 8/3, 27/4, 64/5, and 125/6.
   (d)   The first five terms of their difference are $-1/2, 2-2/3, 6-3/4, 12-4/5$, and $20-5/6$, or $-1/2$, 4/3, 21/4, 56/5, and 115/6.

(e)   The first five terms of their product are $0(1/2)$, $2(2/3)$, $6/(3/4)$, $12(4/5)$, and $20(5/6)$, or 0, $4/3$, $9/2$, $48/5$, and $50/3$.

(f)   The first five terms of their quotient are $0/(1/2)$, $2/(2/3)$, $6(3/4)$, $12/(4/5)$, and $20/(5/6)$, or 0, 3, 8, 15, and 24.

(g)   The first five terms of their convolution are $0(1/2)$, $0(2/3) + 2(1/2)$, $0(3/4) + 2(2/3) + 6(1/2)$, $0(4/5) + 2(3/4) + 6(2/3) + 12(1/2)$, and $0(5/6) + 2(4/5) + 6(3/4) + 12(2/3) + 20(1/2)$, or 0, 1, $13/3$, $23/2$, and $241/10$.

2-2   (a)   For any $\varepsilon > 0$, we want $1/n^p < \varepsilon$, or $n > (1/\varepsilon)^{1/p}$. Thus if $N > (1/\varepsilon)^{1/p}$, then $|1/n^p - 0| < \varepsilon$ for all $n > N$.

(b)   If $\varepsilon > 0$, we want $|(n^2 - 1)/(2n^2 + 3) - 1/2| < \varepsilon$, or $|-5/2(2n^2 + 3)| < \varepsilon$, or $5/2\varepsilon < 2n^2 + 3$, or $n^2 > 5/4\varepsilon - 3/2$. Now if $5/4\varepsilon - 3/2 \leq 0$, the last inequality is true for all n. Otherwise, we have $n > (5/4\varepsilon - 3/2)^{1/2}$. Hence if we let $N = 0$ when $5/4\varepsilon - 3/2 \leq 0$, or $N > (5/4\varepsilon - 3/2)^{1/2}$ when $5/4\varepsilon - 3/2 > 0$, we get $|(n^2 - 1)/(2n^2 + 3) - 1/2| < \varepsilon$ for $n > N$.

(c)   Inasmuch as $|\sin n/n| \leq 1/n$, this proof is the same as in Example 2-6.

2-3   (a)   Since $(n^2 + 2n - 3)/5n^2 = (1 + 2/n - 3/n^2)/5$, and by Problem 2-2a, $\{2/n\} \to 0$ and $\{-3/n^2\} \to 0$ we can apply Theorem 2-1a and conclude that $\{(n^2 + 2n - 3)/5n^2\} \to 1/5$.

(b)   Here

$$\{2n/(n + 1) - (n + 1)/2n\} = \{(4n^2 - (n + 1)^2)/2n(n + 1)\}$$
$$= \{(3n^2 - 2n - 1)/2n(n + 1)\}$$
$$= \{3 - 2/n - 1/n^2\}/\{2(1 + 1/n)\} \to 3/2$$

by Problem 2-2a and Theorem 2-1c.

(c)   We have $0 \leq |(-1)^n/n^2| \leq 1/n^2$, and so $\{|(-1)^n/n^2|\} \to 0$ by Theorem 2-2. Now $-|(-1)^n/n^2| \leq (-1)^n/n^2 \leq |(-1)^n/n^2|$, and so $\{(-1)^n/n^2\} \to 0$ by Theorem 2-2 and the previously established limit. Thus $\{1/n^2 + (-1)^n/n^2\} \to 0$ by Theorem 2-1a.

(d)   Here $0 < (n + 1)^{1/2} - n^{1/2}$
$$= [(n + 1)^{1/2} - n^{1/2}][(n + 1)^{1/2} + n^{1/2}]/[(n + 1)^{1/2} + n^{1/2}]$$
$$= 1/[(n + 1)^{1/2} + n^{1/2}] < 1/n^{1/2}$$
Since $\{1/n^{1/2}\} \to 0$, it follows from Theorem 2-2 that $\{(n + 1)^{1/2} - n^{1/2}\} \to 0$.

(e)   Now $\{1/n\} \to 0$, and $\{(n + 1)/n\} = \{1 + 1/n\} \to 1$. Thus by Theorem 2-1b, $\{(n + 1)/n^2\} = \{1/n\}\{(n + 1)/n\} \to 0 \cdot 1 = 0$

(f)   Here $0 < 1/(n^2 + 1)^{1/2} < 1/n$, and $\{1/n\} \to 0$.

(g)   We have $\{(n^2 + a)/(n + a)/(n^2 + b)/(n + b)\}$
$$= \{(n + b)/(n + a)\}\{(n^2 + a)/(n^2 + b)\}$$
$$= \{(1 + b/n)/(1 + a/n)\}\{(1 + a/n^2)/(1 + b/n^2)\} \to (1/1)(1/1) = 1$$

(h)   Using Problem 2-2b, we have $\{[(n^2 - 1)/(2n^2 + 3)]^3\}$
$$= \{(n^2 - 1)/(2n^2 + 3)\}\{(n^2 - 1)/(2n^2 + 3)\}\{(n^2 - 1)/(2n^2 + 3)\}$$
$$\to (1/2)(1/2)(1/2) = 1/8.$$

Remark   Parts (a), (b), (d), (e), and (g) show that the converses of the various parts of Theorem 2-1 are not true.

2-4 (a) This result follows from the inequality $\big||a_n| - |L|\big| \le |a_n - L|$ and Definition 2-1. (See Problem 1-11f for the inequality.)

(b) We can find N such that if $n > N$, then $|a_n - L| < 1$, or $L - 1 < a_n < L + 1$. Set $A =$ the largest of $a_1, a_2, \ldots, a_n$ and $B =$ the smallest of $a_1, a_2, \ldots, a_n$. Then the smaller of B and $L - 1 \le a_n \le$ the larger of A and $L + 1$, and so $\{a_n\}$ is bounded.

The sequence $\{(-1)^n\}$ shows that the converses of both these statements are false.

2-5 (a) If $a = 1$, clearly $\{a^n\} \to 1$. Take first $0 < a < 1$. Then $a^{n+1} < a^n$, or our sequence decreases. Since it is bounded, it has a limit L. But $\{a^{n-1}\} \to L$ as well, and so $L = aL$, or $L = 0$. If $a > 1$, then $1/a < 1$ and we have $\{1/a^n\} = \{(1/a)^n\} \to 0$. Thus $\{a^n\} \to \infty$ if $a > 1$. Summarizing,

$$\{a^n\} \to \begin{cases} 0 \text{ if } 0 < a < 1 \\ 1 \text{ if } a = 1 \\ \infty \text{ if } a > 1 \end{cases}$$

(b) This can be solved by an analysis similar to that in the solution of part (a). We give an alternate method. If $a = 1$, then $\{a^{1/n}\} \to 1$. Suppose that $a > 1$. We put $a^{1/n} = 1 + b_n$, and it is clear that $b_n > 0$. Then $a = (1 + b_n)^n \ge 1 + nb_n$ by Bernoulli's inequality (see the remark following the solution of Problem 1-29). Thus $0 < b_n \le (a - 1)/n$, and so $\{b_n\} \to 0$, or $\{a^{1/n}\} \to 1$. If $0 < a < 1$, then $1/a > 1$, and $\{1/a^{1/n}\} = \{(1/a)^{1/n}\} \to 1$, or $\{a^{1/n}\} \to 1$. Hence $\{a^{1/n}\} \to 1$ for all $a > 0$.

(c) If we let $a_n = n^2/2^n$, then $a_{n+1}/a_n = (1/2)(1 + 1/n)^2$. Now if $n > 3$, then $1 + 1/n \le 4/3$, and so $(1 + 1/n)^2 \le 16/9 < 2$. Thus for $n \ge 3$, $a_{n+1}/a_n < 1$, and our sequence is decreasing if we neglect the first two terms. Since $a_n > 0$, there must be a limit $L \ge 0$. We now show that $L = 0$. To this end, let $b_n = n^3/2^n$. Then $b_{n+1}/b_n = (1/2)(1 + 1/n)^3$, and if $n \ge 4$, then $1 + 1/n \le 5/4$ and $(1 + 1/n)^3 \le 75/64 < 2$. Hence, except for the first three terms, $\{b_n\}$ is decreasing and, since $b_n > 0$, $\{b_n\} \to l$. Now $a_n = b_n/n$, and so $a_n \to 0 \cdot l = 0$. Thus L is indeed zero. By considering $n^4/2^n$, the same procedure shows that $l = 0$. Indeed, it can be proved by induction that $\{n^k/2^n\} \to 0$ for all positive integers k. It can also be proved similarly that $\{n^k/a^n\} \to 0$ for any $a > 1$ and all positive integers k.

(d) If we put $a = 1/(1 + b)$, then $b > 0$ and

$$a^n = 1/(1 + b)^n = 1 \Big/ \sum_{k=0}^{n} \binom{n}{k} b^k$$

by the binomial theorem. But

$$1 \Big/ \sum_{k=0}^{n} \binom{n}{k} b^k \le 1 \Big/ \binom{n}{2} b^2 = 2/n(n - 1)b^2$$

if $n \ge 2$. Thus $na^n \le 2/b^2(n - 1)$, and since $\{2/b^2(n - 1)\} \to 0$, it follows that $\{na^n\} \to 0$ as well.

(e) Let $a_n = (n^{1/n})^{1/2} = (n^{1/2})^{1/n}$ and put $a_n = 1 + b_n$. Then $n^{1/2} = a_n^n = (1 + b_n)^n \ge 1 + nb_n$ by Bernoulli's inequality. Thus $b_n \le (n^{1/2} - 1)/n < 1/n^{1/2}$, and so $1 \leq n^{1/n} = a_n{}^2 = 1 + 2b_n + b_n{}^2 \leq 1 + 2/n^{1/2} + 1/n$. It can now be seen that $\{n^{1/n}\} \to 1$.

(f) From part (e) and Theorem 2-1b we have $\{(n^m)^{1/n}\} = \{(n^{1/n})^m\} \to 1^m = 1$.

2-6 (a) We have $a_n > Ka_{n-1} > K^2 a_{n-2} > \cdots > K^{n-1} a_1 > 0$. But $\{K^{n-1}\} \to \infty$ by Problem 2-5a since $K > 1$. Thus $\{a_n\} \to \infty$.

(b) Here $a_n < Ka_{n-1} < K^2 a_{n-2} < \cdots < K^{n-1} a_1$, and since $\{K^{n-1}\} \to 0$ by Problem 2-5a because $K < 1$, we have $\{a_n\} \to 0$. This limit would also follow from part (a) by considering $\{b_n\} = \{1/a_n\}$ inasmuch as it would then be the case that $b_n > (1/K)b_{n-1}$.

(c) Suppose first that $l > 1$. Pick ε such that $l - \varepsilon > 1$. Then we can find N such that $a_{n+1}/a_n > l - \varepsilon$ for $n > N$. If we take $m > N$ and put $b_n = a_{m+n}$ and $K = l - \varepsilon$, we then have $b_{n+1} > Kb_n$. By part (a), this last inequality means that $\{b_n\} \to \infty$. Hence $\{a_n\} \to \infty$ as well. If $l < 1$, pick ε such that $l + \varepsilon < 1$. We can then find N such that $a_{n+1}/a_n < l + \varepsilon$ for $n > N$. Take $m > N$ and put $b_n = a_{m+n}$, $K = l + \varepsilon$. Then we have $b_{n+1}/b_n < K$, and part (b) means that $\{b_n\} \to 0$. Thus $\{a_n\} \to 0$ too. The reader should construct examples to show that if $l = 1$, then the sequence $\{a_n\}$ can behave in various ways as far as its limit is concerned.

2-7 (a) If we set $a_n = a^n/n!$, then $a_{n+1}/a_n = a/(n+1)$. Since $\{a/(n+1)\} \to 0$, Problem 2-6c implies that $\{a^n/n!\} \to 0$.

(b) Take $\varepsilon > 0$. From part (a) we can find N such that $(1/\varepsilon)^n/n! < 1$ for $n > N$, or $1/n! < \varepsilon^n$. But the last inequality means that $1/(n!)^{1/n} < \varepsilon$ for $n > N$, and so $\{1/(n!)^{1/n}\} \to 0$, or $\{(n!)^{1/n}\} \to \infty$.

2-8 (a) Put $a_n = n!/n^n$. Then $a_{n+1}/a_n = [n/(n+1)]^n < 1$. Thus $\{a_n\}$ decreases, and since $a_n > 0$, it follows that $\{n!/n^n\}$ converges. (It is not hard to show that the limit is zero.)

(b) If we denote the general term of this sequence by a_n, then $a_{n+1}/a_n = [n/(n+1)][(2n+2)/(2n+1)]^2 = (4n^2 + 4n)/(4n^2 + 4n + 1) < 1$.
Since $a_n > 0$, $\{a_n\}$ is bounded. Hence $\{a_n\} \to L \geq 0$. (We shall see in Chapter 6 that $L = \pi$.)

(c) If a_n denotes the general term of this sequence, then

$$a_{n+1} - a_n = 1/2(n+1) + 1/(2n+1) + \sum_{k=2}^{n} 1/(n+k) - \sum_{k=2}^{n} 1/(n+k) - 1/(n+1)$$

$$= 1/(2n+1) - 1/(2n+2) > 0$$

Thus $\{a_n\}$ is increasing. On the other hand, since $1/(n+k) \leq 1/(n+1)$ for $k = 1, 2, \ldots, n$, we have $a_n \leq n/(n+1) < 1$. Thus $\{a_n\}$ is bounded, and so it has a limit. (The value of this limit will be given in Chapter 7.)

(d) Set $a_n = (1 + 1/n)^n$. We show that $a_{n-1} \leq a_n$. Now this inequality is $[1 + 1/(n-1)]^{-1} \leq \{(1 + 1/n)/[1 + 1/(n-1)]\}^n$, or $(n-1)/n \leq [(n^2 - 1)/n^2]^n$, or $1 - 1/n \leq (1 - 1/n^2)^n$. But $-1/n^2 > -1$ if $n > 1$, and so Bernoulli's inequality gives us $(1 - 1/n^2)^n \geq 1 - n/n^2 = 1 - 1/n$. Inasmuch as this is our last inequality above, and all the steps are reversible, we indeed have $a_{n-1} \leq a_n$.

On the other hand,

$$(1 + 1/n)^n = \sum_{k=0}^{n} \binom{n}{k} 1/n^k = 1 + \sum_{k=1}^{n} (1/k!)n(n-1) \cdots [n-(k-1)]/n^k$$

$$= 1 + \sum_{k=1}^{n} (1/k!)(1 - 1/n)(1 - 2/n) \cdots [1 - (k-1)/n]$$

$$< 1 + \sum_{k=1}^{n} 1/k! < 1 + \sum_{k=0}^{n-1} 1/2^k = 1 + (1 - 1/2^n)/(1 - 1/2)$$

$$= 1 + 2 - 1/2^{n-1} < 3$$

(The binomial theorem and the formula for the sum of a geometric progression have been used in this calculation.) Thus $\{a_n\}$ is bounded, and hence has a limit. [The limit of this sequence is the number e that is the base of the natural logarithms. In other words, $\{(1 + 1/n)^n\} \to e$].

2-9 We show by induction that $\{a_n\}$ is increasing. More precisely, if P_n denotes the proposition $a_{n+1} > a_n$, we establish the truth of P_n by induction. We have first, because $2^{1/2} > 1$, that $2(2)^{1/2} > 2$, or $[2(2)^{1/2}]^{1/2} > 2^{1/2}$. Thus P_1 is true. Next, assume that P_n is true. Then $a_{n+1} > a_n$, or $2a_{n+1} > 2a_n$, or $(2a_{n+1})^{1/2} > (2a_n)^{1/2}$, or $a_{n+2} > a_{n+1}$. Hence P_{n+1} is true if P_n is true. The proposition $a_n < 2$, denoted by Q_n, will also be established by induction. Q_1 is clearly true. If we assume that Q_n is true, then $2a_n < 4$, or $(2a_n)^{1/2} < 2$, or $a_{n+1} < 2$, and so Q_{n+1} is true. Thus $\{a_n\}$ is increasing and bounded, and hence has limit L. Now since $a_n^2 = 2a_{n-1}$, we must have $L^2 = 2L$, or $L = 2$.

2-10 Observe first that $a_{n+1} = 1 + 1/a_n = 1 + 1/(1 + 1/a_{n-1}) = 1 + a_{n-1}/(1 + a_{n-1})$. Let α be the positive root of $x^2 - x - 1 = 0$. Then $\alpha = (1 + 5^{1/2})/2$, and $\alpha = 1 + 1/\alpha$. Thus if $a_1 = \alpha$, then $a_2 = \alpha$, $a_3 = \alpha$, $\ldots$, $a_n = \alpha$, and so $\{a_n\} \to \alpha$. Suppose that $0 < a_1 < \alpha$. Then $a_1^2 - a_1 - 1 < 0$, or $a_1 < 1 + 1/a_1 = a_2$. But $1 + 1/a_1 > 1 + 1/\alpha = \alpha$, and so $a_1 < \alpha < a_2$. Now $a_3 = 1 + 1/a_2 < 1 + 1/\alpha = \alpha$, and $a_3 - a_1 = 1 + a_1/(1 + a_1) - a_1 = (1 + a_1 - a_1^2)/(1 + a_1) > 0$. Thus $a_1 < a_3 < \alpha < a_2$. Next, $a_4 = 1 + 1/a_3 > 1 + 1/\alpha = \alpha$, and $a_4 - a_2 = 1 + a_2/(1 + a_2) - a_2 = (1 + a_2 - a_2^2)/(1 + a_2) < 0$ since $a_2 > \alpha$. We now have $a_1 < a_3 < \alpha < a_4 < a_2$. The same argument gives us $a_3 < a_5 < \alpha < a_6 < a_4$, and it is easy to see that we can use this same procedure to prove by induction that $a_{2n-1} < a_{2n+1} < \alpha < a_{2n+2} < a_{2n}$. In other words, the sequence $\{a_n\}$ is doubly monotone, and we have $\{a_{2n-1}\} \to L_1$, $\{a_{2n}\} \to L_2$ with $0 < L_1 \leq L_2$. By our first observation, we must have $L_1 = 1 + L_1/(1 + L_1)$, and $L_2 = 1 + L_2/(1 + L_2)$. The first of these equations entails $L_1 = (1 + 2L_1)/(1 + L_1)$, or $L_1^2 - L_1 - 1 = 0$, and so, since $L_1 > 0$, $L_1 = \alpha$. Inasmuch as L_2 satisfies the same equation, we must also have $L_2 = \alpha$, and so $\{a_n\} \to \alpha$.

If $a_1 > \alpha$, a similar analysis shows that $\{a_n\}$ is also doubly monotone but that $a_{2n} < a_{2n+2} < \alpha < a_{2n+1} < a_{2n-1}$, and it follows in the same way that $\{a_n\} \to \alpha$.

We have thus established the rather surprising fact that $\{a_n\} \to (1 + 5^{1/2})/2$ regardless of the value of a_1.

Remark If we define a sequence $\{b_n\}$ by $b_1 > 0$, $b_2 > 0$, and $b_{n+1} = b_n + b_{n-1}$ for $n \geq 2$, it is easy to see that if we put $a_n = b_{n+1}/b_n$, we have $a_{n+1} = 1 + 1/a_n$ for $n \geq 1$

with $a_1 = b_2/b_1$. Thus the ratios of the successive terms of any sequence of positive terms defined by the same recurrence relation as Fibonacci's sequence have limit $(1 + 5^{1/2})/2$. In particular, if $b_1 = 1$ and $b_2 = 1$, we get the sequence of Fibonacci numbers, and hence the sequence $1/1, 2/1, 3/2, 5/3, 8/5, 11/8, \dots$ has limit $(1 + 5^{1/2})/2$. The number $(1 + 5^{1/2})/2$ was called the "golden ratio" by the ancient Greeks.

2-11 Since $a_1 < b_1$, we have $a_2 = (a_1 b_1)^{1/2} > (a_1{}^2)^{1/2} = a_1$, and $b_2 = (a_1 + b_1)/2 < (b_1 + b_1)/2 = b_1$. Further, the inequality of the arithmetic-geometric means gives us $a_2 < b_2$. It is clear that the same argument enables us to prove by induction that $a_{n-1} < a_n < b_n < b_{n-1}$. Thus $\{a_n\}$ increases, and $\{b_n\}$ decreases. Since $a_n < b_1$, and $a_1 < b_n$, we have $\{a_n\} \to a$, and $\{b_n\} \to b$ with $0 < a \leq b$. But it must be the case that $b = (a + b)/2$, and so $a = b$.

2-12 It is clear that a_{n+2} is of the form $a_{n+2} = (\alpha_n/2^n)a_1 + (\beta_n/2^n)a_2$ for $n \geq 1$. Now $a_{n+4} = (a_{n+3} + a_{n+2})/2 = (\alpha_{n+1}/2^{n+2})a_1 + (\beta_{n+1}/2^{n+2})a_2 + (\alpha_n/2^{n+1})a_1 + (\beta_n/2^{n+1})a_2 = [(\alpha_{n+1} + 2\alpha_n)/2^{n+2}]a_1 + [(\beta_{n+1} + 2\beta_n)/2^{n+2}]a_2$. Hence $\alpha_{n+2} = \alpha_{n+1} + 2\alpha_n$, and $\beta_{n+2} = \beta_{n+1} + 2\beta_n$. An easy calculation shows that $\alpha_1 = 1$, $\alpha_2 = 1$, and $\beta_1 = 1$, $\beta_2 = 3$. Inasmuch as $\{\alpha_n\}$ and $\{\beta_n\}$ satisfy the same recurrence relation, it follows that $\beta_n = \alpha_{n+1}$. Hence $a_{n+2} = (\alpha_n/2^n)a_1 + (\alpha_{n+1}/2^n)a_2$. Put $\gamma_n = \alpha_n/2^n$. Then $a_{n+2} = \gamma_n a_1 + 2\gamma_{n+1} a_2$, and we also have $\gamma_{n+2} = \alpha_{n+2}/2^{n+2} = (\alpha_{n+1} + 2\alpha_n/2)/2^{n+2} = (\alpha_{n+1}/2^{n+1} + \alpha_n/2^n)/2 = (\gamma_{n+1} + \gamma_n)/2$. Thus $\{\gamma_n\}$ satisfies the same recurrence relation as $\{a_n\}$. Now $\gamma_1 = 1/2$, and $\gamma_2 = 1/4$, and so $\gamma_2 < \gamma_1$. By Problem 1-16a, $\gamma_2 < \gamma_3 < \gamma_1$, and $\gamma_2 < \gamma_4 < \gamma_3 < \gamma_1$. It is clear that we can apply Problem 1-16a over and over to obtain $\gamma_2 < \gamma_4 < \cdots < \gamma_{2n} < \cdots < \gamma_{2n-1} < \cdots < \gamma_3 < \gamma_1$. Therefore $\{\gamma_n\}$ is doubly monotone and, since $\gamma_{2n} < \gamma_1$ and $\gamma_{2n-1} > \gamma_2$ for all n, we have $\gamma_{\{2n-1\}} \to L_1$ and $\{\gamma_{2n}\} \to L_2$ with $L_1 \geq L_2 > 0$. But it follows from $\gamma_{2+2} = (\gamma_{2n+1} + \gamma_{2n})/2$ that we must have $L_2 = (L_1 + L_2)/2$, or $L_1 = L_2 = L$. This means that $\{a_n\} \to La_1 + 2La_2$ for any a_1 and a_2. In particular, if $a_1 = a_2 = a$, we have $a_n = a$ for all n, and so we must have $a = La + 2La$, or $L = 1/3$. Therefore, $\{a_n\} \to (1/3)a_1 + (2/3)a_2$.

Remark If $a_1 > 0$, $a_2 > 0$, and we put (a) $a_n = (a_{n-1}a_{n-2})^{1/2}$ or we put (b) $a_n = 2/(1/a_{n-1} + 1/a_{n-2}) = 2a_{n-1}a_{n-2}/(a_{n-1} + a_{n-2})$, both for $n \geq 3$, it can be shown that in (a) $\{a_n\} \to a_1^{1/3}a_2^{2/3}$, and in (b) $\{a_n\} \to 1/[1/3)(1/a_1) + (2/3)(1/a_2)] = 3a_1a_2/(2a_1 + a_2)$.

2-13 (a) Suppose for definiteness that $\{a_n\}$ is increasing, and consider $A_{n+1} - A_n$

$$= \frac{a_{n+1}}{n+1} + \frac{1}{n+1}\sum_{k=1}^{n} a_k - \frac{1}{n}\sum_{k=1}^{n} a_k = \frac{a_{n+1}}{n+1} + \sum_{k=1}^{n} a_k\left[\frac{1}{n+1} - \frac{1}{n}\right]$$

$$= \frac{a_{n+1}}{n+1} + \sum_{k=1}^{n} \frac{-a_k}{n(n+1)}$$

Now $-a_k \geq -a_{n+1}$ for $k = 1, 2, \dots, n$, and so

$$A_{n+1} - A_n \geq \frac{a_{n+1}}{n+1} - a_{n+1}\sum_{k=1}^{n}\frac{1}{n(n+1)} = 0$$

If $\{a_n\}$ is decreasing, the proof is similar.

(b) Put $a_n = L - b_n$. Then $\{b_n\} \to 0$, and

$$A_n = L - \frac{1}{n}\sum_{k=1}^{n} b_n = L - B_n$$

If $\varepsilon > 0$ is given, we can find an integer m such that for $n > m$ we have $|b_n| < \varepsilon/2$. Thus, by the triangle inequality,

$$|B_n| \le \frac{1}{n}\left|\sum_{k=1}^{m} b_k\right| + \frac{1}{n}\sum_{k=m+1}^{n} |b_k| \le \frac{1}{n}\left|\sum_{k=1}^{m} b_k\right| + \frac{\varepsilon}{2}\frac{n-m}{n} < \frac{1}{n}\left|\sum_{k=1}^{m} b_k\right| + \frac{\varepsilon}{2}$$

for all $n > m$. Since m is fixed we can pick $N > m$ such that

$$\frac{1}{n} < \frac{\varepsilon}{2}\left|\sum_{k=1}^{m} b_k\right|^{-1}$$

for $n > N$. Hence $|B_n| < \varepsilon$ for all $n > N$, or $\{B_n\} \to 0$. Therefore, $\{A_n\} = \{L - B_n\} \to L$.
(c) We have $\{1/a_n\} \to 1/L$, and, by part (b), this means that

$$\left(\frac{1}{n}\sum_{k=1}^{n} \frac{1}{a_k}\right) \to \frac{1}{L}$$

or $\{1/H_n\} \to 1/L$, or $\{H_n\} \to L$.
(d) The inequality referred to is $H_n \le G_n \le A_n$. Inasmuch as $\{A_n\} \to L$ by part (b), and $\{H_n\} \to L$ by part (c), it follows that $\{G_n\} \to L$ as well.

2-14 By Problem 2-13d,

$$\left\{\left[\prod_{k=1}^{n} a_k/a_{k-1}\right]^{1/n}\right\} \to L$$

or $\{a_n^{1/n}/a_0^{1/n}\} \to L$. But $\{a_0^{1/n}\} \to 1$ by Problem 2-5c, and so we can apply Theorem 2-1c and obtain $\{a_n^{1/n}\} \to L$. The reader should note how this result could be applied to the sequence of Problem 2-5e.

CHAPTER 3

3-1 (a) If r is the radius of the circle, we have $A = \pi r^2$. Thus $r = (A/\pi)^{1/2}$, and so $f(A) = (A/\pi)^{1/2}$.
(b) The height h of an equilateral triangle of side s is given by $h = (3)^{1/2}s/2$. The area A is thus $A = sh/2 = s^2(3)^{1/2}/4$, and so $f(s) = s^2(3)^{1/2}/4$.
(c) If $i/10 \le x < (i+1)/10$ where $i = 0, 1, 2, \ldots, 9$, then $f(x) = i$, while $f(1) = 0$.

3-2 In parts (a), (b), and (c) the domain is $x \ge 0$. With this in mind,
(a) $(f+g)(x) = x^2/(x^2+1) + x^{1/2} = (x^{5/2} + x^2 + x^{1/2})/(x^2+1)$
(b) $(f-g)(x) = (x^2 - x^{5/2} - x^{1/2})/(x^2+1)$
(c) $fg(x) = [x^2/(x^2+1)]x^{1/2} = x^{5/2}/(x^2+1)$
(d) $(g/f)(x) = x^{1/2}/[x^2/(x^2+1)] = (x^2+1)/x^{3/2}$ with domain $(0, \infty)$
(e) $f(g)(x) = (x^{1/2})^2/[(x^{1/2})^2 + 1] = x/(x+1)$ with domain $[0, \infty)$
(f) $g(f)(x) = [x^2/(x^2+1)]^{1/2}$ with domain $(-\infty, \infty)$

3-3 (a) Here $f(-x) = 1/[1 + (-x)^2] = 1/(1 + x^2) = f(x)$, and so if f is even.
   (b) Since $f(-x) = -x/[1 + (-x)^2] = -x/(1 + x^2) = -f(x)$, f is odd.
   (c) From trigonometry, $f(-x) = \sin(-x) = -\sin x = -f(x)$, and so f is odd.
   (d) As in part (c), $f(-x) = \cos(-x) = \cos x = f(x)$, and hence f is even.

3-4 (a) Let f and g be odd functions. Then $f(g)(-x) = f(g(-x)) = f(-g(x)) = -f(g(x)) = -f(g)(x)$, and so $f(g)$ is odd.
   (b) If f is even and g is odd, then $f(g)(-x) = f(g(-x)) = f(g(x)) = f(g)(x)$, and so $f(g)$ is even.
   (c) If f is arbitrary and g is even, then $f(g)(-x) = f(g(-x)) = f(g(x)) = f(g)(x)$. $f(g)$ is thus even.

3-5 (a) Let $g(x) = [f(x) + f(-x)]/2$. Then $g(-x) = [f(-x) + f(-(-x))]/2 = [f(-x) + f(x)]/2 = g(x)$.
   (b) We have $f(x) = [f(x) + f(-x)]/2 + [f(-x) - f(-x)]/2$, and we saw in part (a) that g, the first expression on the right-hand side, is even. Let h denote the second expression on the right-hand side. Then $h(-x) = [f(-x) - f(-(-x))]/2 = [f(-x) - f(x)]/2 = -[f(x) - f(-x)]/2 = -h(x)$, and so h is odd and we have $f = g + h$ with g even and h odd.

3-6 (a) If $0 \le x_1 < x_2 \le 1$, then $x_1^{1/2} < x_2^{1/2}$, and so $x_1^{1/2} + 3 < x_2^{1/2} + 3$, or $f(x_1) < f(x_2)$. Now if $y = f(x)$, we have $x = f^{-1}(y)$, and so we can find f^{-1} by solving $y = x^{1/2} + 3$ for x in terms of y. Thus $x = (y - 3)^2$, and we must have $3 \le y \le 4$. Therefore $f^{-1}(x) = (x - 3)^2$ with domain [3, 4].
   (b) Let $2 \le x_1 < x_2 \le 5$. Then $x_1^2 < x_2^2$ and $0 < x_1^2 - 1 < x_2^2 - 1$, or $1/(x_2^2 + 1) < 1/(x_1^2 + 1)$, and so f is strictly decreasing. If $y = 1/(x^2 + 1)$, then $x = [(y + 1)/y]^{1/2}$ with $1/24 \le y \le 1/3$. Hence $f^{-1}(x) = [(x + 1)/x]^{1/2}$ with domain [1/24, 1/3].
   (c) For $1 \le x_1 < x_2 \le 9$, we have $2x_1^2 < 2x_2^2$, and so $2x_1^2 + x_1 - 3 < 2x_2^2 + x_2 - 3$. That is, f is strictly increasing. If $y = 2x^2 + x + 3$, we have $2x^2 + x - (y + 3) = 0$, or $x = (-1 \pm [1 + 8(y + 3)]^{1/2})/4$. Since $x \ge 1$, we can only have $x = (-1 + [25 + 8y]^{1/2})/4$ with $0 \le y \le 168$. Thus $f^{-1}(x) = (-1 + [25 + 8y]^{1/2})/4$ on the domain [0, 168].

3-7 Suppose that $ad - bc > 0$, and take $-d/c < x_1 < x_2$. We want to show that $(ax_1 + b)/(cx_1 + d) < (ax_2 + b)/(cx_2 + d)$, or (since $cx_1 + d$ and $cx_2 + d$ are either both positive, or both negative) $(ax_1 + b)(cx_2 + d) < (ax_2 + b)(cx_1 + d)$, or $acx_1x_2 + bcx_2 + adx_1 + bd < acx_1x_2 + bcx_1 + adx_2 + bd$, or $bcx_2 + adx_1 < bcx_1 + adx_2$, or $0 < ad(x_2 - x_1) - bc(x_2 - x_1) = (ad - bc)(x_2 - x_1)$. Inasmuch as the last inequality is true, and the steps are reversible, we have shown that f is strictly increasing. The proof that f is strictly decreasing if $ad - bc < 0$ is similar. If $ad - bc = 0$, we have (a) $a/b = c/d = k$ if $b \ne 0$, and $d \ne 0$, or (b) $d = b = 0$. If (a) is the case, then $a = bk$ and $c = dk$, or $f(x) = b(kx + 1)/d(kx + 1) = b/d$. If (b) is the case, then $f(x) = ax/bx = a/b$. Thus in either case f is a constant function if $ad - bc = 0$. For $ad - bc \ne 0$, put

$$y = \frac{ax + b}{cx + d} = \frac{[(a/c)(cx + d) - ad/c + b]}{cx + d} = \frac{a}{c} + \frac{1}{c}\frac{bc - ad}{cx + d}$$

or

$$cy - a = \frac{bc - ad}{cx + d}$$

or

$$cx + d = \frac{bc - ad}{cy - a}$$

or

$$x = \frac{1}{c}\frac{bc - ad}{cy - a} - \frac{d}{c} = \frac{(1/c)(bc - ad - cyd + ad)}{cy - a} = \frac{b - dy}{cy - a}$$

Thus the equation of f^{-1} is

$$f^{-1}(x) = \frac{b - dx}{cx - a}$$

and we see that $f = f^{-1}$ when $a = -d$. This shows that there are functions f such that $f = f^{-1}$ other than the obvious example of $f(x) = x$.

Remark   The number $ad - bc$ which appears in this problem is called the *determinant* of a, b, c, and d, and it is usually denoted by

$$\begin{vmatrix} a & b \\ c & d \end{vmatrix}$$

Determinants can be extended to arrays of numbers having n rows and n columns for $n = 2, 3, \ldots$. They have application in many branches of mathematics.

3-8 Let $f(a) \le x_1 < x_2 \le f(b)$. Then $x_1 = f(f^{-1}(x_1))$ and $x_2 = f(f^{-1}(x_2))$, and so $f(f^{-1}(x_1)) < f(f^{-1}(x_2))$. Since f is strictly increasing, the last inequality can be true only if $f^{-1}(x_1) < f^{-1}(x_2)$. In other words, f^{-1} is strictly increasing. Needless to say, it can be proved in the same way that the inverse of a strictly decreasing function is also strictly decreasing.

3-9 Let $x_1 < x_2$, with both in the domain. Then $f(x_1) \le f(x_2)$, and $g(x_1) \le g(x_2)$. These inequalities may be added to give $f(x_1) + g(x_1) \le f(x_2) + g(x_2)$, or $(f + g)(x_1) \le (f + g)(x_2)$. Thus $f + g$ is increasing.
To show that fg need not be increasing when f and g are, let $f(x) = x$ and $g(x) = x - 1$ for $0 \le x \le 1$. It is clear that f and g are increasing. We have $fg(x) = x^2 - x$. Now $0 < 1/2$, and $fg(1/2) = 1/4 - 1/2 = -1/4 < 0 = fg(0)$, and so fg is not increasing. On the other hand, if f and g are both positive, that is, $f(x) \ge 0$ and $g(x) \ge 0$ for all x in their domain, as well as both increasing, then, if $x_1 < x_2$, $f(x_1)g(x_1) \le f(x_1)g(x_2) \le f(x_2)g(x_2)$, or $fg(x_1) \le fg(x_2)$, and so fg is increasing.

3-10 Suppose that $a \le x_1 < x_2 \le b$. Then $g(x_1) \le g(x_2)$, and so $f(g(x_1)) \le f(g(x_2))$. Thus $f(g)$ is indeed increasing.

3-11 We must have $f^{-1}(f(x) + f(y)) = f^{-1}(f(xy)) = xy$. Put $f(x) = u$ and $f(y) = v$. Then $x = f^{-1}(u)$, $y = f^{-1}(v)$, and $f^{-1}(u + v) = f^{-1}(u)f^{-1}(v)$. The last equation is the relation we seek.

3-12 (a) Solving for y, we have $y = (1 - x^2)/x$ for $x \neq 0$. The restrictions of this function is the family of implicit functions.
(b) We must have $y^2 = 1 - x_2$ for $|x| \leq 1$. Now this equation leads to $y = (1 - x^2)^{1/2}$ and $y = -(1 - x^2)^{1/2}$, and these two functions determine the family of implicit functions.
(c) It is clear we must have $|x| \leq 1$ and $|y| = 1 - |x|$. If $0 \leq x \leq 1$, $|x| = x$; while if $-1 \leq x < 0$, $|x| = -x$. Thus $|y| = 1 - x$ if $0 \leq x \leq 1$, and $|y| = 1 + x$ if $-1 \leq x < 0$. Therefore, if $y \geq 0$, we have

$$y = \begin{cases} 1 + x \text{ for } -1 \leq x < 0 \\ 1 - x \text{ for } 0 \leq x \leq 1 \end{cases}$$

and if $y < 0$,

$$y = \begin{cases} -1 - x \text{ for } -1 \leq x < 0 \\ -1 + x \text{ for } 0 \leq x \leq 1 \end{cases}$$

These two functions determine the family of implicit functions.
(d) We have $y^{2/3} = 1 - x^{2/3}$, or $y = \pm(1 - x^{2/3})^{3/2}$ for $|x| \leq 1$, and this gives us the two functions $y = (1 - x^{2/3})^{3/2}$, and $y = -(1 - x^{2/3})^{3/2}$ which both have domain $[-1, 1]$. These two functions generate the family of implicit functions.

3-13 (a) Since $|x^p - 0| = |x|^p$, we see that if $\varepsilon > 0$ is given and we let $\delta = \varepsilon^{1/p}$, then $|x^p - 0| < \varepsilon$ for all x such that $0 < x < \delta$.
(b) We have $|ax + b - (ac + b)| = |a|\,|x - c|$. For $\varepsilon > 0$, let $\delta = \varepsilon/|a|$. Then if $0 < |x - c| < \delta$, we have $|a|\,|x - c| < \varepsilon$, or $|ax - b - (ac + b)| < \varepsilon$.
(c) We are interested in $|x^2 - a^2| = |x + a|\,|x - a|$. Since we are only concerned with values of x that are in some sense "close" to a, we can restrict ourselves to x such that $x \neq a$, and $a - 1 < x < a + 1$. This means that $2a - 1 < x + a < 2a + 1$, and if we let M be the larger of $|2a - 1|$ and $|2a + 1|$, then $|x + a| < M$. Thus if $\varepsilon > 0$ and we let δ be the smaller of ε/M and 1, then $|x^2 - a^2| < \varepsilon$ when $0 < |x - a| < \delta$.
(d) Here, for $x \neq 0$, $|1/x - 1/a| = |x - a|/|ax|$. Let $b > 0$ be such that the interval $(a - b, a + b)$ does not contain zero, and let M be the larger of $1/|a + b|$ and $1/|a - b|$. Then if $x \neq a$ and $a - b < x < a + b$, we have $1/|x| < M$. For any $\varepsilon > 0$, let δ be the smaller of $\varepsilon\,|a|/M$ and b. Then $|1/x - 1/a| < \varepsilon$ when $0 < |x - a| < \delta$.
(e) We have $|x^{1/2} - a^{1/2}| = |x - a|/(x^{1/2} + a^{1/2})$. Let $b > 0$ be such that $0 < a - b$. Then if $x \neq a$, and $a - b < x < a + b$, we have $(a - b)^{1/2} + a^{1/2} < x^{1/2} + a^{1/2} < (a + b)^{1/2} + a^{1/2}$, or $1/(x^{1/2} + a^{1/2}) < 1/[(a - b)^{1/2} + a^{1/2}] = M$. Given $\varepsilon > 0$, let δ be the smaller of ε/M and b. Then $|x^{1/2} - a^{1/2}| < \varepsilon$ for all x such that $0 < |x - a| < \delta$.

3-14 (a) From Problem 3-13a, $0 = \lim_{y \to 0+} y^p = \lim_{y \to 0+} 1/(1/y)^p$, and so $\lim_{x \to \infty} 1/x^p = 0$ by Exercise 3-14.
(b) From Problem 3-13b, we have $\lim_{x \to a} x = a$. Thus from Theorem 3-1b, $\lim_{x \to a} x^n = a^n$.

(c) From Problem 3-13c, $\lim\limits_{x \to a} 1/x = 1/a$, and so from Theorem 3-1b,

$\lim\limits_{x \to a} 1/x^n = \lim\limits_{x \to a} (1/x)^n = (1/a)^n = 1/a^n$.

(d) From part (b) and Theorem 3-1a, we have

$$\lim_{x \to a} \sum_{k=0}^{n} c_k x^k = \sum_{k=0}^{n} \lim_{x \to a} c_k x^k = \sum_{k=0}^{n} c_k \lim_{x \to a} x^k = \sum_{k=0}^{n} c_k a^k$$

(e) From part (d) and Theorem 3-1c, we have

$$\lim_{x \to a} \left[\sum_{k=0}^{n} c_k x^k \Big/ \sum_{k=0}^{m} b_k x^k \right] = \lim_{x \to a} \sum_{k=0}^{n} c_k x^k \Big/ \lim_{x \to a} \sum_{k=0}^{m} b_k x^k$$

$$= \sum_{k=0}^{n} c_k a^k \Big/ \sum_{k=0}^{m} b_k a^k$$

since $\sum\limits_{k=0}^{m} b_k a^k \neq 0$.

(f) Here

$$\sum_{k=0}^{n} a_k x^k \Big/ \sum_{k=0}^{n} b_k x^k = \sum_{k=0}^{n} a_k / x^{n-k} \Big/ \sum_{k=0}^{n} b_k / x^{n-k}$$

and we have from part (a) $\lim\limits_{x \to \infty} 1/x^{n-k} = 0$ for $k = 0, 1, \ldots, n-1$. Thus it follows from Theorem 3-1a and part (c) that

$$\lim_{x \to \infty} \left[\sum_{k=0}^{n} a_k x^k \Big/ \sum_{k=0}^{n} b_k x^k \right] = \frac{a_n}{b_n}$$

(g) We have, from parts (b) and (f) and Theorem 3-3, that

$$\lim_{x \to \infty} \left[\sum_{k=1}^{n} a_k x^k \Big/ \sum_{k=1}^{n} b^k x^k \right]^m = \left[\lim_{x \to \infty} \left(\sum_{k=1}^{n} a_k x^k \Big/ \sum_{k=1}^{n} b_k x^k \right) \right]^m = (a_n/b_n)^m$$

(h) It follows from part (d) and Theorem 3-3 that

$$\lim_{x \to a} \sum_{k=0}^{n} a_k \left[\sum_{j=0}^{m} b_j x^j \right]^k = \sum_{k=0}^{n} a_k \left[\lim_{x \to a} \sum_{j=0}^{m} b_j x^j \right]^k = \sum_{k=0}^{n} a_k \left[\sum_{j=0}^{m} b_j a^j \right]^k$$

(i) It follows from Exercise 3-17, Problem 3-13c, part (d), and Theorem 3-3 that

$$\lim_{x \to a} \left[\left| \sum_{k=0}^{n} c_k x^k \right| \right]^{1/2} = \left[\lim_{x \to a} \left| \sum_{k=0}^{n} c_k x^k \right| \right]^{1/2}$$

$$= \left[\left| \lim_{x \to a} \sum_{k=0}^{n} c_k x^k \right| \right]^{1/2} = \left[\left| \sum_{k=0}^{n} c_k a^k \right| \right]^{1/2}$$

3-15 (a) Since $|\sin (1/x^{1/2})| \leq 1$, we have $-x \leq x \sin (1/x^{1/2}) \leq x$ for $x > 0$, and it follows from Theorem 3-2 that $\lim\limits_{x \to 0+} x \sin (1/x^{1/2}) = 0$.

(b) If $4/10 \leq x < 5/10$, we have $f(x) = 4$, and so $\lim\limits_{x \to 1/2-} f(x) = 4$.

(c) We have $(x^2 - a^2)^{1/2}/(x - a)^{1/2} = (x + a)^{1/2}$, and so

$$\lim_{x \to a+} (x^2 - a^2)^{1/2}/(x - a)^{1/2} = \lim_{x \to a+} (x + a)^{1/2} = (2a)^{1/2}$$

The last relation follows from Problem 3-13e.

3-16 (a) Since $x^{1/4} = (x^{1/2})^{1/2}$, it follows from Problem 3-13e and Theorem 3-3 that $\lim_{x \to a} x^{1/4} = (\lim_{x \to a} x^{1/2})^{1/2} = (a^{1/2})^{1/2} = a^{1/4}$. This function is thus continuous on its domain.

(b) It follows from Problem 3-13b and d that this function is continuous for all $x \neq 0$. Now $\lim_{x \to 0-} x = 0$, and $\lim_{x \to 0+} 1/x = \infty$, and so f has an infinite discontinuity at zero.

(c) If $x > 0$, then $|x|/x = 1$; while if $x < 0$, then $|x|/x = -1$. It is clear from this that f is continuous for all $x \neq 0$, and that $\lim_{x \to 0+} f(x) = 1$ and $\lim_{x \to 0-} f(x) = -1$. Hence f has a simple discontinuity at zero.

(d) If $a < 2$, it follows from Problem 3-13c that $\lim_{x \to a} (x^2/2 - 2) = a^2/2 - 2$; and if $a > 2$, it follows from Problem 3-14c that $\lim_{x \to a} (2 - 8/x^2) = 2 - 8/a^2$.

These relations mean that f is continuous for all $x \neq 2$. Now $f(2+) = 0 = f(2-)$, and so, inasmuch as $f(2) = 1 \neq 0$, this means that f has a removable discontinuity at $x = 2$. We can remove the discontinuity by putting $f(2) = 0$ instead of 1.

3-17 For $n = 1, 2, \ldots$, let $a_n = 1/2\pi n$, and $b_n = 2/\pi(4n + 1)$. Then $a_n > 0$, $b_n > 0$, $\{a_n\} \to 0$, and $\{b_n\} \to 0$. Further, $f(a_n) = \sin(2\pi n) = 0$, and $f(b_n) = \sin[(4n + 1)\pi/2] = 1$, and so $\{f(a_n)\} \to 0$ and $\{f(b_n)\} \to 1$. Thus $f(0+)$ does not exist, and so $\lim_{x \to 0} f(x)$ does not exist either. Hence f is not continuous at zero, and inasmuch as $f(0+)$ does not exist, the discontinuity cannot be simple. But $|\sin 1/x| \leq 1$, or $|f(x)| \leq 1$, and hence f cannot have an infinite discontinuity at zero. Therefore, f must have an oscillatory discontinuity at zero. If the reader will sketch the graph of this function near zero, he will see that the term "oscillatory discontinuity" is well taken.

3-18 Assume for definiteness that f is increasing and let $a < c \leq b$. Let $\{a_n\}$ be an increasing sequence such that $a_n < c$ for all n, and $\{a_n\} \to c$. Then $\{f(a_n)\}$ is also an increasing sequence, and $f(a_n) \leq f(c)$ for all n, or $\{f(a_n)\}$ is bounded. Thus $\{f(a_n)\}$ must have a limit. Suppose that $f(c-)$ fails to exist. Then we can find increasing sequences $\{a_n\}$ and $\{b_n\}$ such that $a_n < c$, $b_n < c$ for all n, $\{a_n\} \to c$, $\{b_n\} \to c$, and $\{f(a_n)\} \to L_1, \{f(b_n)\} \to L_2$ with $L_1 < L_2$. Pick $\varepsilon > 0$ such that $L_1 + \varepsilon < L_2 - \varepsilon$. We can then find N such that $f(a_n) < L_1 + \varepsilon$ and $L_2 - \varepsilon < f(b_n)$ for all $n > N$. Let m be a fixed integer $> N$ and consider $b_m < c$. Since $\{a_n\} \to c$, we can find a_k with $k > N$ such that $b_m < a_k < c$. We then have $f(a_k) < L_1 + \varepsilon < L_2 - \varepsilon < f(b_m)$. But this inequality contradicts the hypothesis that f is increasing, and so $f(c-)$ must exist. If $a \leq c < b$, a similar argument shows that $f(c+)$ exists. The proof when f is decreasing is also similar.

This problem means that a monotone function can have only simple discontinuities.

3-19   We must have $f(x) = f(x + 0) = f(x) + f(0)$, and so $f(0) = 0$. Thus $0 = f(0) = f(x - x) = f(x) + f(-x)$, or $f(x) = -f(-x)$. Hence f is odd and we can concentrate on $x \geq 0$. Now $f(2x) = f(x + x) = f(x) + f(x) = 2f(x)$, and it is easy to see that we can prove by induction that $f(nx) = nf(x)$ for all positive integers n. Put $c = f(1)$. Then $f(n) = cn$, and $c = f(1) = f(n/n) = nf(1/n)$. In other words, $f(1/n) = c/n$ for all positive integers n. If r is a positive rational number, then $r = m/n$, where m and n are positive integers, and the above results imply that $f(r) = f(m/n) = mf(1/n) = m(c/n) = cr$. Now let x be positive and irrational. We can select a sequence of positive rational numbers $\{r_k\}$ such that $\{r_k\} \to x$. Because f is continuous, $\{f(r_k)\} \to f(x)$. On the other hand, $\{f(r_k)\} = \{cr_k\} \to cx$. Therefore, $f(x) = cx$ (because a sequence cannot have two limits) for all $x \geq 0$. The fact that f is odd now means that $f(x) = cx$ for all x.

Remark   The equation $f(x + y) = f(x) + f(y)$ is an example of a *functional equation*. Another example is the equation of Problem 3-11. Many of the numerous trigonometric identities are also functional equations. As in this problem, it is often the case that a functional equation determines the function satisfying it to a large extent.

CHAPTER 4

4-1   By the binomial theorem,

$$(x + h)^n = \sum_{k=0}^{n} \binom{n}{k} x^k h^{n-k}$$

Thus,

$$(x + h)^n - x^n = h \sum_{k=0}^{n-1} \binom{n}{k} x^k h^{n-1-k}$$

and so

$$\frac{dx^n}{dx} = \lim_{h \to 0} \frac{(x + h)^n - x^n}{h} = \lim_{h \to 0} \sum_{k=0}^{n-1} \binom{n}{k} x^k h^{n-1-k} = \binom{n}{n-1} x^{n-1} = nx^{n-1}$$

(b)   We have $[(x + h)^{1/2} - x^{1/2}]/h$

$$= \frac{(x + h)^{1/2} - x^{1/2}}{h} \frac{(x + h)^{1/2} + x^{1/2}}{(x + h)^{1/2} + x^{1/2}} = \frac{h}{h[(x + h)^{1/2} + x^{1/2}]}$$

and so

$$\frac{dx^{1/2}}{dx} = \lim_{h \to 0} \frac{1}{(x + h)^{1/2} + x^{1/2}} = \frac{1}{2x^{1/2}}$$

(c)   Here

$$\frac{1/(x + h)^{1/2} - 1/x^{1/2}}{h} = \frac{1}{h} \frac{x^{1/2} - (x + h)^{1/2}}{x^{1/2}(x + h)^{1/2}} \frac{(x + h)^{1/2} + x^{1/2}}{(x + h)^{1/2} + x^{1/2}}$$

$$= \frac{1}{h} \frac{-h}{x^{1/2}(x + h)^{1/2}[x^{1/2} + (x + h)^{1/2}]}$$

Thus

$$\frac{d1/x^{1/2}}{dx} = \lim_{h \to 0} \frac{-1}{x^{1/2}(x+h)^{1/2}[x^{1/2}+(x+h)^{1/2}]} = \frac{-1}{2x^{3/2}}$$

4-2 All solutions to these make use of the chain rule.   With this in mind, we have:

(a) $f'(x) = p(ax+b)^{p-1}a$

(b) $f'(x) = -p(ax^2+bx+c)^{-(p+1)}(2ax+b)$

(c) $f'(x) = p\left(\sum_{k=0}^{n} c_k x^k\right)^{p-1}\left(\sum_{k=1}^{n} kc_k x^{k-1}\right)$

(d) $f'(x) = p(\tan x)^{p-1}\sec^2 x$

(e) $f'(x) = p(\sin^{-1} x)^{p-1}\dfrac{1}{(1-x^2)^{1/2}}$

(f) $f'(x) = p(e^x + \log x)^{p-1}(e^x + 1/x)$

(g) $f'(x) = p(\sin x^p)^{p-1}(\cos x^p)px^{p-1}$

(h) $f'(x) = \dfrac{1}{1+(x^p)^2}px^{p-1}$

4-3 $\dfrac{dy}{dx} = -am\sin(mx) + bm\cos(mx)$

$\dfrac{d^2y}{dx^2} = -am^2\cos(mx) - bm^2\sin(mx) = -m^2y$

4-4 $\dfrac{dy}{dx} = m(\tan x + \sec x)^{m-1}(\sec^2 x + \sec x \tan x)$

$= m\sec x(\tan x + \sec x)^m = my\sec x$

4-5 $\dfrac{d}{dx}[\tan^{-1}(u/v)] = \dfrac{1}{1+(u/v)^2}\dfrac{v\, du/dx - u\, dv/dx}{v^2} = \dfrac{v\, du/dx - u\, dv/dx}{u^2+v^2}$

4-6 $\dfrac{d}{dx}e^{e^{e^x}} = e^{e^{e^x}}\dfrac{d}{dx}e^{e^x} = e^{e^{e^x}}e^{e^x}\dfrac{de^x}{dx} = e^{e^{e^x}}e^{e^x}e^x$

$\dfrac{d}{dx}\log[\log(\log x)] = \dfrac{1}{\log(\log x)}\dfrac{d}{dx}\log(\log x)$

$= \dfrac{1}{\log(\log x)}\dfrac{1}{\log x}\dfrac{d}{dx}\log x = \dfrac{1}{x\log x\log(\log x)}$

4-7 If we denote $\log_{h(x)}f(x)$ by $l(x)$, we have $f(x) = h(x)^{l(x)}$, or $\log f(x) = l(x)\log h(x)$, and so $l(x) = \log f(x)/\log h(x)$.   Thus,

$$\frac{d}{dx}[\log_{h(x)}f(x)] = \frac{\log h(x)f'(x)/f(x) - \log f(x)h'(x)/h(x)}{\log^2 h(x)}$$

4-8 If we denote $f(x)^{g(x)}$ by $h(x)$, then $\log h(x) = g(x)\log f(x)$, and so

$$\frac{d}{dx}[\log h(x)] = \frac{d}{dx}[g(x)\log f(x)]$$

or

$$\frac{h'(x)}{h(x)} = g(x)\frac{f'(x)}{f(x)} + g'(x)\log f(x)$$

Thus,

$$h'(x) = f(x)^{g(x)}\left[\frac{g(x)}{f(x)}f'(x) + g'(x)\log f(x)\right]$$

In particular,

$$\frac{dx^x}{dx} = x^x[1 + \log x]$$

and so forth.

Remark The function h'/h appearing above is called the *logarithmic derivative* of h. Notice that if we know the logarithmic derivative of f, then f' can be found by simply multiplying it by f. It sometimes happens, particularly when we are dealing with a product or quotient of functions, or, as above, with exponentiation of functions, that it is easier to find the logarithmic derivative than to directly compute the derivative. As an illustration, consider the following problem.

Supplementary Problem 4-8 Find f' if

$$f(x) = \frac{x^2(3-x)^{1/3}e^{\tan^{-1}x}(\cos x)^{1/2}}{(1-x^2)^5(3+x)^{2/3}}$$

Solution $\log|f(x)| = 2\log|x| + \frac{1}{3}\log|3-x| + \tan^{-1}x + \frac{1}{2}\log|\cos x|$

$$- 5\log|1-x^2| - \frac{2}{3}\log|3+x|$$

and so

$$\frac{f'(x)}{f(x)} = \frac{2}{x} - \frac{1}{3}\frac{1}{3-x} + \frac{1}{1+x^2} - \frac{1}{2}\frac{\sin x}{\cos x} + 10\frac{x}{1-x^2} - \frac{2}{3}\frac{1}{3+x}$$

Multiplication of the right-hand side of this equation by $f(x)$ will yield $f'(x)$.

4-9 If f is even, then $f(x) = f(-x)$. If we differentiate each side of this identity, we have $Df(x) = Df(-x) = -Df(-x)$. In other words, Df is odd. Similarly, if g is odd, then $g(x) = -g(-x)$, and $Dg(x) = -Dg(-x) = Dg(-x)$. Hence, Dg is even.

4-10 (a) We prove Leibniz' rule by induction. When $n = 1$ it is simply the familiar formula for the derivative of a product. Assume that it is true for some integer n. Then

$$D^{n+1}(fg) = D\,[D^n(fg)] = D\left[\sum_{k=0}^{n}\binom{n}{k} D^k f D^{n-k}g\right]$$

$$= \sum_{k=0}^{n}\binom{n}{k} D\,[D^k f D^{n-k}g]$$

$$= \sum_{k=0}^{n}\binom{n}{k}[D^k f D^{n+1-k}g + D^{k+1}f D^{n-k}g]$$

$$= \sum_{k=0}^{n}\binom{n}{k} D^k f D^{n+1-k}g + \sum_{k=0}^{n}\binom{n}{k} D^{k+1}f D^{(n+1)-(k+1)}g$$

$$= D^0 f D^{n+1}g + \sum_{k=1}^{n}\left[\binom{n}{k} + \binom{n}{k-1}\right] D^k f D^{n+1-k}g + D^{n+1}f D^0 g$$

$$= D^0 f D^{n+1}g + \sum_{k=1}^{n}\binom{n+1}{k} D^k f D^{n+1-k}g + D^{n+1}f D^0 g$$

[we have used Eq. (1 18)]

$$= \sum_{k=0}^{n+1}\binom{n+1}{k} D^k f D^{n+1-k}g$$

Thus, if Leibniz' rule is true for some integer n, it is true for $n+1$. The induction is complete.

(b) The computation of the first three derivatives is a routine matter; we shall accordingly compute $D^n x^3 \cos x$ for $n \geq 4$. Observe first that $Dx^3 = 3x^2$, $D^2 x^3 = 6x$, $D^3 x^3 = 6$, and $D^n x^3 = 0$ for $n \geq 4$. At the same time, we have $D \cos x = -\sin x$, $D^2 \cos x = -\cos x$, $D^3 \cos x = \sin x$, and $D^4 \cos x = \cos x$. It is clear that the higher derivatives of cosine will repeat cyclically according to the formula

$$D^n \cos x = \begin{cases}(-1)^{n/2} \cos x & \text{if } n \text{ is even} \\ (-1)^{(n+1)/2} \sin x & \text{if } n \text{ is odd}\end{cases}$$

Consequently, if $n \geq 4$, then

$$D^n x^3 \cos x = \sum_{k=0}^{3}\binom{n}{k} D^k x^3 D^{n-k}\cos x$$

$$= \begin{cases}x^3(-1)^{n/2}\cos x + 3nx^2(-1)^{n/2}\sin x + 3n(n-1)(-1)^{(n-2)/2}\cos x \\ \quad + n(n-1)(n-2)(-1)^{(n-2)/2}\sin x \text{ if } n \text{ is even} \\ x^3(-1)^{(n+1)/2}\sin x + 3nx^2(-1)^{(n-1)/2}\cos x + 3n(n-1)(-1)^{(n-1)/2}\sin x \\ \quad + n(n-1)(n-2)(-1)^{(n-3)/2}\cos x \text{ if } n \text{ is odd}\end{cases}$$

(c) It is easy to see that $D^k x^m = m(m-1)\cdots(m-k+1)x^{m-k} = [m!/(m-k)!]x^{m-k}$ if $k \leq m$, and that $D^k x^m = 0$ if $k > m$. Our formula follows by incorporating these results into Leibniz' rule.

4-11   Implicit differentiation results in

$$px^{p-1}y^q + qx^p y^{q-1}(dy/dx) = (p+q)(x+y)^{p+q-1}[1 + dy/dx]$$

Thus

$$[qx^p y^{q-1} - (p+q)(x+y)^{p+q}]\frac{dy}{dx} = (p+q)(x+y)^{p+q-1} - px^{p-1}q$$

or

$$\left[qx^p y^{q-1} - \frac{(p+q)x^p y^q}{x+y}\right]\frac{dy}{dx} = \frac{(p+q)x^p y^q}{x+y} - px^{p-1}y^q$$

or

$$x^p y^{q-1}\left[\frac{q(x+y) - (p+q)y}{x+y}\right]\frac{dy}{dx} = x^{p-1}y^q\left[\frac{(p+q)x - p(x+y)}{x+y}\right]$$

or

$$x[qx - py]\,dy/dx = y[qx - py], \quad \text{or} \quad x\,dy/dx = y$$

4-12   We have

$$\frac{1}{1 + (x/y)^2}\frac{y - x\,dy/dx}{y^2} + \frac{1}{(x^2 + y^2)^{1/2}}\frac{x + y\,dy/dx}{(x^2 + y^2)^{1/2}} = 0$$

or

$$\frac{y - x\,dy/dx + x + y\,dy/dx}{x^2 + y^2} = 0$$

or

$$x + y + (y - x)\,dy/dx = 0 \quad \text{or} \quad dy/dx = (x+y)/(x-y)$$

4-13 $e^{x+y}(1 + dy/dx) = e^x + e^y\,dy/dx$, or $(e^{x+y} - e^y)\,dy/dx = e^x - e^{x+y}$, or $e^x\,dy/dx = -e^y$, or $dy/dx = -e^{y-x}$.

4-14   We have to examine the existence of $\lim_{h\to 0}[|0 + h| - |0|]/h = \lim_{h\to 0}|h|/h$. But if $h > 0$, $|h|/h = 1$; while if $h < 0$, $|h|/h = -1$. Thus the limit does not exist.

4-15   We have $|f(x) - f(0)| = |f(x)| = |x|\,|\sin(1/x)| \le |x|$. Thus if $|x| < \varepsilon$, $|f(x) - f(0)| < \varepsilon$, and so f is continuous at zero. On the other hand, $f'(0) = \lim_{h\to 0}[f(0 + h) - f(0)]/h = \lim_{h\to 0}h\sin(1/h)/h = \lim_{h\to 0}\sin(1/h)$, and this limit does not exist (see Problem 3-17).

Remark   While a function which is differentiable at a given value is necessarily continuous there, Problems 4-14 and 4-15 show that the converse need not be true. Indeed, Problem 4-15 shows that there need not be right-hand or left-hand derivatives there.

4-16   If $x \ne 0$, we have $f'(x) = 2x\sin(1/x) - \cos(1/x) + a$, while $f'(0) = \lim_{h\to 0}f(h)/h = \lim_{h\to 0}[h\sin(1/h) + a] = a$ [since $|\sin(1/h)| \le 1$]. Thus as $x \to 0$, $f'(x)$

oscillates between $a + 1$ and $a - 1$, and so $\lim_{h \to 0} f'(x)$ does not exist. Therefore f' is discontinuous at zero.

Remark This example can be modified to illustrate another point if we take $0 < a < 1$. Then $a - 1 < 0$ and it is possible to find values of x as close to zero as desired for which $f'(x) < 0$. This means that, even though $f'(0) = a > 0$, there is no interval including zero throughout which f is an increasing function. What *is* true if $f'(x_0) > 0$ is that there is an interval containing x_0 such that if $x_1 > x_0 > x_2$, then $f(x_1) > f(x_2)$. As we have seen, if $x_1 > x_2$ and either $x_2 > x_0$, or $x_1 < x_0$, then it need not be true that $f(x_1) > f(x_2)$.

4-17 We have $D[x - \sin x] = 1 - \cos x$. Now $1 - \cos x \geq 0$. Further, $1 - \cos x = 0$ only if $x = \pm 2k\pi$ for $k = 0, 1, 2, \ldots$. Thus, $1 - \cos x > 0$ if $(2k - 1)\pi < x < 2k\pi$, or $-2k\pi < x < -(2k - 1)\pi$. This means that f is strictly increasing on abutting intervals. But it is easy to see that if a function is strictly increasing on the intervals $[a, b]$ and $[b, c]$, then it is strictly increasing on $[a, c]$. In other words, the function $f(x) = x - \sin x$ is strictly increasing for all x.

As for g, $Dg(x) = \alpha - \cos x$. Since $\alpha - \cos x > 0$ if $\alpha > 1$, and $\alpha - \cos x < 0$ if $\alpha < -1$, g is strictly increasing if $\alpha > 1$, strictly decreasing if $\alpha < -1$. We saw above that g is strictly increasing if $\alpha = 1$, and it can be shown similarly that g is strictly decreasing if $\alpha = -1$. Inasmuch as $\alpha - \cos x$ changes sign if $|\alpha| < 1$, g cannot be either strictly increasing or strictly decreasing if $|\alpha| < 1$. To sum up, g is strictly increasing if $\alpha \geq 1$, strictly decreasing if $\alpha \leq -1$.

Remark Since $f(0) = 0$, it follows that if $x > 0$, then $f(x) = x - \sin x > 0$. We thus have the inequality, especially interesting for x near zero, $x < \sin x$ for $x > 0$.

4-18 Since f is continuous on $[0, \pi/2]$ (recall that $\lim_{x \to 0} \sin x/x = 1$), if we can show that $\dfrac{d}{dx} [x/\sin x] > 0$ for $0 < x < \pi/2$, the assertion follows. Now $\dfrac{d}{dx} [x/\sin x] = [\sin x - x \cos x]/\sin^2 x$, and the right-hand side is positive if and only if $\sin x - x \cos x > 0$, or, since $0 < \cos x < 1$ for $0 < x < \pi/2$, if and only if $\tan x - x > 0$. Since $\tan 0 - 0 = 0$, if we can show that $\tan x - x$ increases strictly for $0 < x < \pi/2$, it will follow that $\tan x - x > 0$ for such x. But $\dfrac{d}{dx} [\tan x - x] = \sec^2 x - 1 = \tan^2 x > 0$ for $0 < x < \pi/2$, and so $\dfrac{d}{dx} [x/\sin x] > 0$ for $0 < x < \pi/2$.

4-19 (a) We have

$$D \frac{f}{g} = \frac{(f'g - g'f)}{g^2} = \frac{f'}{g} - \frac{g'f}{g_2} = \frac{f'g'}{g'g} - \frac{g'f}{gg} = \left(\frac{f'}{g'} - \frac{f}{g} \right) \frac{g'}{g}$$

Now $g' > 0$, and so g is strictly increasing; since $g(a) = 0$, this means that $g(x) > 0$ for $a < x < b$. Hence the sign of $Df/g(x)$ is the same as the sign of $f'(x)/g'(x) - f(x)/g(x)$. But

$$\frac{f(x)}{g(x)} = \frac{f(x) - f(a)}{g(x) - g(a)} = \frac{f'(z)}{g'(z)}$$

where $a < z < x$. Inasmuch as f'/g' is increasing, we can conclude that

$$\frac{f'(x)}{g'(x)} - \frac{f(x)}{g(x)} \geq 0$$

or $Df/g(x) \geq 0$ for $a < x \leq b$. Therefore, f/g is increasing on $(a, b]$. It can be seen from the proof that if f'/g' is strictly increasing, then so is f/g. Further, if $\lim_{x \to a+} f(x)/g(x)$ exists and is put equal to $f/g(a)$, then f/g is increasing on $[a, b]$.

(b)   When $g(x) = x$, we must put $a = 0$ in part (a). Then $f(x)/x$ will be increasing on $(0, b]$ if f is continuous on $[0, b]$ and differentiable on $(0, b)$; if $f(0) = 0$; and if f' is increasing.

(c)   Put $f(x) = x$, and $g(x) = \sin x$ in part (a). Then f and g are continuous on $[0, \pi/2]$ and differentiable on $(0, \pi/2)$; $f(0) = 0 = g(0)$; $g'(x) = \cos x > 0$ for $0 < x < \pi/2$; and $f'(x)/g'(x) = 1/\cos x = \sec x$ is an increasing function on $(0, \pi/2)$. Hence the function $x/\sin x$ increases on $(0, \pi/2]$, and inasmuch as $\lim_{x \to 0} x/\sin x = 1$, it also increases on $[0, \pi/2]$ as we observed at the conclusion of the solution of part (a).

(d)   Observe that $h(x) = \sin \pi x/x(1-x)$ is symmetric about $x = 1/2$; that is, $h(1/2 + \delta) = h(1/2 - \delta)$ for $0 < \delta < 1/2$ [because $\sin (\pi/2 + \pi\delta) = \cos \pi\delta$, and $\sin (\pi/2 - \pi\delta) = \cos (-\pi\delta) = \cos \pi\delta$]. Thus if we can show that h is strictly increasing on $(0, 1/2]$, it will follow that h is strictly decreasing on $[1/2, 1)$. Consequently, $h(1/2) = 4$ will be the maximum of h on $(0, 1)$, and $h(0) = h(1/2) = \pi$ (because $\lim_{x \to 0} \sin \pi x/x = \pi$) will be the minimum of h on $(0, 1)$. In other words, we will have $\pi < h(x) \leq 4$ for $0 < x < 1$.

To show that h is strictly increasing on $(0, 1/2]$, we put $f(x) = \sin \pi x$ and $g(x) = x(1-x)$, both for $0 \leq x \leq 1/2$. Clearly, f and g fulfill the conditions of part (a). Therefore, all we have to do is show that $f'(x)/g'(x) = \pi \cos \pi x/(1 - 2x)$ is strictly increasing on $(0, 1/2)$. To this end,

$$D\left[\frac{\pi \cos \pi x}{1 - 2x}\right] = \frac{-\pi^2(1 - 2x) \sin \pi x + 2\pi \cos \pi x}{(1 - 2x)^2}$$

and so f'/g' is strictly increasing on $(0, 1/2)$ if $l(x) = 2\pi \cos \pi x - \pi^2(1 - 2x) \sin \pi x > 0$ on $(0, 1/2)$. But

$$l'(x) = -2\pi^2 \sin \pi x + 2\pi^2 \sin \pi x - \pi^3(1 - 2x) \cos \pi x = -\pi^3(1 - 2x) \cos \pi x$$

Since $-\pi^3(1 - 2x) \cos \pi x < 0$ for $0 < x < 1/2$, l must be strictly decreasing on $(0, 1/2)$. Thus $l(x) > l(1/2) = 2\pi \cos \pi/2 = 0$, or f'/g' is strictly increasing on $(0, 1/2)$.

4-20   Notice first that this function is defined only when $x^2 - 2x - 8 \geq 0$. That is, when $(x + 2)(x - 4) \geq 0$, which is $x \geq 4$ or $x \leq -2$. For these x, we have

$$f'(x) = (1/2)(x^2 - 2x - 8)^{-1/2}(2x - 2) = (1/2)(2x - 2)/(x^2 - 2x - 8)^{1/2}$$

Thus $f'(x) = 0$ only if $2x - 2 = 0$, or $x = 1$. Therefore, since $x = 1$ is not in the domain of f, f has no horizontal tangent.

4-21   (a)   We have $f'(x) = (x-1)^{1/2} + (1/2)x/(x-1)^{1/2}$, and

$$f''(x) = (1/2)[3(x-1)^{1/2} - (3x-2)/2(x-1)^{1/2}]/(x-1)$$

$$= (1/4)(3x-4)/(x-1)^{3/2}$$

We can see that $f''(x) < 0$ if $1 < x < 3/4$, and $f''(x) > 0$ if $4/3 < x$. Thus the graph of f is concave upward if $4/3 < x$, and concave downward if $1 < x < 4/3$, and so $(4/3, f(4/3))$ is the only point of inflection.

(b)   Here $f'(x) = [2x(x^2-9) - (x^2-4)2x]/(x^2-9)^2 = -10x/(x^2-9)^2$, and

$$f''(x) = \frac{-10(x^2-9)^2 + 40x^2(x^2-9)}{(x^2-9)^4}$$

$$= \frac{30(x^2+3)}{(x^2-9)^4} = \frac{30(x^2+3)}{(x^2-9)^4(x+3)(x-3)}$$

From the last equation, we see that $f''(x) > 0$ if $|x| > 3$, and $f''(x) < 0$ if $|x| < 3$. In other words, the graph f is concave upward if $|x| > 3$, and concave downward if $|x| < 3$. Since 3 and -3 are not in the domain of f, there are no points of inflection.

(c)   For this function $f'(x) = -2xe^{-x^2}$, and $f''(x) = -2e^{-x^2} + 4x^2e^{-x^2} = 2e^{-x^2}[2x^2 - 1]$. Thus $f''(x) < 0$ if $|x| < 1/2^{1/2}$, and $f''(x) > 0$ if $|x| > 1/2^{1/2}$. Hence the graph of f is concave downward if $|x| < 1/2^{1/2}$, and concave upward if $|x| > 1/2^{1/2}$. Thus the points $(1/2^{1/2}, e^{-1/2})$ and $(-1/2^{1/2}, e^{-1/2})$ are points of inflection.

4-22   (a)   Let a and b belong to I, and suppose that $a < b$. We will show that if $f''(x) > 0$ on I, then not only is the midpoint of the chord joining $(a, f(a))$ and $(b, f(b))$ above the graph of f, but the entire chord, excepting of course the endpoints, is above the graph of f. The equation of this chord is

$$y = \frac{x-a}{b-a}[f(b) - f(a)] + f(a)$$

for $a \le x \le b$. Now the chord will be above the graph if

$$\frac{x-a}{b-a}[f(b) - f(a)] + f(a) > f(x)$$

for $a < x < b$, or

$$\frac{f(b) - f(a)}{b-a} > \frac{f(x) - f(a)}{x-a}$$

Let $F(x) = f(x) - f(a)$, and $G(x) = x - a$ for $a \le x \le b$. Then F and G are continuous on $[a, b]$, differentiable on $(a, b]$, $F(a) = 0 = G(a)$, $G'(x) = 1 > 0$ for $a < x \le b$, and $F'(x)/G'(x) = f'(x)/1$ is strictly increasing because $f''(x) > 0$ on I. In other words, F and G satisfy the conditions of Problem 4-19a, and so F/G is strictly increasing on $(a, b]$, or

$$\frac{f(b) - f(a)}{b-a} > \frac{f(x) - f(a)}{x-a}$$

if $a < x < b$.

(b) We have $D^2x^p = p(p-1)x^{p-2} > 0$ for all $x > 0$ because $p > 1$; $D^2e^x = e^x > 0$ for all x; and $D^2x \log x = 1/x > 0$ for all $x > 0$. The stated inequalities thus follow from part (a).

4-23 We have $dy/dx = 4kx(x^2 - 3)$, and

$$d^2y/dx^2 = 4k(x^2 - 3) + 8kx^2 = 12k(x^2 - 1)$$

Thus $d^2y/dx^2 = 0$ if $x = \pm 1$, and, since $d^2y/dx^2 > 0$ if $x < -1$, $d^2y/dx^2 < 0$ if $|x| < 1$, and $d^2y/dx^2 > 0$ if $x > 1$, the points $(-1, 4k)$ and $(1, 4k)$ are points of inflection. The slopes of the normals at these points are $-1/8k$ and $1/8k$ respectively, and so the equations of the normal lines are $(y - 4k)/(x + 1) = -1/8k$ and $(y - 4k)/(x - 1) = 1/8k$ respectively. For these lines to pass through $(0, 0)$, we must have $-4k/-1 = 1/8k$ and $-4k/1 = -1/8k$. Both of these equations give $32k^2 = 1$, or $k = 1/4(2)^{1/2}$.

4-24 By implicit differentiation, we have $mx^{m-1}y^n + nx^my^{n-1}(dy/dx) = 0$, or $dy/dx = -my/nx$, and so the slope of the tangent line at (x_1, y_1) is $-my_1/nx_1$. The equation of the tangent line is thus $(y - y_1)/(x - x_1) = -my_1/nx_1$, or $my_1(x - x_1) + nx_1(y - y_1) = 0$. It intersects the x axis at $((m + n)x_1/n, 0)$, the y axis at $(0, (m + n)y_1/m)$. Thus the lengths of the line segments in question are

$$L_1 = ([(m + n)x_1/n - x_1]^2 + y_1{}^2)^{1/2} = (m^2x_1{}^2/n^2 + y_1{}^2)^{1/2}$$
$$= (1/|n|)(m^2x_1{}^2 + n^2y_1{}^2)^{1/2}$$

and

$$L_2 = (x_1{}^2 + [(m + n)y_1/m - y_1]^2)^{1/2} = (x_1{}^2 + n^2y_1{}^2/m^2)^{1/2}$$
$$= (1/|m|)(m^2x_1{}^2 + n^2y_1{}^2)^{1/2}$$

Hence

$$L_1/L_2 = |m|/|n|$$

4-25 Let (x_1, y_1) be any point on the curve. Since $(2/3)x^{-1/3} + (2/3)y^{-1/3} \, dy/dx = 0$, or $dy/dx = -y^{1/3}/x^{1/3}$, the equation of the tangent line at (x_1, y_1) is $(y - y_1)/(x - x_1) = -y_1^{1/3}/x_1^{1/3}$. The points at which this line intersects the axes are

$$([y_1x_1^{1/3} + y_1^{1/3}x_1]/y_1^{1/3}, 0)$$

which is $(x_1^{1/3}a^{2/3}, 0)$, and $(0, x_1^{2/3}y_1^{1/3} + y_1)$ which is $(0, y_1^{1/3}a^{2/3})$. The distance between these points is then

$$(x_1^{2/3}a^{4/3} + y_1^{2/3}a^{4/3})^{1/2} = [a^{4/3}(x_1^{2/3} + y_1^{2/3})]^{1/2} = (a^{4/3}a^{2/3})^{1/2} = a$$

4-26 (a) We have $f'(x) = 3x^2 - 18x + 24 = 3(x - 4)(x - 2)$. Hence 2 and 4 are the only critical numbers of f. Inasmuch as $f'(x) > 0$ if $0 < x < 2$, $f'(x) < 0$ if $x < 2 < 4$, and $f'(x) > 0$ if $2 < x < 6$, $f(2)$ is a maximum and $f(4)$ is a minimum. Therefore, the absolute maximum is the largest of $f(0), f(2)$, and $f(6)$; while the absolute minimum is the smallest of the numbers $f(0), f(4)$, and $f(6)$. Now $f(0) = 0$, $f(2) = 20$, $f(4) = 16$, and $f(6) = 36$. Thus $f(0) = 0$ is the absolute minimum, and $f(6) = 36$ is the absolute maximum.
(b) Here $f'(x) = (2/3)x^{-1/3}$. Hence 0 is the only critical number of f. Since $f(0) = 0$ and $f(1) = f(-1) = 1$, $f(0)$ is the absolute minimum and $f(1) = 1$ is the absolute maximum.

(c) Now $f'(x) = \log x + 1$. Therefore, the only critical number of f is $1/e$. Since $f'(x) < 0$ if $0 < x < 1/e$, and $f'(x) > 0$ if $1/e < x < 1$, it follows that $f(1/e) = -1/e$ is the absolute minimum. Further, it can be seen that f decreases on $(0, 1/e)$ and increases on $(1/e, 1)$, and so the absolute maximum of f must be $\lim_{x \to 0-} f(x) = 0$, or $\lim_{x \to 0+} f(x) \leq 0$ [$\lim_{x \to 0+} f(x)$ exists because $f(x) < 0$ and f is decreasing on $(0, 1/e)$]. In other words, 0 is the absolute maximum.

4-27 Suppose first that $m > 1$ and $n > 1$. Regardless of the oddness or evenness of m and n, we have

$$f'(x) = m(x-a)^{m-1}(x-b)^n + n(x-a)^m(x-b)^{n-1}$$
$$= (x-a)^{m-1}(x-b)^{n-1}[m(x-b) + n(x-a)]$$
$$= (x-a)^{m-1}(x-b)^{n-1}[(m+n)x - (mb+na)]$$

From this calculation it can be seen that the critical numbers are a, b, and $(mb + na)/(m + n)$. Let us put $c = (mb + na)/(m + n)$. Observe that $a < c < b$ because $na/(m+n) < nb/(m+n)$, and so

$$mb/(m+n) + na/(m+n) < mb/(m+n) + nb/(m+n) = b$$

and $ma/(m+n) < mb/(m+n)$, which gives us $a = ma/(m+n) + na/(m+n) < mb/(m+n) + na/(m+n)$.

(a) If m and n are both even, then $m - 1$ and $n - 1$ are both odd. Hence $(x-a)^{n-1}$ and $(x-b)^{n-1}$ change sign at $x = a$ and $x = b$ respectively. Thus $f'(x) < 0$ if $x < a$, $f'(x) > 0$ if $a < x < c$, $f'(x) < 0$ if $c < x < b$, and $f'(x) > 0$ if $b < x$. In other words, $f(a)$ and $f(b)$ are minimums, and $f(c)$ is a maximum.

(b) If m and n are both odd, then $m - 1$ and $n - 1$ are both even, and so $(x-a)^{m-1}$ and $(x-b)^{n-1}$ do not change sign at $x = a$ and $x = b$ respectively. Hence $f'(x) \leq 0$ if $x < c$, and $f'(x) \geq 0$ if $x > 0$, or $f(c)$ is a minimum.

(c) If m is odd and n is even, then $(x-a)^{m-1}$ does not change sign at $x = a$, while $(x-b)^{n-1}$ changes sign at $x = b$. Thus $f'(x) \geq 0$ if $x < c$, $f'(x) < 0$ if $c < x < b$, and $f'(x) > 0$ if $b < x$, and so $f(c)$ is a maximum, and $f(b)$ is a minimum.

(d) In the same way as in part (c) it can be shown that if m is even and n is odd, then $f(a)$ is a maximum, and $f(c)$ is a minimum.

When $m = 1$, or $n = 1$, or both, the problem can be solved in a similar way. Since the solution is somewhat simpler in these cases, we leave the details to the reader and just summarize the results. For $m = n = 1$, $f((a + b)/2)$ is a minimum. If $m = 1$ and $n > 1$, then $f((na + b)/(n + 1))$ is a maximum and $f(b)$ is a minimum when n is even; when n is odd, $f((na + b)/(n + 1))$ is a minimum. Finally, if $n = 1$ and $m > 1$ is even, then $f(a)$ is a maximum and $f((mb + a)/(m + 1))$ is a minimum; if m is odd, then $f((mb + a)/(m + 1))$ is a minimum.

4-28 (a) Put $f(x) = \tan x - x$ for $0 \leq x < \pi/2$. Then $f'(x) = \sec^2 x - 1 = \tan^2 x > 0$ if $0 < x < \pi/2$, and so f is strictly increasing on $[0, \pi/2)$. Therefore, if $0 < x < \pi/2$, then $f(x) > f(0) = 0$, or $\tan x - x > 0$.

(b) We will prove this inequality by induction. Put $f(x) = e^x - 1 - x$ for $x > 0$. Then $f'(x) = e^x - 1$, and since $e^x > 1$ if $x > 0$, we see that $f'(x) > 0$

if $x > 0$, and so f is strictly increasing on $[0, \infty)$. Inasmuch as $f(0) = 0$, we have $e^x - 1 - x > 0$, or $e^x > 1 + x$ if $x > 0$. Assume now that our inequality is true for some integer n. Put

$$g(x) = e^x - \sum_{k=0}^{n+1} \frac{x^k}{k!}$$

for $x \geq 0$. Then

$$g'(x) = e^x - \sum_{k=0}^{n+1} \frac{kx^{k-1}}{k!} = e^x - \sum_{k=0}^{n} \frac{x^m}{m!}$$

But our induction hypothesis now implies that $g'(x) > 0$ if $x > 0$, and so g is strictly increasing on $[0, \infty)$. Thus

$$g(x) = e^x - \sum_{k=0}^{n+1} \frac{x^k}{k!} > g(0) = 0$$

if $x > 0$, and so the induction is complete.

(c)   If $f(x) = x^p(a - x)^q$ for $0 \leq x \leq a$, then

$$f'(x) = px^{p-1}(a - x)^q - qx^p(a - x)^{q-1}$$
$$= x^{p-1}(a - x)^{q-1}[p(a - x) - qx]$$
$$= x^{p-1}(a - x)^{q-1}[pa - (p + q)x]$$

It can be seen that $f'(x) > 0$ if $0 < x < pa/(p + q)$, and $f'(x) < 0$ if $pa/(p + q) < x < a$. Hence the absolute maximum of this function is $f(pa/(p + q))$. That is, if $0 \leq x \leq a$, then

$$x^p(a - x)^q \leq \left(\frac{pa}{p + q}\right)^p \left(a - \frac{pa}{p + q}\right)^q = \left(\frac{pa}{p + q}\right)^p \left(\frac{qa}{p + q}\right)^q$$
$$= \frac{p^p q^q a^{p+q}}{(p + q)^{p+q}}$$

(d)   Here put $f(x) = a^2 \sec^2 x + b^2 \csc^2 x$ for $0 < x < \pi/2$. Then $f'(x) = 2a^2 \sec^2 x \tan x - 2b^2 \csc^2 x \cot x$, and the critical numbers are the roots of $f'(x) = 0$. That is, we must solve the equation $a^2 \sec^2 x \tan x = b^2 \csc^2 x \cot x$. If we write it in terms of sines and cosines, it becomes $a^2 \sin^4 x = b^2 \cos^4 x$, or $\tan^4 x = b^2/a^2$, or $x = \tan^{-1}(b/a)^{1/2}$. Inasmuch as $\lim_{x \to \pi/2-} f(x) = \infty = \lim_{x \to 0+} f(x)$, and there is only one critical number in $(0, \pi/2)$, it must be the case that $f(\tan^{-1}(b/a)^{1/2})$ is the absolute minimum of f. Therefore, if $0 < x < \pi/2$, then $a^2 \sec^2 x + b^2 \csc^2 x \geq a^2 \sec^2 [\tan^{-1}(b/a)^{1/2}] + b^2 \csc^2 [\tan^{-1} (b/a)^{1/2}] = a^2[(a + b)^{1/2}/a^{1/2}]^2 + b^2[(a + b)^{1/2}/b^{1/2}]^2 = a(a + b) + b(a + b) = (a + b)^2$.

Remark   In the course of part (d) we had occasion to evaluate $\sec [\tan^{-1}(b/a)^{1/2}]$ and $\csc [\tan^{-1}(b/a)^{1/2}]$. This can be done analytically along the lines indicated when we evaluated $\cos (\sin^{-1} x)$ in Example 4-7, but the easiest way is to consider the right triangle in Figure 4-3. The indicated angle obviously has its tangent equal to

$(b/a)^{1/2}$. Therefore, we can interpret it as $\tan^{-1}(b/a)^{1/2}$. By the Pythagorean formula the length of the hypotenuse of this triangle must be $(a+b)^{1/2}$. Since the secant of the indicated angle is the length of the hypotenuse divided by the length of the adjacent side, we have $\sec[\tan^{-1}(b/a)^{1/2}] = (a+b)^{1/2}/a^{1/2}$. In the same way, $\mathrm{cosec}[\tan^{-1}(b/a)^{1/2}] = (a+b)^{1/2}/b^{1/2}$.

By means of such simple diagrams it is possible to evaluate geometrically the composition of a trigonometric function by an inverse trigonometric function.

4-29 Let us call either a maximum or minimum of f an extremum, and let us first suppose that $a^2 \le 3b$ and examine $df(x)/dx = 3x^2 + 2ax + b$. Since the derivative exists for all x, f can have extremum only if $3x^2 + 2ax + b = 0$. The roots of this equation are $[-2a \pm (4a^2 - 12b)^{1/2}]/6$ and there are real roots, and hence possible extrema, only if $a^2 - 3b \ge 0$. Thus if $a^2 - 3b < 0$, f can have no extremum. If $a^2 - 3b = 0$, then $df(x)\,dx = 3x^2 + 2ax + a^2/3 = 3(x + a/3)^2 \ge 0$, and so f is increasing for all x and thus can have no extremum.

Conversely, let us now assume that f has no extremum. If α and β are distinct real roots of $3x^2 + 2ax + b = 0$, with $\alpha < \beta$, then $df(x)/dx = 3(x - \alpha)(x - \beta)$. Thus if $x < \alpha$, then $df(x)/dx > 0$, if $\alpha < x < \beta$, then $df(x)/dx < 0$, and if $x > \beta$, then $df(x)/dx > 0$, and so f has a maximum at α and a minimum at β. Our hypothesis thus means that $3x^2 + 2ax + b = 0$ must have either a double real root, or no real roots. As above, we must therefore have $a^2 - 3b \le 0$.

4-30 We have

$$f'(x) = \sum_{k=1}^{n} -2(a_k - x) = -2\sum_{k=1}^{n} a_k + 2nx$$

Now $f'(x) = 0$ when

$$x = \frac{1}{n}\sum_{k=1}^{n} a_k$$

and since $f''(x) = 2n > 0$, this value must minimize f.

Remark This problem is of some interest in statistics. It shows that the "variance" of $a_1, a_2, \ldots, a_n$ is the smallest "second central moment" of $a_1, a_2, \ldots, a_n$.

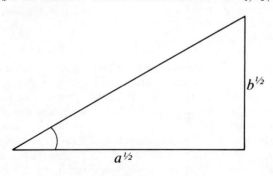

$b^{1/2}$

$a^{1/2}$

4-3

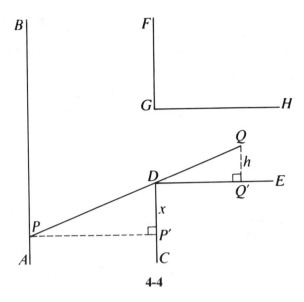

4-4

4-31   If we let $f(x) = x^p - px + p - 1$ for $x \geq 0$, then $f'(x) = px^{p-1} - p$, and $f'(x) = 0$ only if $x = 1$.   Since $f''(x) = p(p-1)x^{p-2}$, we have $f''(1) = p(p-1) > 0$, and so $f(1)$ is the minimum.   In other words, $x^p - px + p - 1 \geq f(1) = 0$.

Remark   This simple inequality has far-reaching consequences.   Consult E. F. Beckenbach, R. Bellman: *Inequalities*, Chapter 1, Sections 14, 17, and 18.

4-32   If s represents the side of the square, and r the radius of the circle, then $k = 4s + 2\pi r$, and the sum in question is $A = s^2 + \pi r^2$.   The first equation gives r as a function of s, and so we can regard the second as a function of s for $0 \leq s \leq k/4$ and differentiate it with respect to s to obtain $dA/ds = 2s + 2\pi r \, dr/ds$.   Differentiation of the first equation gives us $0 = 4 + 2\pi \, dr/ds$, or $dr/ds = -2/\pi$.   Thus $dA/ds = 2s - 4r$, and this is zero when $2r = s$.   If $s < 2r$, then $dA/ds < 0$, while if $0 > 2r$, then $dA/ds > 0$.   Thus A is decreasing for $0 \leq s \leq 2r$ and increasing for $2r \leq s \leq k/4$, or A is minimized when $s = 2r$.   Also, A assumes its maximum at $s = 0$, or $s = k/4$.   Inasmuch as $A(0) = k^2/4\pi > k^2/16 = A(k/4)$, the sum of the areas is a maximum when the wire is left uncut and bent into a circle.

4-33   If PQ represents the pipe in Figure 4-4, the conditions of the problem mean that if P slides along wall AB while PQ slides along corner D, then Q will just touch wall GH at some time during the turn.   The hall is thus as wide as the maximum distance from Q to DE.   In other words, if Q' is the projection of Q on DE, we must find the maximum of QQ'.

Accordingly, let $h = QQ'$, P' be the projection of P on DC, and $x = DP'$.   Then $PP' = m$, $PQ = n$, and the triangles $PP'D$ and $DQ'Q$ are similar.   Hence $x/PD = h/DQ$.   But $PD = (m^2 + x^2)^{1/2}$, and so $DQ = n - (m^2 + x^2)^{1/2}$.   Thus

$$x/(m^2 + x^2)^{1/2} = h/[n - (m^2 + x^2)^{1/2}]$$

and we can express h as a function of x by $h(x) = nx/(m^2 + x^2)^{1/2} - x$ for $0 \le x \le (n^2 - m^2)^{1/2}$. Now

$$\frac{dh}{dx} = \frac{n(m^2 + x^2)^{1/2} - nx[x/(m^2 + x^2)^{1/2}]}{m^2 + x^2} = nm^2/(m^2 + x^2)^{3/2} - 1$$

and $dh/dx = 0$ when $x = m^{2/3}(n^{2/3} - m^{2/3})^{1/2}$. Inasmuch as

$$d^2h/dx^2 = -3nm^2x/(m^2 + x^2)^{5/2}$$

we see that $d^2h/dx^2 < 0$ for $x > 0$, and so $m^{2/3}(n^{2/3} - m^{2/3})^{1/2}$ maximizes h. Thus $h[m^{2/3}(n^{2/3} - m^{2/3})^{1/2}] = (n^{2/3} - m^{2/3})^{3/2}$ is the width of the hall in feet.

4-34 If θ is the angle between a and b, then $0 \le \theta \le \pi$, and the area A is given by $A = (1/2)ab \sin \theta$. Now A is maximized when $\sin \theta = 1$, or when $\theta = \pi/2$. Thus a right triangle has the greatest area under the given conditions, and the third side giving the greatest areas is the hypotenuse and thus has length $(a^2 + b^2)^{1/2}$.

Remark This problem shows that maximum and minimum problems can occasionally be solved without the aid of calculus.

4-35 In Figure 4-5, let AC be the beam, BD be the intervening wall, EF the wall to be braced, and put $\theta = \angle BAD$. Then $BD = a$, $DE = b$, and the length l of the beam is given by $l = AC = AB + BC$. Now $BC = b/\cos \theta$, and $AB = a/\sin \theta$. Thus

$l = b/\cos \theta + a/\sin \theta$ for $0 < \theta < \pi/2$, and so

$$\frac{dl}{d\theta} = \frac{b \sin \theta}{\cos^2 \theta} - \frac{a \cos \theta}{\sin^2 \theta} = \frac{1}{\sin^2 \theta \cos^2 \theta} [b \sin^3 \theta - a \cos^3 \theta]$$

Hence $dl/d\theta = 0$ if and only if $\tan^3 \theta = a/b$. Since $dl/d\theta < 0$ if $\tan^3 \theta < a/b$, and $dl/d\theta > 0$ if $\tan^3 \theta > a/b$, we see that l is minimized if $\theta = \tan^{-1}(a/b)^{1/3} = \theta_0$. The

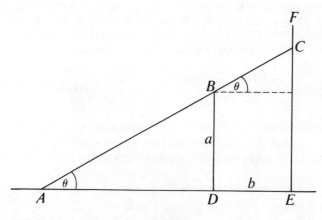

4-5

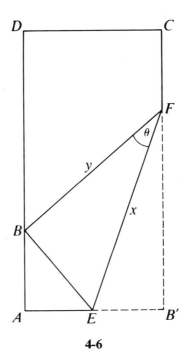

4-6

shortest beam required thus has length

$$l = \frac{b}{\cos \theta_0} + \frac{a}{\sin \theta_0} = \frac{b(a^{2/3} + b^{2/3})^{1/2}}{b^{1/3}} + \frac{a(a^{2/3} + b^{2/3})^{1/2}}{a^{1/3}}$$

$$= (a^{2/3} + b^{2/3})^{3/2}$$

4-36   In Figure 4-6 let B' be the position of B before the paper was folded.   Denote $\angle BFE$ by θ, EF by x, and BF by y.   Then $\angle EFB' = \theta$, $\angle BFB' = 2\theta$, and $\angle BFB' = \angle DBF$ because $AD \parallel B'C$.   From these relations we get $a = y \sin 2\theta = 2y \sin \theta \cos \theta$, and $x = y/\cos \theta = a/2 \sin \theta \cos^2 \theta$ for $0 < \theta < \pi/2$.   Hence

$$\frac{dx}{d\theta} = \frac{-a}{2} \frac{\cos^3 \theta - 2 \sin^2 \theta \cos \theta}{\sin^2 \theta \cos^4 \theta} = \frac{a}{2 \sin^2 \theta \cos^3 \theta} [2 \sin^2 \theta - \cos^2 \theta]$$

and so $dx/d\theta = 0$ if and only if $\tan^2 \theta = 1/2$.   Inasmuch as $dx/d\theta < 0$ if $\tan^2 \theta < 1/2$, and $dx/d\theta > 0$ if $\tan^2 \theta > 1/2$, x is minimized when $\theta = \tan^{-1} (1/2^{1/2}) = \theta_0$.   Since $\sin \theta_0 = 1/3^{1/2}$, and $\cos \theta_0 = (2/3)^{1/2}$, the minimum value of x is $3(3)^{1/2}a/4$.

4-37   Let C be the center of the circle, T the point at which the circle intersects the x axis, and denote $\angle QCT$ by θ (see Figure 4-7).   Then $OT = \overset{\frown}{QT} = \theta$, and $PT = \sin \theta$.   Thus $x = OP = OT - PT = \theta - \sin \theta$.   But $\theta = 2\pi t$, and so $x = 2\pi t - \sin 2\pi t$.   Thus $v = dx/dt = 2\pi - 2\pi \cos 2\pi t$, and $a = d^2x/dt^2 = 4\pi^2 \sin 2\pi t$.

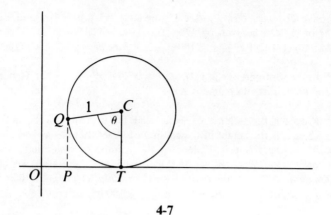

4-7

4-38 We have $v = ds/dt = (1/2)1/(t+1)^{1/2}$, and $a = d^2s/dt^2 = (-1/4)1/(t+1)^{3/2} = -2[1/2(t+1)^{1/2}]^3 = -2v^3$.

4-39 Implicit differentiation gives $2v \, dv/dt = -2b/s \, ds/dt$, or $2va = (-ab/s^2)v$, and so $a = -b/s^2$.

4-40 In Figure 4-8, let AC represent the ladder, BD the fence, and $CE \perp AD$. Then $BD = 8$, $AC = 10$, and we are interested in $CE = h$. If we let $x = AD$, then $dx/dt = 2$, and $AB = (x^2 + 64)^{1/2}$. Thus, since $\triangle ADB$ is similar to $\triangle AEC$, $h/10 = 8/(x^2 + 64)^{1/2}$, or $h = 80/(x^2 + 64)^{1/2}$, and so

$$dh/dt = [-80x/(x^2 + 64)^{3/2}] \, dx/dt = -160x/(x^2 + 64)^{1/2}$$

When the upper end of the ladder reaches the top of the fence, $x = 6$ and consequently $dh/dt = -.96$ ft/sec.

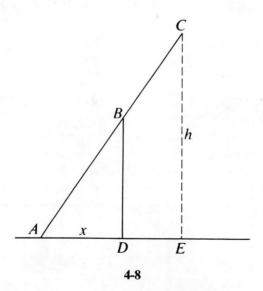

4-8

4-41   We have $OA = x$, $OB = y$ (see Figure 4-9), $x^2 + y^2 = 100$ and $dx/dt = k$.
The area M of $\triangle BOA$ is given by $M = (1/2)x(100 - x^2)^{1/2} = (1/2)(100x^2 - x^4)^{1/2}$,
and so $dM/dt = (1/4)[200x - 4x^3]/(100x^2 - x^4)^{1/2} \, dx/dt = k(50 - x^2)/(100 - x^2)^{1/2}$
units/min.

Since P has coordinate $(x/2, y/2)$, $OP = (x^2/4 + y^2/4)^{1/2} = 5$. Hence OP is
constant, and so its rate of change is zero.

4-42   Let W denote the volume of water that has run out of the tank.   Then
$W = \pi 1^2 S$, where S is the height of the column of water that has leaked out.   Thus
$dW/dt = \pi \, dS/dt = \pi v$.   But if V denotes the volume of water left in the tank
we clearly have $dV/dt = -dW/dt$, and inasmuch as $V = 100\pi h$ we also have
$dV/dt = 100\pi \, dh/dt$.   Thus $100\pi \, dh/dt = -\pi V$, or $(dh/dt)(1/v) = -1/100$.   Finally,
$2v \, dv/dt = 2g \, dh/dt$, or $dv/dt = g(dh/dt)(1/v) = -g/100$ in./sec^2.

4-43   These are all indeterminate forms of the 0/0 type.   We evaluate them by
using L'Hospital's rule.

(a) $\displaystyle \lim_{x \to 0} \frac{a^x - b^x}{x} = \lim_{x \to 0} \frac{a^x \log a - b^x \log b}{1} = \log a - \log b = \log \frac{a}{b}$

(b) $\displaystyle \lim_{x \to 1} \frac{1 - 4 \sin^2 (\pi x/6)}{1 - x^2} = \lim_{x \to 1} \frac{(-4\pi/3) \sin (\pi x/6) \cos (\pi x/6)}{-2x}$

$$= \frac{(-4\pi/3)(1/2)(3^{1/2}/2)}{-2} = \pi(3^{1/2})/6$$

(c) $\displaystyle \lim_{x \to 0} \frac{e^x - e^{-x} - 2x}{x - \sin x} = \lim_{x \to 0} \frac{e^x + e^{-x} - 2}{1 - \cos x} = \lim_{x \to 0} \frac{e^x - e^{-x}}{\sin x}$

$$= \lim_{x \to 0} \frac{e^x + e^{-x}}{\cos x} = \frac{2}{1} = 2$$

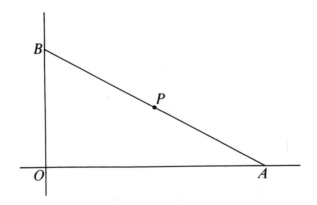

4-9

4-44 We apply L'Hospital's rule to evaluate these ∞/∞ indeterminate forms.

(a) $\displaystyle \lim_{x\to\infty} \frac{\log x}{x^{\alpha}} = \lim_{x\to\infty} \frac{1/x}{\alpha x^{\alpha-1}} = \lim_{x\to\infty} \frac{1}{\alpha x^{\alpha}} = 0$

(b) $\displaystyle \lim_{x\to 0+} \frac{\log x}{\operatorname{cosec} x} = \lim_{x\to 0+} \frac{1/x}{-\operatorname{cosec} x \cot x} = \lim_{x\to 0+} \frac{-\sin^2 x}{x \cos x}$

To this 0/0 form we apply the rule again. Thus

$$\lim_{x\to 0+} \frac{-\sin^2 x}{x \cos x} = \lim_{x\to 0+} \frac{-2\sin x \cos x}{\cos x - x \sin x} = \frac{0}{1} = 0$$

(c) $\displaystyle \lim_{x\to 0+} \frac{\log(\sin x)}{\log(\tan x)} = \lim_{x\to 0+} \frac{\cos x/\sin x}{\sec^2 x/\tan x} = \lim_{x\to 0+} \cos^2 x = 1$

Remark Parts (b) and (c) illustrate the desirability of simplification when applying L'Hospital's rule. Indeed, without the simplification carried out in part (b), repeated applications of the rule would keep resulting in ∞/∞ forms.

4-45 We reduce these $\infty - \infty$ forms to ones to which L'Hospital's rule may be applied.

(a) $\displaystyle \lim_{x\to 1} \left[\frac{x}{x-1} - \frac{1}{\log x} \right] = \lim_{x\to 1} \frac{x\log x - (x-1)}{(x-1)\log x}$

$$= \lim_{x\to 1} \frac{\log x + 1 - 1}{\log x + (x-1)/x} = \lim_{x\to 1} \frac{1/x}{1/x + 1/x^2} = \frac{1}{2}$$

(b) If we let $x = 1/y$, then

$\displaystyle \lim_{x\to\infty} [x(x^2 + a^2)^{1/2} - x^2]$

$$= \lim_{y\to 0+} \left[\frac{1}{y}(1/y^2 + a^2)^{1/2} - \frac{1}{y^2} \right] = \lim_{y\to 0+} \frac{(1 + a^2 y^2)^{1/2} - 1}{y^2}$$

$$= \lim_{y\to 0+} \frac{a^2 y/(1 + a^2 y)^{1/2}}{2y} = \lim_{y\to 0+} \frac{a^2}{2(1 + a^2 y^2)^{1/2}} = \frac{a^2}{2}$$

4-46 We reduce these $0 \cdot \infty$ forms to ones to which L'Hospital's rule may be applied.

(a) $\displaystyle \lim_{x\to 0+} x^{\alpha} \log x = \lim_{x\to 0+} \frac{\log x}{1/x^{\alpha}} = \lim_{x\to 0+} \frac{1/x}{-\alpha/x^{\alpha+1}} = \lim_{x\to 0+} \frac{-x^{\alpha}}{\alpha} = 0$

(b) $\displaystyle \lim_{x\to\infty} x \sin \frac{a}{x} = \lim_{x\to\infty} \frac{\sin(a/x)}{1/x} = \lim_{x\to\infty} \frac{(-a/x^2)\cos(a/x)}{-1/x^2}$

$$= \lim_{x\to\infty} a \cos \frac{a}{x} = a$$

4-47 Each of these forms, one of the 1^{∞}, one of the 0^0, and one of the ∞^0 type is reduced to one to which L'Hospital's rule may be applied by considering its logarithm.

(a) $\log\left[\lim_{y\to\infty}\left(1+\dfrac{x}{y}\right)^y\right]=\lim_{y\to\infty}y\log\left(1+\dfrac{x}{y}\right)$

$$=\lim_{y\to\infty}\frac{\log(1+x/y)}{1/y}=\lim_{y\to\infty}\frac{-x/(1+x/y)y^2}{-1/y^2}$$

$$=\lim_{y\to\infty}\frac{x}{1+x/y}=x$$

Thus $\lim_{y\to\infty}\left(1+\dfrac{x}{y}\right)^y=e^x$

(b) $\log[\lim_{x\to0+}(\sin x)^x]=\lim_{x\to0+}x\log(\sin x)$

$$=\lim_{x\to0+}\frac{\log(\sin x)}{1/x}=\lim_{x\to0+}\frac{\cos x/\sin x}{-1/x^2}=\lim_{x\to0+}\frac{-x^2\cos x}{\sin x}$$

$$=\lim_{x\to0+}\frac{-2x\cos x+x^2\sin x}{\cos x}=\frac{0}{1}=0$$

and so $\lim_{x\to0+}(\sin x)^x=e^0=1$.

(c) $\log\left[\lim_{x\to0+}\left(\dfrac{1}{x}\right)^{\sin x}\right]=\lim_{x\to0+}\sin x\log\left(\dfrac{1}{x}\right)$

$$=\lim_{x\to0+}\frac{-\log x}{1/\sin x}=\lim_{x\to0+}\frac{-1/x}{\cos x/\sin^2 x}$$

$$=\lim_{x\to0+}\frac{\sin^2 x}{x\cos x}$$

$$=\lim_{x\to0+}\frac{2\sin x\cos x}{\cos x-x\sin x}=\frac{0}{1}=0$$

Hence $\lim_{x\to0+}\left(\dfrac{1}{x}\right)^{\sin x}=e^0=1$

4-48   If we apply L'Hospital's rule, we get

$$\lim_{x\to0+}\frac{e^{-1/x}}{x}=\lim_{x\to0+}\frac{(1/x^2)e^{-1/x}}{1}=\lim_{x\to0+}\frac{e^{-1/x}}{x^2}$$

In other words, an application of L'Hospital's rule leads to another indeterminate form. Although we have encountered such outcomes before, this one is different in that, as is easy to see, the nth application of the rule leads to $\lim_{x\to0}e^{-1/x}/n!\,x^{n+1}$, which is another indeterminate form. Thus, no matter how many times the rule is applied, the situation will not change.

However, if we modify the problem slightly, L'Hospital's rule becomes effective. Indeed, let $y=1/x$. Then $\lim_{x\to0+}e^{-1/x}/x=\lim_{y\to\infty}y\,e^{-y}=\lim_{y\to\infty}y/e^y=\lim_{y\to\infty}1/e^y=0$.

4-49 $$\lim_{x\to 0}\frac{x^2 \sin (1/x)}{\sin x}=\lim_{x\to 0}\frac{x \sin (1/x)}{\sin x/x}=\frac{\lim_{x\to 0} x \sin (1/x)}{\lim_{x\to 0} \sin x/x}=\frac{0}{1}=0$$

(Recall the fundamental limit $\lim_{x\to 0} \sin x/x = 1$.) An attempt to use L'Hospital's rule leads to the expression

$$\frac{2x \sin (1/x) - \cos (1/x)}{\cos x}$$

which has no limit as $x \to 0$. Thus f/g may tend to a limit when f'/g' does not, and L'Hospital's rule is seen to be a sufficient but not a necessary condition.

4-50 There is an incorrect application of L'Hospital's rule because $(6x - 4)/(2x - 2)$ is not an indeterminate form at $x = 2$. The correct statement is

$$\lim_{x\to 2}\frac{3x^2 - 4x - 4}{x^2 - 2x}=\lim_{x\to 2}\frac{6x - 4}{2x - 2}=\frac{8}{2}=4$$

4-51 The given conditions permit us to use L'Hospital's rule and we have

$$\lim_{x\to \infty}\frac{xf'(x)}{f(x)}=\lim_{x\to \infty}\frac{f'(x) + xf''(x)}{f'(x)}$$

$$=\lim_{x\to \infty}\frac{2f''(x) + xf'''(x)}{f''(x)}=\lim_{x\to \infty}\left[2 + \frac{xf'''(x)}{f''(x)}\right]=2 + k$$

4-52 Let $\theta = \angle POA$, $PP_1 \perp OA$, and $PP_2 \perp MA$ (see Figure 4-10). Then $\widehat{AP}^2 = r\theta = AM$, and, because $\triangle MAB$ is similar to $\triangle MP_2 P_1$, $AB/AM = PP_2/P_2 M$. But $PP_2 = OA - OP_1 = r - r \cos \theta$, and $P_2 M = AM - P_2 A = AM - P_1 P = r\theta - r \sin \theta$. Thus

$$AB = AM(PP_2/P_2 M) = r\theta[(r - r \cos \theta)/(r\theta - r \sin \theta)] = r[(\theta - \theta \cos \theta)/(\theta - \sin \theta)]$$

As P approaches A, $\theta \to 0$. If we let $\theta \to 0$ in our expression for AB, we get an indeterminate form. We evaluate it by applying L'Hospital's rule. Thus

$$\lim_{P\to A} AB = r \lim_{\theta\to 0}\frac{\theta - \theta \cos \theta}{\theta - \sin \theta}=r \lim_{\theta\to 0}\frac{1 - \cos \theta + \theta \sin \theta}{1 - \cos \theta}$$

$$=r \lim_{\theta\to 0}\frac{2 \sin \theta + \theta \cos \theta}{\sin \theta}=r \lim_{\theta\to 0}\frac{3 \cos \theta - \theta \sin \theta}{\cos \theta}=r\frac{3}{1}=3r$$

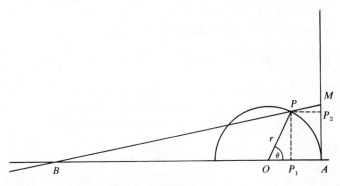

4-10

4-53 Since $\tan^{-1} x = \tan^{-1} x - \tan^{-1} 0$, we have, by the Mean Value Theorem, $\tan^{-1} x = x/(1 + \xi^2)$, where $0 < \xi < x$.   But then $1 < 1 + \xi^2 < 1 + x^2$, or $1/(1 + x^2) < 1/(1 + \xi^2) < 1$.   Thus, since $x > 0$, $x/(1 + x^2) < \tan^{-1} x < x$.

4-54 By the Mean Value Theorem, $e^x - e^a = (x - a)e^\xi$, where $a < \xi < x$.   But $e^a < e^\xi < e^x$, and so $e^a(x - a) < e^x - e^a < e^x(x - a)$.

4-55 The Mean Value Theorem gives us $\sin x - \sin y = (x - y)\sin \xi$, where ξ is between x and y.   Thus $|\sin x - \sin y| = |x - y| |\sin \xi| \leq |x - y|$.

4-56 Applying the Mean Value Theorem, we have $\log (x + 1) = \log (x + 1)$ $- \log 1 = x/(1 + \xi)$, where $1 + \xi$ is between 1 and $x + 1$.   If $x > 0$, $1 < 1 + \xi < 1 + x$, or $1/(1 + x) < 1/(1 + \xi) < 1$; while if $-1 < x < 0$, $0 < 1 + x < 1 + \xi < 1$, or $1 < 1/(1 + \xi) < 1/(1 + x)$.   In the first case we have $x/(x + 1) < x/(1 + \xi) < x$, while in the second case we have $x/(x + 1) < x/(1 + \xi) < x$ because $x < 0$.   Thus in either case

$$\frac{x}{x + 1} < \log (x + 1) < x$$

4-57 If we let

$$f(x) = \frac{a_n}{n + 1} x^{n+1} + \frac{a_{n-1}}{n} x^n + \cdots + \frac{a_1}{2} x^2 + a_0 x$$

then f is continuous and differentiable on any interval (f is a polynomial), and $f(0) = f(1) = 0$.   Thus we can apply Rolle's theorem and conclude that there is at least one number r such that $0 < r < 1$, and $0 = f'(r) = a_n r^n + a_{n-1} r^{n-1} + \cdots + a_1 r + a_0$.   The root we seek is r.

CHAPTER 5

5-1 (a) Let $u = \log (1 + x^2)$.   Then $du = 2x/(1 + x^2)$, and so

$$\int \frac{x \log (1 + x^2)}{(1 + x^2)}\, dx = \frac{1}{2} \int u\, du = \frac{u^2}{4} + C = \frac{1}{4} \log^2 (1 + x^2) + C$$

(b) If $u = 1 + x^{2/3}$, then $du = (2/3)x^{-1/3}\, dx$, and

$$\int \frac{1}{x^{1/3}(1 + x^{2/3})}\, dx = \frac{3}{2} \int \frac{1}{u}\, du = \frac{3}{2} \log |u| + C = \frac{3}{2} \log (1 + x^{2/3}) + C$$

(c) If $u = (1 - x^2)/x^2$, then $du = -2/x^3\, dx$, and

$$\int \frac{1}{x^3} \left(\frac{1 - x^2}{x^2}\right)^{10} dx = -\frac{1}{2} \int u^{10} du = -\frac{1}{22} u^{11} + C = -\frac{1}{22} \left(\frac{1 - x^2}{x^2}\right)^{11} + C$$

(d) Let $u = x^2$.   Then $du = 2x\, dx$, and so

$$\int \frac{1}{x(x^4 -- 1)^{1/2}}\, dx = \frac{1}{2} \int \frac{2x}{x^2(x^4 - 1)^{1/2}}\, dx = \frac{1}{2} \int \frac{du}{u(u^2 - 1)^{1/2}}$$

$$= \frac{1}{2} \sec^{-1} u + C = (1/2) \sec^{-1} x^2 + C$$

(e)   Here   put $u = (x - a)/(x - b)$.   Then $du = (a - b)/(x - b)^2 \, dx$,   and

$$\int (x - a)^{p-1}(x - b)^{-p-1} \, dx = \int \left(\frac{x - a}{x - b}\right)^{p-1} \frac{1}{(x - b)^2} \, dx$$

$$= \frac{1}{a - b} \int u^{p-1} \, du$$

$$= u^p/p(a - b) + C = [1/p(a - b)][(x - a)/(x - b)]^p + C$$

5-2   (a)   If $u = \cos x$, then $du = -\sin x \, dx$, and

$$\int \frac{\sin x}{\cos^{1/2} x} \, dx = -\int u^{-1/2} \, du = -2u^{1/2} + C = -2\cos^{1/2} x + C$$

(b) $\displaystyle\int (\sec x + \tan x)^2 \, dx = \int (\sec^2 x + 2\sec x \tan x + \tan^2 x) \, dx$

$$= \int \sec^2 x \, dx + 2 \int \sec x \tan x \, dx + \int (\sec^2 x - 1) \, dx$$

$$= 2 \int \sec^2 x \, dx + 2 \int \sec x \tan x \, dx - \int dx$$

$$= 2 \tan x + 2 \sec x - x + C$$

(c)   If $u = \tan 3x$, then $du = 3 \sec^2 3x \, dx$, and

$$\int \sec^2 3x \tan^5 3x \, dx = (1/3) \int u^5 \, du = (1/18)u^6 + C = \tan^6 3x/18 + C$$

(d) $\displaystyle\int \frac{1}{1 + \cos x} \, dx = \int \frac{1 - \cos x}{1 - \cos^2 x} \, dx = \int \frac{1}{\sin^2 x} \, dx - \int \frac{\cos x}{\sin^2 x} \, dx$

$$= \int \operatorname{cosec}^2 x \, dx - \int \operatorname{cosec} x \cot x \, dx = -\cot x + \operatorname{cosec} x + C$$

(e) $\displaystyle\int \sin^{1/2} x \cos^3 x \, dx = \int \sin^{1/2} x [1 - \sin^2 x] \cos x \, dx$

$$= \int \sin^{-1/2} x \cos x \, dx - \int \sin^{5/2} x \cos x \, dx$$

$$= (2/3) \sin^{3/2} x - (2/7) \sin^{7/2} x + C$$

(f) $\displaystyle\int \sin^2 x \cos^2 x \, dx = \int [(1 - \cos 2x)/2][(1 + \cos 2x)/2] \, dx$

$$= (1/4) \int (1 - \cos^2 2x) \, dx = (1/4) \int \sin^2 2x \, dx$$

$$= (1/8) \int (1 - \cos 4x) \, dx = x/8 - \sin 4x/32 + C$$

(g) $\displaystyle\int \sec^5 x \tan^3 x \, dx = \int \sec^4 x \tan^2 x \sec x \tan x \, dx$

$$= \int \sec^4 x \, (\sec^2 x - 1) \sec x \tan x \, dx$$

$$= \int (\sec^6 x - \sec^4 x) \sec x \tan x \, dx$$

Thus if $u = \sec x$, then $du = \sec x \tan x \, dx$, and

$$\int \sec^5 x \tan^3 x \, dx$$

$$= \int (u^6 - u^4) \, du = (1/7)u^7 - (1/5)u^5 + C$$

$$= \sec^7 x/7 - \sec^5 x/5 + C$$

(h) $\displaystyle\int \tan^2 x \sec^4 x \, dx = \int \tan^2 x[1 + \tan^2 x] \sec^2 x \, dx$

$$= \int (\tan^2 x + \tan^4 x) \sec^2 x \, dx, \text{ and if we let } u = \tan x$$

then $du = \sec^2 x \, dx$, and so $\displaystyle\int (\tan^2 x + \tan^4 x) \sec^2 x \, dx$

$$= \int (u^2 + u^4) \, du = u^3/3 + u^5/5 + C = \tan^3 x/3 + \tan^5 x/5 + C$$

(i) $\displaystyle\int \cos ax \cos bx \, dx = (1/2) \int [\cos (a + b)x + \cos (a - b)x] \, dx$

$$= \sin (a + b)x/2(a + b) + \sin (a - b)x/2(a - b) + C$$

provided that $a \neq b$. But if $a = b$, the antiderivative is $\int \cos^2 ax \, dx$, which we know how to evaluate (see Example 5-4). Similar identities permit evaluation of $\int \cos ax \, dx$, and $\int \sin ax \sin bx \, dx$ in the same way.

(j) Here

$$\int \frac{1}{a^2 \sin^2 x + b^2 \cos^2 x} \, dx = \int \frac{\sec^2 x}{a^2 \tan^2 x + b^2} \, dx$$

If $u = \tan x$, then $du = \sec^2 x \, dx$, and so

$$\int \frac{\sec^2 x}{a^2 \tan^2 x + b^2} \, dx = \int \frac{du}{a^2 u^2 + b^2} = \frac{1}{a^2} \int \frac{du}{u^2 + (b/a)^2}$$

$$= \frac{1}{a^2} \frac{a}{b} \tan^{-1} \frac{a}{b} u + C = \frac{1}{ab} \tan^{-1} \left(\frac{a}{b} \tan x\right) + C$$

5-3 (a) Since $d \tan x = \sec^2 x \, dx$, we have $\displaystyle\int \sec^2 x \, e^{\tan x} \, dx = \int e^u \, du = e^u + C$, where $u = \tan x$. Thus

$$\int \sec^2 x \, e^{\tan x} \, dx = e^{\tan x} + C$$

(b) Here

$$\int \frac{e^{2x} - 1}{e^{2x} + 1} \, dx = \int \frac{e^x - e^{-x}}{e^x + e^{-x}} \, dx$$

Thus if $u = e^x + e^{-x}$, then $du = (e^x + e^{-x}) \, dx$, and so

$$\int \frac{e^{2x} - 1}{e^{2x} + 1} \, dx = \int \frac{1}{u} \, du = \log |u| + C = \log (e^x + e^{-x}) + C$$

(c) If $u = e^x$, then $du = e^x \, dx$, and

$$\int e^{x + e^x} \, dx = \int e^{e^x} e^x \, dx = \int e^u \, du = e^u + C = e^{e^x} + C$$

(d) We have

$$\int a^x b^{2x} e^{3x} \, dx = \int e^{x \log a} e^{2x \log b} e^{3x} \, dx$$

$$= \int e^{(\log a + 2\log b + 3)x} \, dx$$

$$= e^{(\log a + 2\log b + 3)x} / (\log a + 2 \log b + 3) + C$$

$$= a^x b^{2x} e^{3x} / (\log a + 2 \log b + 3) + C$$

5-4 (a) If $u = \tan x$, then $du = \sec^2 x \, dx$, and

$$\int \frac{\sec^2 x}{9 + \tan^2 x} \, dx = \int \frac{1}{3^2 + u^2} \, du = \frac{1}{3} \tan^{-1} \frac{u}{3} + C = \frac{1}{3} \tan^{-1} \left(\frac{\tan x}{3} \right) + C$$

(b) If $u = \log x$, then $du = (1/x) \, dx$, and

$$\int \frac{1}{x(4 - \log^2 x)^{1/2}} \, dx = \int \frac{1}{(2^2 - u^2)^{1/2}} \, du$$

$$= \sin^{-1} \frac{u}{2} + C = \sin^{-1} \left(\frac{\log x}{2} \right) + C$$

(c) Let $u = e^x$. Then $du = e^x \, dx$, and

$$\int \frac{1}{(e^{2x} - 16)^{1/2}} \, dx = \int \frac{1}{u(u^2 - 4^2)^{1/2}} \, du = \frac{1}{2} \sec^{-1} \frac{u}{2} + C = \frac{1}{2} \sec^{-1} \frac{e^x}{2} + C$$

(d) Completion of the square gives $8 + 2x - x^2 = 8 - (x^2 - 2x) = 9 - (x^2 - 2x + 1) = 9 - (x - 1)^2$, and so

$$\int \frac{1}{(8 + 2x - x^2)^{1/2}} \, dx = \int \frac{1}{(3^2 - (x - 1)^2)^{1/2}} \, dx = \sin^{-1} \frac{x - 1}{3} + C$$

The completion of the square is often useful when dealing with the expression $ax^2 + bx + c$.

5-5 (a) Let $u = \tan^{-1} x$ and $dv = dx$. Then $du = dx/(1 + x^2)$ and $v = x$. Hence

$$\int \tan^{-1} x \, dx = x \tan^{-1} x - \int [x/(1 + x^2)] \, dx$$

$$= x \tan^{-1} x - (1/2) \log (1 + x^2) + C$$

(b) Let $u = \log x$ and $dv = x \, dx$. Then $du = dx/x$ and $v = x^2/2$, and so $\int x \log x \, dx = x^2 \log x/2 - \int (x^2/2x) \, dx = x^2 \log x/2 - x^2/4 + C$.

(c) If $u = x^2$, $dv = e^x \, dx$, then $du = 2x \, dx$, $v = e^x$, and $\int x^2 e^x \, dx = x^2 e^x - 2 \int x \, e^x \, dx$. We can integrate the last antiderivative by parts again, and get

$$\int x e^x \, dx = x e^x - \int e^x \, dx = x e^x - e^x + C$$

Hence

$$\int x^2 e^x \, dx = x^2 e^x - 2xe^x + 2e^x + C$$

(d) Let $u = e^x$, $dv = \sin x \, dx$. Then $du = e^x \, dx$, $v = -\cos x$, and so $\int e^x \sin x \, dx = -e^x \cos x + \int e^x \cos x \, dx$. It seems that we are making no progress, but, as above, $\int e^x \cos x \, dx = e^x \sin x - \int e^x \sin x \, dx$. Thus $2 \int e^x \sin x \, dx = e^x \sin x - e^x \cos x + C$ and so

$$\int e^x \sin x \, dx = (e^x \sin x - e^x \cos x)/2 + C$$

(e) If $u = x$, $dv = dx/(2x + 1)^{1/2}$, then $du = dx$, $v = (2x + 1)^{1/2}$, and

$$\int [x/(2x + 1)^{1/2}] \, dx = x(2x + 1)^{1/2} - \int (2x + 1)^{1/2} \, dx$$
$$= x(2x + 1)^{1/2} - (2x + 1)^{3/2}/3 + C$$

(f) Let $u = x^2$ and $dv = x \, dx/(1 + x^2)^{1/2}$. Then $du = 2x \, dx$, $v = (1 + x^2)^{1/2}$, and so

$$\int [x^3/(1 + x^2)^{1/2}] \, dx = x^2(1 + x^2)^{1/2} - \int 2x(1 + x^2)^{1/2} \, dx$$
$$= x^2(1 + x^2)^{1/2} - 2(1 + x^2)^{3/2}/3 + C$$

(g) Let $u = x^2$ and $dv = xe^{-x^2}$. Then $du = 2x \, dx$, $v = -e^{-x^2}/2$, and $\int x^3 e^{-x^2} \, dx = -x^2 e^{-x^2}/2 + \int xe^{-x^2} \, dx = -x^2 e^{-x^2}/2 - e^{-x^2}/2 + C$.

(h) If $u = \sec x$, $dv = \sec^2 x \, dx$, then $du = \sec x \tan x \, dx$, $v = \tan x$, and so

$$\int \sec^3 x \, dx = \sec x \tan x - \int \sec x \tan^2 x \, dx$$
$$= \sec x \tan x - \int \sec x [\sec^2 x - 1] \, dx$$
$$= \sec x \tan x + \int \sec x \, dx - \int \sec^3 x \, dx$$
$$= \sec x \tan x + \log |\sec x + \tan x| - \int \sec^3 x \, dx$$

Thus $\int \sec^3 x \, dx = \sec x \tan x/2 + \log |\sec x + \tan x|/2 + C$.

5-6 (a) Let $x = u - 1$. Then $dx = du$, and

$$\int x(x + 1)^{1/3} \, dx = \int (u - 1)u^{1/3} \, du$$
$$= \int (u^{4/3} - u^{1/3}) \, du$$
$$= 3u^{7/3}/7 - 3u^{4/3}/4 + C$$
$$= 3(x + 1)^{7/3}/7 - 3(x + 1)^{4/3}/4 + C$$

(b) Let $x = (u-1)^2$. Then $dx = 2(u-1)\,du$, $u = 1 + x^{1/2}$, and

$$\int [1/(1 + x^{1/2})]\,dx = 2\int [(u-1)/u]\,du = 2\int (1 - 1/u)\,du$$
$$= 2u - 2\log |u| + C'$$
$$= 2(1 + x^{1/2}) - 2\log (1 + x^{1/2}) + C'$$
$$= 2x^{1/2} - 2\log (1 + x^{1/2}) + C$$

(c) If $x = a\tan\theta$, then $dx = a\sec^2\theta\,d\theta$, and

$$\int \frac{1}{(x^2 + a^2)^{1/2}}\,dx = \int \frac{\sec^2\theta}{a(\tan^2\theta + 1)^{1/2}}\,d\theta = \int \sec\theta\,d\theta$$
$$= \log |\sec\theta + \tan\theta| + C' = \log \left| \sec\left(\tan^{-1}\frac{x}{a}\right) + \tan\left(\tan^{-1}\frac{x}{a}\right)\right| + C'$$
$$= \log \left| \frac{(x^2 + a^2)^{1/2}}{a} + \frac{x}{a}\right| + C' = \log |x + (x^2 + a^2)^{1/2}| + C$$

(d) If $x = a\sec\theta$, then $dx = a\sec\theta\tan\theta\,d\theta$, and

$$\int \frac{1}{(x^2 - a^2)^{1/2}}\,dx = \int \frac{a\sec\theta\tan\theta}{a(\sec^2\theta - 1)^{1/2}}\,d\theta = \int \sec\theta\,d\theta$$
$$= \log |\sec\theta + \tan\theta| + C' = \log \left| \sec\left(\sec^{-1}\frac{x}{a}\right) + \tan\left(\sec^{-1}\frac{x}{a}\right)\right| + C'$$
$$= \log \left| \frac{x}{a} + \frac{(x^2 - a^2)^{1/2}}{a}\right| + C' = \log |x + (x^2 - a^2)^{1/2}| + C$$

(e) Let $x = a\sin\theta$. Then $dx = a\cos\theta\,d\theta$, and

$$\int \frac{(a^2 - x^2)^{1/2}}{x^2}\,dx = \int \frac{a(1 - \sin^2\theta)^{1/2}}{a^2\sin^2\theta}\, a\cos\theta\,d\theta$$
$$= \int \cot^2\theta\,d\theta = \int (\mathrm{cosec}^2\,\theta - 1)\,d\theta = -\cot\theta - \theta + C$$
$$= -\cot\left(\sin^{-1}\frac{x}{a}\right) - \sin^{-1}\frac{x}{a} + C = -\frac{(a^2 - x^2)^{1/2}}{x} - \sin^{-1}\frac{x}{a} + C$$

(f) $\int \dfrac{1}{(x^2 - 4x + 5)^2}\,dx = \int \dfrac{1}{[(x-2)^2 + 1]^2}\,dx = \int \dfrac{du}{(u^2 + 1)^2}$

where $u = x - 2$. Now let $u = \tan\theta$. Then $du = \sec^2\theta\,d\theta$, and

$$\int du/(u^2 + 1)^2 = \int [\sec^2\theta/(\tan^2\theta + 1)^2]\,d\theta = \int \cos^2\theta\,d\theta$$
$$= (1/2)\int [1 + \cos 2\theta]\,d\theta = \theta/2 + \sin 2\theta/4 + C$$
$$= \tan^{-1} u/2 + \sin (\tan^{-1} u)\cos (\tan^{-1} u)/2 + C$$
$$= \tan^{-1} u/2 + (1/2)[u/(u^2 + 1)^{1/2}][1/(u^2 + 1)^{1/2}] + C$$
$$= \tan^{-1} u/2 + u/2(u^2 + 1) + C$$

Hence

$$\int \frac{1}{(x^2 - 4x + 5)^2}\, dx = \frac{\tan^{-1}(x-2)}{2} + \frac{1}{2}\frac{x-2}{x^2 - 4x + 5} + C$$

(g) Let $x - a = (b - a)\sin^2\theta$. Then $dx = 2(b - a)\sin\theta\cos\theta\, d\theta$, $b - x = (b - a)(1 - \sin^2\theta) = (b - a)\cos^2\theta$, and so

$$\int \frac{1}{[(x-a)(b-x)]^{1/2}}\, dx = \int \frac{2(b-a)\sin\theta\cos\theta}{[(b-a)\sin^2\theta\,(b-a)\cos^2\theta]^{1/2}}\, d\theta$$

$$= 2\int d\theta = 2\theta + C = 2\sin^{-1}\left(\frac{x-a}{b-a}\right)^{1/2} + C$$

Supplementary Problem 5-6   For more practice with these techniques, the reader may want to show that

(a) $\int (x^2 - a^2)^{1/2}\, dx = (x/2)(x^2 - a^2)^{1/2} - (a^2/2)\log|x + (x^2 - a^2)^{1/2}| + C$

(b) $\int (a^2 - x^2)^{1/2}\, dx = (x/2)(a^2 - x^2)^{1/2} + (a^2/2)\sin^{-1}(x/a) + C$

(c) $\int (x^2 + a^2)^{1/2}\, dx = (x/2)(x^2 + a^2)^{1/2} + (a^2/2)\log|x + (x^2 + a^2)^{1/2}| + C$

5-7 (a)   The partial fraction expansion in this case is

$$\frac{1}{x^2 - a^2} = \frac{A}{x+a} + \frac{B}{x-a} = \frac{(A+B)x + a(B-A)}{x^2 - a^2}$$

Thus $A + B = 0$, and $a(B - A) = 1$, or $B = -A$ and $2aB = 1$. Hence $B = 1/2a$ and $A = 1/2a$, and so

$$\int [1/(x^2 - a^2)]\, dx = (1/2a)\int [1/(x-a)]\, dx - (1/2a)\int [1/(x+a)]\, dx$$

$$= (1/2a)\log|x-a| - (1/2a)\log|x+a| + C$$

$$= (1/2a)\log|(x-a)/(x+a)| + C$$

(b)   Since the degree of the numerator exceeds that of the denominator, we must first divide before seeking the partial fraction expansion. Thus $x^3/(x^2 - 2x - 3) = x + 2 + (7x + 6)/(x^2 - 2x + 3)$, and we want

$$\frac{7x+6}{x^2 - 2x - 3} = \frac{A}{x-3} + \frac{B}{x-1} = \frac{A(x+1) + B(x-3)}{x^2 - 2x - 3}$$

Hence $7x + 6 = A(x + 1) + B(x - 3)$, and if we let $x = 3$, and $x = -1$, we find that $A = 27/4$, and $B = 1/4$ respectively. We now have

$$\int \frac{x^3}{x^2 - 2x - 3}\, dx = \int x\, dx + 2\int dx + \frac{27}{4}\int \frac{dx}{x-3} + \frac{1}{4}\int \frac{dx}{x+1}$$

$$= \frac{x^2}{2} + 2x + \frac{27}{4}\log|x-3| + \frac{1}{4}\log|x+1| + C$$

(c) Here

$$\frac{2x^3 + x^2 + 5x + 4}{x^4 + 8x^2 + 16} = \frac{2x^3 + x^2 + 5x + 4}{(x^2 + 4)^2} = \frac{1}{x^2 + 4}\left[\frac{2x^3 + x^2 + 5x + 4}{x^2 + 4}\right]$$

$$= \frac{1}{x^2 + 4}\left[2x + 1 - \frac{3x}{x^2 + 4}\right] = \frac{2x + 1}{x^2 + 4} - \frac{3x}{(x^2 + 4)^2}$$

and so

$$\int \frac{2x^3 + x^2 + 5x + 4}{x^4 + 8x^2 + 16}\,dx = \int \frac{2x}{x^2 + 4}\,dx + \int \frac{1}{x^2 + 4}\,dx - 3\int \frac{x}{(x^2 + 4)^2}\,dx$$

$$= \log(x^2 + 4) + \frac{1}{2}\tan^{-1}\frac{x}{2} + \frac{3}{2}\frac{1}{x^2 + 4} + C$$

(d) We want

$$\frac{4x^2 - 3x}{(x + 2)(x^2 + 1)} = \frac{A}{x + 2} + \frac{Bx + C}{x^2 + 1} = \frac{A(x^2 + 1) + (Bx + C)(x + 2)}{(x + 2)(x^2 + 1)}$$

and so $4x^2 - 3x = A(x^2 + 1) + (Bx + C)(x + 2)$. In this equation put $x = -2$. Then it becomes $22 = 5A$, or $A = 22/5$. Next let $x = 0$. Then $0 = 22/5 + 2C$, or $C = -11/5$. Finally, let $x = 1$. Then $1 = 44/5 + 3B - 33/5$, and so $B = -2/5$. Thus

$$\int \frac{4x^2 - 3x}{(x + 2)(x^2 + 1)}\,dx = \frac{22}{5}\int \frac{dx}{x + 2} - \frac{2}{5}\int \frac{x}{x^2 + 1}\,dx - \frac{11}{5}\int \frac{dx}{x^2 + 1}$$

$$= \frac{22}{5}\log|x + 2| - \frac{1}{5}\log(x^2 + 1) - \frac{11}{5}\tan^{-1}x + C$$

5-8 (a) If we first make the substitution $x = \sin\theta$, $dx = \cos\theta\,d\theta$, then $\int x^2 \sin^{-1}x\,dx = \int \theta \sin^2\theta \cos\theta\,d\theta$. Now put $u = \theta$ and $dv = \sin^2\theta \cos\theta\,d\theta$, or $du = d\theta$ and $v = \sin^3\theta/3$, and integrate by parts. Thus

$$\int \theta \sin^2\theta \cos\theta\,d\theta = \theta \sin^3\theta/3 - (1/3)\int \sin^3\theta\,d\theta$$

$$= \theta \sin^3\theta/3 - (1/3)\int (1 - \cos^2\theta)\sin\theta\,d\theta$$

$$= \theta \sin^3\theta/3 + \cos\theta/3 - \cos^3\theta/9 + C$$

In other words,

$$\int x^2 \sin^{-1}x\,dx = x^3 \sin^{-1}x/3 + \cos(\sin^{-1}x)/3 - \cos^3(\sin^{-1}x)/9 + C$$

$$= x^3 \sin^{-1}x/3 + (1 - x^2)^{1/2}/3 - (1 - x^2)^{3/2}/9 + C$$

(b) Let $u = \log x$ and $dv = x/(1 - x^2)^{1/2}\,dx$, or $du = (1/x)\,dx$ and $v = -(1 - x^2)^{1/2}$, and integrate by parts.

$$\int \frac{x\log x}{(1 - x^2)^{1/2}}\,dx = -(1 - x^2)^{1/2}\log x + \int \frac{(1 - x^2)^{1/2}}{x}\,dx$$

In the last antiderivative make the substitution $x = \sin \theta$, $dx = \cos \theta \, d\theta$.   Then

$$\int [(1 - x^2)^{1/2}/x] \, dx = \int [\cos^2 \theta / \sin \theta] \, d\theta = \int [(1 - \sin^2 \theta)/\sin \theta] \, d\theta$$

$$= \int \cosec \theta \, d\theta - \int \sin \theta \, d\theta$$

$$= \log |\cosec \theta - \cot \theta| + \cos \theta + C$$

$$= \log |\cosec (\sin^{-1} x) - \cot (\sin^{-1} x)| + \cos (\sin^{-1} x) + C$$

$$= \log |1/x - (1 - x^2)^{1/2}/x| + (1 - x^2)^{1/2} + C$$

Therefore,

$$\int [x \log x/(1 - x^2)^{1/2}] \, dx = -(1 - x^2)^{1/2} \log |x| + \log |1 - (1 - x^2)^{1/2}|$$

$$- \log |x| + (1 - x^2)^{1/2} + C$$

(c)   Here

$$\int x^{1/2}(1 + x^2)^2 \, dx = \int x^{1/2}(1 + 2x^2 + x^4) \, dx$$

$$= \int x^{1/2} \, dx + 2 \int x^{5/2} \, dx + \int x^{9/2} \, dx$$

$$= 2x^{3/2}/3 + 4x^{7/2}/7 + 2x^{11/2}/11 + C$$

(d)   Put $u = \log^2 x$ and $dv = x^4 \, dx$.   Then $du = (2 \log x/x) \, dx$, $v = x^5/5$, and $\int x^4 \log^2 x \, dx = x^5 \log^2 x/5 - (2/5) \int x^4 \log x \, dx$.   Integrate the last antiderivative by parts again.

$$\int x^4 \log x \, dx = x^5 \log x/5 - x^5/25 + C$$

Thus

$$\int x^4 \log^2 x \, dx = x^5 \log^2 x/5 - 2x^5 \log x/25 + 2x^5/125 + C$$

(e)   Let $u = 3x^2 - 2 \cos 3x$.   Then $du = (6x + 6 \sin 3x) \, dx$, and so

$$\int \frac{x + \sin 3x}{3x^2 - 2 \cos 3x} \, dx = \frac{1}{6} \int \frac{du}{u} = \frac{1}{6} \log |u| + C = \frac{1}{6} \log |3x^2 - 2 \cos 3x| + C$$

(f)   We have

$$\int \frac{x^{1/2} + 1}{x^{1/2}(x + 1)} \, dx = \int \left[\frac{1}{x + 1} + \frac{1}{x^{1/2}(x + 1)} \right] dx$$

$$= \log |x + 1| + \int [1/x^{1/2}(x + 1)] \, dx$$

In the last antiderivative let $x = u^2$, $dx = 2u \, du$.   Then

$$\int [1/x^{1/2}(x + 1)] \, dx = \int [2u/u(u^2 + 1)] \, du = 2 \int du/(u^2 + 1) = 2 \tan^{-1} u + C$$

Thus

$$\int \frac{x^{1/2}+1}{x^{1/2}(x+1)} \, dx = \log|x+1| + 2\tan^{-1} x^{1/2} + C$$

(g) Put $x = 2\sec\theta$, $dx = 2\sec\theta\tan\theta \, d\theta$. Then

$$\int 1/x^3(x^2-4)^{1/2} \, dx = \int [2\sec\theta\tan\theta/8\sec^2\theta \, 2\tan\theta] \, d\theta$$

$$= (1/8)\int [1/\sec^2\theta] \, d\theta = (1/8)\int \cos^2\theta \, d\theta$$

$$= (1/16)\int (1+\cos 2\theta) \, d\theta = \theta/16 + \sin 2\theta/32 + C$$

$$= \theta/16 + \sin\theta\cos\theta/16 + C = \sec^{-1}(x/2)/16$$
$$+ \sin(\sec^{-1}(x/2))\cos(\sec^{-1}(x/2))/16 + C$$
$$= \sec^{-1}(x/2)/16 + (x^2-4)^{1/2}/8x^2 + C$$

(h) $\displaystyle\int \frac{1}{(x+3)^{1/2}-(x+2)^{1/2}} \, dx = \int \frac{(x+3)^{1/2}+(x+2)^{1/2}}{(x+3)-(x+2)} \, dx$

$$= \int (x+3)^{1/2} \, dx + \int (x+2)^{1/2} \, dx$$

$$= 2(x+3)^{3/2}/3 + 2(x+2)^{3/2}/3 + C$$

(i) We have $\displaystyle\int \sin 2x e^{\sin^2 x} \, dx = 2\int \sin x e^{\sin^2 x} \cos x \, dx = \int 2u e^{u^2} \, du$, where $u = \sin x$, $du = \cos x \, dx$. Thus $\displaystyle\int \sin 2x e^{\sin^2 x} \, dx = e^{u^2} + C = e^{\sin^2 x} + C$

(j) Note first that $\displaystyle\int [1/(e^x - 16e^{-x})] \, dx = \int \{e^x/[(e^x)^2 - 16]\} \, dx$. If we put $u = e^x$ and $du = e^x \, dx$, then

$$\int [1/(e^x - 16e^{-x})] \, dx = \int [1/(u^2 - 4^2)] \, du = (1/8)\log|(u-4)/(u+4)| + C$$

(from Problem 5-7a) $= (1/8)\log|(e^x-4)/(e^x+4)| + C$.

(k) Let $[1+(1+x)^{1/2}]^{1/2} = u$, or $x = (u^2-1)^2 - 1$. Then

$$dx = 4u(u^2-1) \, du, \quad \text{and} \quad \int [1+(1+x)^{1/2}]^{-1/2} \, dx$$

$$= \int [4u(u^2-1)/u] \, du = 4\int (u^2-1) \, du = 4u^3/3 - 4u + C$$

$$= 4[1+(1+x)^{1/2}]^{3/2}/3 - 4[1+(1+x)^{1/2}]^{1/2} + C$$

(l) Here let $x = e^u$, $dx = e^u \, du$. Then $\displaystyle\int \sin(\log x) \, dx = \int e^u \sin u \, du = (e^u \sin u - e^u \cos u)/2 + C$ (by Problem 5-5d) $= [x\sin(\log x) - x\cos(\log x)]/2 + C$.

(m)   If we put $x = 2 \tan \theta$ and $dx = 2 \sec^2 \theta \, d\theta$, then

$$\int [(x^2 + 4)^{1/2}/x^4] \, dx = (1/4) \int (\sec^3 \theta/\tan^4 \theta) \, d\theta$$

$$= (1/4) \int (\cos \theta/\sin^4 \theta) \, d\theta$$

$$= (-1/12)(1/\sin^3 \theta) + C$$

$$= (-1/12) \operatorname{cosec}^3 (\tan^{-1} (x/2)) + C$$

$$= -(x^2 + 4)^{3/2}/12x^3 + C$$

(n)   Let $x = \tan \theta$.   Then $dx = \sec^2 \theta \, d\theta$, and

$$\int [\tan^{-1} x/(1 + x^2)^{3/2}] \, dx = \int (\theta \sec^2 \theta/\sec^3 \theta) \, d\theta$$

$$= \int \theta \cos \theta \, d\theta = \theta \sin \theta - \int \sin \theta \, d\theta$$

(we have integrated by parts)

$$= \theta \sin \theta + \cos \theta + C = x \tan^{-1} x/(1 + x^2)^{1/2} + 1/(1 + x^2)^{1/2} + C$$

(o)   We have

$$\int \frac{x + \sin x}{1 + \cos x} \, dx = \int \frac{\sin x}{1 + \cos x} \, dx + \int \frac{x - x \cos x}{1 - \cos^2 x} \, dx$$

$$= \int \frac{\sin x}{1 + \cos x} \, dx + \int x \operatorname{cosec}^2 x \, dx - \int x \operatorname{cosec} x \cot x \, dx$$

Now $\int [\sin x/(1 + \cos x)] \, dx = -\log |1 + \cos x|$, and integrating by parts,

$$\int x \operatorname{cosec}^2 x \, dx = -x \cot x + \int \cot x \, dx$$

$$= -x \cot x + \log |\sin x|$$

and

$$- \int x \operatorname{cosec} x \cot x \, dx = x \operatorname{cosec} x - \int \operatorname{cosec} x \, dx$$

$$= x \operatorname{cosec} x - \log |\operatorname{cosec} x - \cot x|$$

Hence

$$\int \frac{x + \sin x}{1 + \cos x} \, dx$$

$$= x \operatorname{cosec} x - x \cot x + \log \left| \frac{\sin x}{(\operatorname{cosec} x - \cot x)(1 + \cos x)} \right| + C$$

$$= x/\sin x - x \cos x/\sin x + \log |\sin^2 x/(1 - \cos x)(1 + \cos x)| + C$$

$$= x(1 - \cos x)/\sin x + \log |\sin^2 x/\sin^2 x| + C$$

$$= x(1 - \cos^2 x)/\sin x(1 + \cos x) + \log 1 + C$$

$$= x \sin x/(1 + \cos x) + C$$

Alternately, recalling Problem 5-2d, let $u = x + \sin x$, $dv = 1/(1 + \cos x)\ dx$. Then $du = (1 + \cos x)\ dx$,

$$v = \operatorname{cosec} x - \cot x, \text{ and } \int [(x + \sin x)/(1 + \cos x)]\ dx$$

$$= (x + \sin x)(\operatorname{cosec} x - \cot x) - \int (\operatorname{cosec} x - \cot x)(1 + \cos x)\ dx$$

$$= x(\operatorname{cosec} x - \cot x) + (1 - \cos x) - \int [(1 - \cos x)(1 + \cos x)/\sin x]\ dx$$

$$= x(\operatorname{cosec} x - \cot x) + (1 - \cos x) - \int \sin x\ dx$$

$$= x(\operatorname{cosec} x - \cot x) + (1 - \cos x) + \cos x + C'$$

$$= x(\operatorname{cosec} x - \cot x) + C$$

(p)   The partial fraction expansion is

$$\frac{x + 1}{(x + 2)^2(x + 3)} = \frac{A}{x + 3} + \frac{B_1}{x + 2} + \frac{B_2}{(x + 2)^2}$$

$$= \frac{A(x + 2)^2 + B_1(x + 2)(x + 3) + B_2(x + 3)}{(x + 2)^2(x + 3)}$$

Hence we must have $x + 1 = A(x + 2)^2 + B_1(x + 2)(x + 3) + B_2(x + 3)$.   In this equation put $x = -2$.   Then we get $-1 = B_2$.   If we put $x = -3$, we get $-2 = A$.   Hence $x + 1 = -2(x + 2)^2 + B_1(x + 2)(x + 3) - (x + 3)$.   Now put $x = 0$ in this equation.   Then we get $1 = -8 + 6B_1$, or $B_1 = 2$.   Consequently,

$$\int [(x + 1)/(x + 2)^2(x + 3)]\ dx$$

$$= \int [-2/(x + 3) + 2/(x + 2) - 1/(x + 2)^2]\ dx$$

$$= -2 \log |x + 3| + 2 \log |x + 2| + 1/(x + 2) + C$$

(q)   Here we must have

$$\frac{x^3 + x^2 + x - 1}{x^2(x^2 + 1)} = \frac{A_1}{x} + \frac{A_2}{x^2} + \frac{Bx + C}{x^2 + 1}$$

This   entails $x^3 + x^2 + x + 1 = A_1x(x^2 + 1) + A_2(x^2 + 1) + (Bx + C)x^2 = (A_1 + B)x^3 + (A_2 + C)x^2 + A_1x + A_2$.   Hence $A_1 = 1$, $A_2 = -1$.   Inasmuch as $A_1 + B = 1$, and $A_2 + C = 1$, we get $B = 0$, $C = 2$.   Therefore,

$$\int [(x^3 + x^2 + x - 1)/x^2(x^2 + 1)]\ dx = \int [1/x - 1/x^2 + 2/(x^2 + 1)]\ dx$$

$$= \log |x| + 1/x + 2 \tan^{-1} x + C$$

(r)

$$\int (e^x - e^{-2x})^2\ dx = \int (e^{2x} - 2e^{-x} + e^{-4x})\ dx$$

$$= e^{2x}/2 + 2e^{-x} - e^{-4x}/4 + C$$

(s)   Put $(1 + x)^{1/2} = u$, or $x = u^2 - 1$, and $dx = 2u\,du$.   Then

$$\int \{1/x[(1 + x)^{1/2} - 2]\}\,dx = \int [2u/(u^2 - 1)(u - 2)]\,du$$

Now we must have $2u/(u^2 - 1)(u - 2) = A/(u - 1) + B/(u + 1) + C/(u - 2)$, and   this   entails $2u = A(u + 1)(u - 2) + B(u - 1)(u - 2) + C(u - 1)(u + 1)$. If we let $u = 1, 2$, and -1 in this equation, we find that $A = -1$, $C = 4/3$, and $B = -1/3$ respectively.   Hence

$$\int [2u/(u^2 - 1)(u - 2)]\,du = \int [-1/(u - 1) - 1/3(u + 1) + 4/3(u - 2)]\,du$$
$$= -\log |u - 1| - \log |u + 1|/3 + 4\log |u - 2|/3 + C$$

In other words,

$$\int \{1/x[(1 + x)^{1/2} - 2]\}\,dx = -\log |(1 + x)^{1/2} + 1|$$
$$-\log |(1 + x)^{1/2} + 1|/3 - 4\log |(1 + x)^{1/2} + 1|/3 + C$$

(t)   We have

$$\int [(7x - 2)/(7 - 2x^2)^{1/2}]\,dx = \int [7x/(7 - 2x^2)^{1/2}]\,dx - \int [2/(7 - 2x^2)^{1/2}]\,dx$$
$$= (-7/4) \int [-4x/(7 - 2x^2)^{1/2}]\,dx$$
$$- 2^{1/2} \int [1/(7/2 - x^2)^{1/2}]\,dx$$
$$= (-7/2)(7 - 2x^2)^{1/2} - 2^{1/2} \sin^{-1} [(2/7)^{1/2}x] + C$$

(u) $\int \sin^6 \theta\,d\theta = (1/8) \int (1 - \cos 2\theta)^3\,d\theta$

$$= (1/8) \int (1 - 3 \cos 2\theta + 3 \cos^2 \theta - \cos^3 2\theta)\,d\theta$$
$$= \theta/8 - 3 \sin^2 \theta/16 + (3/8) \int \cos^2 2\theta\,d\theta - (1/8) \int \cos^3 2\theta\,d\theta$$
$$= \theta/8 - 3 \sin 2\theta/16 + (3/8) \int (1 + \cos 4\theta)/2\,d\theta - (1/8)$$
$$\times \int (1 - \sin^2 2\theta) \cos 2\theta\,d\theta = \theta/8 - 3 \sin 2\theta/16 + 3\theta/16$$
$$+ 3 \sin 4\theta/64 - \sin 2\theta/16 + \sin^3 2\theta/48 + C$$
$$= 5\theta/16 - \sin 2\theta/4 + \sin^3 2\theta/48 + 3 \sin 4\theta/64 + C$$

(v)   Put $x - 4 = u$, $dx = du$; then $\int [x/(x - 4)^{1/2}]\,dx = \int [(u + 4)/u^{1/2}]\,du$

$= \int [u^{1/2} + 4/u^{1/2}]\,du = \frac{2}{3}u^{3/2} + 8u^{1/2} + C = \frac{2}{3}(x - 4)^{3/2} + 8(x - 4)^{1/2} + C.$

(w)   Since the degree of the numerator exceeds the degree of the denominator we must first divide.   This gives us $(x^3 - 3x^2 - 5x + 5)/(x^2 + 2x - 3) = x - 5 + (8x - 10)/(x + 3)(x - 1)$.   We now have $(8x - 10)/(x + 3)(x - 1) = A/(x + 3) + B/(x - 1)$, or $8x - 10 = A(x - 1) + B(x + 3)$.   If we put $x = 1$ and $x = -3$ in the last identity, we find that $A = 17/2$, $B = -1/2$.   Thus

$$\int [(x^3 - 3x^2 - 5x + 5)/(x^2 + 2x - 3)]\, dx$$

$$= \int [x - 5 + 17/2(x + 3) - 1/2(x - 1)]\, dx$$

$$= x^2/2 - 5x + 17 \log |x + 3|/2 - \log |x - 1|/2 + C$$

(x)   If we put $x = u^2$, $dx = 2u\, du$, then

$$\int (1 + x) \cos x^{1/2}\, dx = \int (1 + u^2) \cos u\, 2u\, du$$

$$= 2 \int u \cos u\, du + 2 \int u^3 \cos u\, du$$

Two straightforward integrations by parts give

$$2 \int u \cos u\, du = 2u \sin u + 2 \cos u + C_1$$

and $\quad 2 \int u^3 \cos u\, du = 2u^3 \sin u + 6u^2 \cos u - 12u \sin u + C_2$.

Hence

$$\int (1 + x) \cos x^{1/2}\, dx = 2x^{3/2} \sin x^{1/2} + 6x \cos x^{1/2} - 10(x^{1/2}) \sin x^{1/2}$$

$$- 10 \cos x^{1/2} + C$$

(y)   We have

$$\int [\sec^3 x/\tan x]\, dx = \int (\sec^2 x/\tan^2 x) \sec x \tan x\, dx$$

$$= \int [\sec^2 x\, (\sec^2 x - 1)] \sec x \tan x\, dx$$

If we put $u = \sec x$, $du = \sec x \tan x\, dx$, then

$$\int [\sec^3 x/\tan x]\, dx = \int [u^2/(u^2 - 1)]\, du = \int [1 + 1/(u^2 - 1)]\, du$$

$$= u + (1/2) \log |(u - 1)/(u + 1)| + C$$

(we have used Problem 5-7a)

$$= \sec x + (1/2) \log |(\sec x - 1)/(\sec x + 1)| + C$$

(z)   If we let $x = \tan^{2/3} \theta$,

$$dx = [2 \sec^2 \theta/3 \tan^{1/3} \theta]\, d\theta$$

then

$$\int [(1 + x^3)^{1/2}/x]\, dx = (2/3) \int [(1 + \tan^2 \theta)^{1/2} \sec^2 \theta/\tan \theta]\, d\theta$$

$$= (2/3) \int [\sec^3 \theta/\tan \theta]\, d\theta = (2/3) \sec \theta$$

$$+ (1/3) |(\sec \theta - 1)/(\sec \theta + 1)| + C$$

[from part (y)]

$$= (2/3) \sec (\tan^{-1} x^{3/2}) + (1/3) \log |(\sec (\tan^{-1} x^{3/2}) - 1)/$$

$$\times (\sec (\tan^{-1} x^{3/2}) + 1)| + C = 2(1 + x^3)^{1/2}/3$$

$$+ (1/3) \log |[(1 + x^3)^{1/2} - 1]/[(1 + x^3)^{1/2} + 1]| + C$$

Alternately, let $1 + x^3 = u^2$, or $x = (u^2 - 1)^{1/3}$, and $dx = [2u/3(u^2 - 1)^{2/3}]\,du$. Then $\int [(1 + x^3)^{1/2}/x]\,dx$

$$= \int \frac{u}{(u^2 - 1)^{1/3}} \frac{2}{3} \frac{u}{(u^2 - 1)^{2/3}}\,du = \frac{2}{3} \int \frac{u^2}{u^2 - 1}\,du$$

$= 2u/3 + (1/3) \log |(u-1)/(u+1)|$ [from the solution of part (y)] $= (2/3)(1 + x^3)^{1/2} + (1/3) \log |[(1 + x^3)^{1/2} - 1]/[(1 + x^3)^{1/2} + 1]| + C.$

5-9   Let $u = \tan(x/2)$, or $x = 2 \tan^{-1} u$.   Then

$$dx = 2\,du/(1 + u^2), \ \sin x = 2 \sin(x/2) \cos(x/2) = 2 \tan(x/2) \cos^2(x/2)$$

$$= 2 \tan(x/2)/\sec^2(x/2) = 2 \ \tan(x/2)/[1 + \tan^2(x/2)] = 2u/(1 + u^2),$$

and

$$\cos x = \cos^2(x/2) - \sin^2(x/2) = [1 - \tan^2(x/2)]/\sec^2(x/2)$$

$$= [1 - \tan^2(x/2)]/[1 + \tan^2(x/2)] = (1 - u^2)/(1 + u^2)$$

Thus $\int [1/(2 + 3 \cos x)]\,dx$

$$= \int \frac{1}{2 + 3(1 - u^2)/(1 + u^2)} \frac{2}{1 + u^2}\,du = \int \frac{2}{5 - u^2}\,du = -2 \int \frac{1}{u^2 - 5}\,du$$

$$= (-1/5^{1/2}) \log |(u - 5^{1/2})/(u + 5^{1/2})| + C$$

(from the result of Problem 5-7a)

$$= (1/5^{1/2}) \log |(5^{1/2} + u)/(5^{1/2} - u)| + C$$

$$= (1/5^{1/2}) \log |[5^{1/2} + \tan(x/2)]/[5^{1/2} - \tan(x/2)]| + C$$

Remark   This substitution reduces any rational function of $\sin x$ and $\cos x$ to a rational function and u which can, in theory, be antidifferentiated.

Supplementary Problem 5-9   Use the substitution in the solution of Problem 5-9 to find

(a) $\int \dfrac{dx}{3 + 2 \cos x}$

(b) $\int \dfrac{\sin x}{1 + \sin x + \cos x}\,dx$

(c) $\int \dfrac{1 + \sin x}{\sin x\,(2 + \cos x)}\,dx$

Solution   (a) $\int \dfrac{dx}{3 + 2 \cos x} = \int \dfrac{1}{3 + 2(1 - u^2)/(1 + u^2)} \dfrac{2}{1 + u^2}\,du$

$$= \int \frac{2}{5 + u^2}\,du = \frac{2}{5^{1/2}} \tan^{-1} \frac{u}{5^{1/2}} + C = \frac{2}{5^{1/2}} \tan^{-1} \left(\frac{1}{5^{1/2}} \tan \frac{x}{2} \right) + C$$

(b)   Here

$$\int \frac{\sin x}{1 + \sin x + \cos x}\, dx = \int \frac{2u/(1 + u^2)}{1 + 2u/(1 + u^2) + (1 - u^2)/(1 + u^2)} \frac{2}{1 + u^2}\, du$$

$$= \int \frac{4u}{1 + u^2 + 2u + 1 - u^2} \frac{1}{1 + u^2}\, du$$

$$= \int \frac{2u}{(1 + u^2)(1 + u)}\, du$$

$$= \int \left[\frac{u + 1}{u^2 + 1} - \frac{1}{u + 1} \right] du$$

(this partial fraction expansion is easy to determine; we leave the details to the reader)

$$= (1/2) \log (u^2 + 1) + \tan^{-1} u - \log |u + 1| + C$$
$$= (1/2) \log (\tan^2 (x/2) + 1) + x/2 - \log |\tan^2 (x/2) + 1| + C$$

(c)   We have

$$\int \frac{1 + \sin x}{\sin x\, (2 + \cos x)}\, dx = \int \frac{1 + 2u/(1 + u^2)}{[2u/(1 + u^2)][2 + (1 - u^2)/(1 + u^2)]} \frac{2}{1 + u^2}\, du$$

$$= \int \frac{(1 + u)^2/(1 + u^2)}{2u(3 + u^2)/(1 + u^2)^2} \frac{2}{1 + u^2}\, du$$

$$= \int \frac{(1 + u)^2}{u(u^2 + 3)}\, du$$

$$= \int \left[\frac{1}{3u} + \frac{(2/3)u + 2}{u^2 + 3} \right] du$$

(we once again leave the details of the partial fraction expansion to the reader)

$$= (1/3) \log u + (1/3) \log (u^2 + 3) + (2/3^{1/2}) \tan^{-1}(u/3^{1/2}) + C$$
$$= (1/3) \log |\tan(x/2)| + (1/3) \log [\tan^2 (x/2) + 3]$$
$$+ (2/3^{1/2}) \tan^{-1} [(1/3^{1/2}) \tan (x/2)] + C$$

5-10   Let $(x - 1)/(x + 1) = u^2$.   Then $x = (1 + u^2)/(1 - u^2)$,

$$dx = \{[(1 - u^2)2u + (1 + u^2)2u]/(1 - u^2)^2\}\, du = [4u/(1 - u^2)^2]\, du$$

and so

$$\int \left(\frac{x - 1}{x + 1} \right)^{1/2} \frac{dx}{x} = \int u \frac{1 - u^2}{1 + u^2} \frac{4u}{(1 - u^2)^2}\, du = \int \frac{4u^2}{1 - u^4}\, du$$

Now the partial fraction expansion of $4u^2/(1 - u^4)$ is found to be $4u^2/(1 - u^4) = 1/(1 + u) + 1/(1 - u) - 2/(1 + u^2)$.   Thus

$$\int \frac{4u^2}{1 - u^4}\, du = \log |1 + u| - \log |1 - u| - 2 \tan^{-1} u + C$$

and so

$$\int\left(\frac{x-1}{x+1}\right)^{1/2}\frac{dx}{x}=\log\left|\frac{1+[(x-1)/(x+1)]^{1/2}}{1-[(x-1)/(x+1)]^{1/2}}\right|-2\tan^{-1}\left(\frac{x-1}{x+1}\right)^{1/2}+C$$

Remark   This problem illustrates a standard technique for handling antiderivatives of algebraic functions.   It consists of finding a substitution that turns the algebraic function into a rational function which can, at least in theory, be dealt with.   Consider the following problem for further practice with this technique.

Supplementary Problem 5-10

(a)   Find $\int(x/[(1+x)^{1/2}-(1+x)^{1/3}])\,dx$

(b)   Find a substitution that will rationalize $\int dx/x(3x^2+2x-1)^{1/2}$.   Also give the antiderivative that results

(c)   Do the same as part (b) for

$$\int dx/[(x+a)^{3/2}+(x-a)^{3/2}]$$

Solution   (a)   If we let $1+x=u^6$, then $dx=6u^5\,du$, and

$$\int(x/[(1+x)^{1/2}-(1+x)^{1/3}])\,dx=\int[(u^6-1)6u^5/(u^3-u^2)]\,du$$

$$=\int 6[u^3(u^6-1)/(u-1)\,du$$

$$=\int 6(u^5+u^4+u^3+u^2+u+1)u^3\,du$$

$$=6u^9/9+6u^8/8+6u^7/7+6u^6/6$$
$$+6u^5/5+6u^4/4+C$$

$$=2(1+x)^{3/2}/3+3(1+x)^{4/3}/4$$
$$+6(1+x)^{7/6}/7+(1+x)$$
$$+6(1+x)^{5/6}/5+3(1+x)^{2/3}/2+C$$

(b)   We have $x(3x^2+2x-1)^{1/2}=x(x+1)[(3x-1)/(x+1)]^{1/2}$.   We seek a rational function r such that $(3x-1)/(x+1)=r^2(u)$, or $x=[1+r^2(u)]/[3-r^2(u)]$. Thus r can be any rational function. Let us take $r(u)=u$.   Then $x=(1+u^2)/(3-u^2)$, and $dx=\{[(3-u^2)2u+(1+u^2)2u]/(3-u^2)^2\}\,du=[8u/(3-u^2)]du$ and so

$$\int dx/x(3x^2+2x-1)^{1/2}$$

$$=\int\frac{1}{[(1+u^2)/(3-u^2)][(1+u^2)/(3-u^2)+1]u}\frac{8u}{(3-u^2)^2}\,du$$

$$=2\int\frac{1}{1+u^2}\,du$$

(c)   We seek rational functions r_1 and r_2 such that $x+a=r_1{}^2(u)$ and $x-a=r_2{}^2(u)$. This entails $r_1{}^2(u)-a=r_2{}^2(u)+a$. If we let $r_1=p_1+q_1$ and $r_2=p_2+q_2$,

the last equation becomes $p_1{}^2(u) + 2p_1(u)q_1(u) + q_1{}^2(u) - a = p_2{}^2(u) + 2p_2(u)q_2(u)$
$+ q_2{}^2(u)$ If we put $2\,p_1(u)q_1(u) = a$, and $2p_2(u)q_2(u) = -a$, we get $p_1{}^2(u) + q_1{}^2(u) =$
$p_2{}^2(u) + q_2{}^2(u)$. In this last equation put $p_1{}^2(u) = p_2{}^2(u) = au^2/2$, and $q_1{}^2(u) = q_2{}^2(u) =$
$a/2u^2$. Then $r_1(u) = (a/2)^{1/2}(u + 1/u)$, and $r_2(u) = (a/2)^{1/2}(u - 1/u)$. Hence $r_1{}^2(u) =$
$(a/2)(u^2 + 2 + 1/u^2)$, and $r_2{}^2(u) = (a/2)(u^2 - 2 + 1/u^2)$, or $x = (a/2)(u^2 + 2 + 1/u^2)$
$- a = (a/2)(u^2 + 1/u^2)$. Note the last equation is consistent inasmuch as $x =$
$(a/2)(u^2 - 2 + 1/u^2) + a = (a/2)(u^2 + 1/u^2)$. Since $dx = (a/2)(2u - 2/u^3)\,du$, we have
$\int dx/[(x + a)^{3/2} + (x - a)^{3/2}]$

$$= \int \frac{a(u - 1/u^3)\,du}{(a/2)^{3/2}[(u + 1/u)^3 + (u - 1/u)^3]} = \left(\frac{2}{a}\right)^{3/2} \int \frac{u^4 - 1}{u^6 + 3u^2}\,du$$

For a more systematic investigation of rationalization, as well as other matters connected with the problem of antidifferentiatiation, consult *Integration of Functions of a Single Variable* by G. H. Hardy, Cambridge Tract No. 2.

5-11 We have $2u\,du = (1 - 1/x^2)\,dx = [(x^2 - 1)/x^2]\,dx$, and so

$$\int \frac{x-1}{x+1} \frac{1}{[x(x^2 + x + 1)]^{1/2}}\,dx = \int \frac{x-1}{x+1} \frac{1}{x} \frac{1}{(x + 1 + 1/x)^{1/2}}\,dx$$

$$= \int \frac{x^2 - 1}{x^2} \frac{1}{(x + 1)^2/x} \frac{1}{(x + 1 + 1/x)^{1/2}}\,dx$$

$$= \int \frac{x^2 - 1}{x^2} \frac{1}{x + 1 + 1/x + 1} \frac{1}{(x + 1 + 1/x)^{1/2}}\,dx$$

$$= \int \frac{2u}{(u^2 + 1)u}\,du = 2 \int \frac{du}{1 + u^2} = 2\,\tan^{-1} u + C$$

$$= 2\,\tan^{-1}\left([x + 1 + 1/x]^{1/2}\right) + C$$

5-12 (a) If $q = r/s$, r and s integers, let $x = u^s$. Then $su^{s-1}\,du = dx$, and

$$\int (1 + x)^p x^{r/s}\,dx = \int s(1 + u^s)^p u^r u^{s-1}\,du$$

$$= s \int u^{r+s-1} \left[\sum_{k=0}^{p} \binom{p}{k} u^{ks}\right]\,du = s \sum_{k=0}^{p} \binom{p}{k} \int u^{(k+1)s+r-1}\,du$$

$$= \sum_{k=0}^{p} \binom{p}{k} \frac{s}{(k+1)s+r} u^{(k+1)s+r} + C$$

$$= \sum_{k=0}^{p} \binom{p}{k} \frac{s}{(k+1)s+r} (x^{1/s})^{(k+1)s+r} + C$$

$$= x^q \sum_{k=0}^{p} \binom{p}{k} \frac{s}{(k+1)s+r} x^{k+1} + C$$

(b)   If $p = r/s$, r and s integers, let $1 + x = u^s$. Then $su^{s-1}\, du = dx$, and

$$\int (1 + x)^{r/s} x^q\, dx = \int su^r(u^s - 1)^q u^{s-1}\, du$$

$$= s \int u^{r+s-1} \left[\sum_{k=0}^{n} (-1)^{q-k} \binom{q}{k} u^{ks} \right] du$$

$$= \sum_{k=0}^{q} (-1)^{q-k} \binom{q}{k} s \int u^{(k+1)s+r-1}\, du$$

$$= \sum_{k=0}^{q} (-1)^{q-k} \binom{q}{k} \frac{s}{(k+1)s+r}\, u^{(k+1)s+r} + C$$

$$= \sum_{k=0}^{q} (-1)^{q-k} \binom{q}{k} \frac{s}{(k+1)s+r}\, [(1 + x)^{1/s}]^{(k+1)s+r} + C$$

$$= (1 + x)^p \sum_{k=0}^{q} (-1)^{q-k} \binom{q}{k} \frac{s}{(k+1)s+r}\, (1 + x)^{k+1} + C$$

(c)   If $p = r/s$, $q = l/m$, and $p + q = n$, where r, s, l, m, and n are integers, let $(1 + x)/x = u^s$, or $x = 1/(u^s - 1)$ and $dx = [-su^{s-1}/(u^s - 1)^2]\, du$.   Thus

$$\int (1 + x)^p x^q\, dx = \int \left(\frac{1+x}{x} \right)^p x^{p+q}\, dx = \int \left(\frac{1+x}{x} \right)^p x^n\, dx$$

$$= -\int u^r \frac{1}{(u^2 - 1)^n} \frac{su^{s-1}}{(u^s - 1)^2}\, du = -s \int \frac{u^{r+s-1}}{(u^s - 1)^{n+2}}\, du$$

It can be seen that we have rationalized $\int (1 + x)^p x^q\, dx$.

5-13 $I_n = \int \tan^{n-2} x\, [\sec^2 x - 1]\, dx = \int \tan^{n-2} x \sec^2 x\, dx - \int \tan^{n-2} x\, dx$

$= \tan^{n-1} x/(n - 1) - I_{n-2}$

5-14   Integrating by parts, we have

$$I_{m,n} = \int x^{m-1} \frac{x}{(1 + x^2)^n}\, dx = -\frac{x^{m-1}}{2(n - 1)(1 + x^2)^{n-1}} + \frac{m-1}{2(n-1)} \int \frac{x^{m-2}}{(1 + x^2)^{n-1}}\, dx$$

$$= -\frac{x^{m-1}}{2(n - 1)(1 + x^2)^{n-1}} + \frac{m-1}{2(n-1)} I_{m-2,\,n-1}$$

Remark   Problems 5-13 and 5-14 are examples of *reduction formulas*: Eq. 5-33 is another one.   Such expressions are especially useful when the parameters are positive integers because if a particular antiderivative of the type under consideration can be found, then all succeeding ones can be found.

5-15   If we integrate by parts, we have $\int f''(x)F(x)\, dx = f'(x)F(x) - \int f'(x)F'(x)\, dx$. Integrating by parts again, we have $\int f''(x)F(x)\, dx = f'(x)F(x) - f(x)F'(x) + \int f(x)F''(x)\, dx$.   More generally, it is easy to see that by repeated integrations by parts (or mathematical induction), we have

$$\int f^{(n)}(x)F(x)\,dx = f^{(n-1)}(x)F(x) - f^{(n-2)}(x)F'(x) + \cdots + (-1)^n \int f(x)F^{(n)}(x)\,dx$$

An interesting instance of this formula is given when $f(x) = e^x$, a case we have already encountered, with $n = 2$, in the solution of Problem 5-5d.

5-16   We know that, under the stated condition, there is a partial fraction expansion of the form

$$\frac{P(x)}{Q(x)} = \sum_{k=1}^{n} \frac{A_k}{x - \alpha_k}$$

Now

$$\frac{Q(x)}{x - \alpha_k} = \frac{Q(x) - Q(\alpha_k)}{x - \alpha_k} \to Q'(\alpha_k)$$

as $x \to \alpha_k$.   Since

$$(x - \alpha_k)\frac{P(x)}{Q(x)} = \frac{x - \alpha_k}{x - \alpha_1} A_1 + \cdots + A_k + \cdots + \frac{x - \alpha_k}{x - \alpha_n} A_n$$

we have that

$$\lim_{x \to \alpha_k} P(x)\frac{(x - \alpha_k)}{Q(x)} = \frac{P(\alpha_k)}{Q'(\alpha_k)} = \lim_{x \to \alpha_k}\left[\frac{x - \alpha_k}{x - \alpha_1} A_1 + \cdots + A_k + \cdots + \frac{x - \alpha_k}{x - \alpha_n} A_k\right] = A_k$$

Hence

$$\int \frac{P(x)}{Q(x)}\,dx = \int\left[\sum_{k=1}^{n}\frac{P(\alpha_k)}{Q'(\alpha_k)}\frac{1}{x - \alpha_k}\right]dx = \sum_{k=1}^{n}\frac{P(\alpha_k)}{Q'(\alpha_k)}\log|x - \alpha_k| + C$$

5-17   To say that $(-1/2)\cos 2x + C = \int \sin 2x\,dx = \sin^2 x + C$ means only that $D(-1/2)\cos 2x = \sin 2x = D(\sin^2 x)$.   Now we know that any two functions which have the same derivative differ by a constant.   Thus $\sin^2 x + \cos 2x/2 = k$ for all x.   In particular, $\sin^2 0 + \cos 0/2 = k$, and so $k = 1/2$.   In other words, we have reestablished the familiar trigonometric identity $\cos 2x = 1 - 2\sin^2 x$.

5-18   We have $y' = 2xy$, or $y'/y = 2x$.   Thus $\int y'/y\,dx = \int 2x\,dx$, or $\log y = x^2 + k$, or $y = e^{x^2 + k} = e^k e^{x^2}$.   Now $1 = e^k e^4$, and so $e^k = e^{-4}$, and the equation is $y = e^{-4}e^{x^2}$.

5-19   Since we may take $s_0 = 0$, we have $v = at + 88$, and $s = at^2/2 + 88t$.   We want $300 = at^2/2 + 88t$, and so the time t_1 required to stop the truck is given by $t_1 = [-88 \pm (88^2 + 600a)^{1/2}]/a$.   When $t = t_1$, $v = 0$, and so $0 = at_1 + 88 = -88 \pm (88^2 + 600a)^{1/2} + 88$.   Then $(88^2 + 600a)^{1/2} = 0$, or $a = -88^2/600$ ft/sec².

5-20   Using the approximation $g = 32$ ft/sec², we have, since s_0 can be taken to be zero, $v = 32t + v_0$, and $s = 16t^2 + v_0 t$.   Now the height s_1 of the building is given by $s_1 = 16(2.5)^2 + (2.5)v_0$, which gives us $v_0 = [s_1 - 16(6.25)]/2.5$.   But the velocity with which the object strikes the ground is 110 ft/sec.   Thus $110 = 32(2 \cdot 5) + [s_1 - 16(6.25)]/2.5$, or $s_1 = 175$ ft.

5-21 Here $s = gt^2/2 + v_0 t = -16t^2 + v_0 t$ because we can take $s_0 = 0$. If s_1 denotes the height in question, then $-16(4)^2 + 4v_0 = s_1 = -16(9)^2 + 9v_0$, or $v_0 = 208$ ft/sec. Thus $s_1 = -256 + 832 = 576$ ft.

5-22 Let a be the unknown acceleration. For the first 60 feet, we have $s = at^2/2$. Let t_1 be the elapsed time for the first 60 feet. Then $60 = at_1^2/2$, and the velocity v_1 at the end of 60 feet is $v_1 = at_1$. Now for $t_1 \le t \le 10$ we have, because the acceleration is zero, $s = 60 + v_1(t - t_1) = 60 + at_1 t - at_1^2 = at_1 t - 60$. But $300 = 10at_1 - 60$, or $360 = 10at_1 = 10a(120/a)^{1/2}$. Thus $36 = (120a)^{1/2}$, and so $a = (36)^2/120$ ft/sec^2 .

5-23 We have $(1/y^{1/2})\, dy/dt = -0.025$, and so $\int y'/y^{1/2}\, dt = \int -0.025\, dt$, or $2(y)^{1/2} = k - 0.025\, t$. If $y = 9$ when $t = 0$, then $2(9)^{1/2} = k$, and so $2(y)^{1/2} = 6 - 0.025t$. Hence $y = 0$ when $0 = 6 - 0.025t$, or $t = 6/0.025 = 240$ sec $= 4$ min.

5-24 Here $dv/dt = -k(v)^{1/2}$, or $(1/v^{1/2})\, dv/dt = -k$, and so $2(v)^{1/2} = -kt + C_1$. Now we have $C_1 = 2(v_0)^{1/2}$ when $t = 0$, and so $v = [(v_0)^{1/2} - kt/2]^2$. Thus $\int (ds/dt)\, dt = \int [(v_0)^{1/2} - kt/2]^2\, dt$, or $s = (-2/3k)[(v_0)^{1/2} - kt/2]^3 + C_2$, and we have $0 = (-2/3k)v_0^{3/2} + C_2$ when $t = 0$, or $0 = (2/3k)v_0^{3/2} - (2/3k)[(v_0)^{1/2} - kt/2]^3$. We must have $3^{3/8} = (2/3k)(225)^{3/2}$, or $k = 2000/3$. Thus $s = v_0^{3/2}/1000 - (1/1000)[(v_0)^{1/2} - 1000t/3]^3$, and $v = [(v_0)^{1/2} - 1000t/3]^2$. Therefore, when $v_0 = 225$, $v = 0$ when $t = 3(15)/1000 = 0.045$ sec. If $v_0 = 400$, the time required for the bomb to come to rest is $3(20)/1000 = 0.06$ sec, and the depth is $(20)^3/1000 = 8$ ft.

5-25 If we regard v as a function of x , and x as a function of t , we can write

$$\frac{dv}{dt} = \frac{dv}{dx}\frac{dx}{dt} = v\frac{dv}{dx}$$

and so we have $v\, dv/dx = -\omega^2 x$. Thus $\int v(dv/dx)\, dx = -\omega^2 \int x\, dx$, or $v^2/2 = -\omega^2 x^2/2 + C$. This equation can be written as $v^2 = \omega^2(2c/\omega^2 - x^2)$, and we see that $x^2 \le 2c/\omega^2$ because $v^2 \ge 0$. For simplicity, let $2c/\omega^2 = r^2$. We then have to deal with the equation $(dx/dt)^2 = \omega^2(r^2 - x^2)$. Since the cases $x = r$, or $x = -r$ for all t can be discounted, we can take $x^2 \ne r^2$ and obtain

$$\frac{\pm 1}{(r^2 - x^2)^{1/2}}\frac{dx}{dt} = \omega$$

or

$$\pm \int \frac{1}{(r^2 - x^2)^{1/2}}\frac{dx}{dt}\, dt = \omega \int dt$$

Hence $\mp\cos^{-1}(x/r) = \omega t + k$, or $x/r = \cos[\pm(\omega t + k)] = \cos(\omega t + k)$ (cosine is an even function). The law of motion is thus $x = r\cos(\omega t + k)$, and we can recognize this as a simple harmonic motion (see Example 4-15) by observing that if a particle P moves around the circle $x^2 + y^2 = r^2$ with a constant angular velocity ω , then the projection of P on the x axis has x coordinate $x = r\cos(\omega t + k)$, where k is determined by the initial position of P .

Remark The trick of writing the acceleration in the form $v\,dv/dx$, as we did above, is very useful in cases where the acceleration depends on the position.

CHAPTER 6

6-1 The partition is $-1 + 0/n,\ -1 + 4/n,\ \ldots,\ -1 + 4k/n,\ \ldots,\ -1 + 4n/n$, and the associated Riemann sum R_n is, in the notation of the remark following the solution of Problem 1-5,

$$R_n = \sum_{k=1}^{n} [(-1 + 4k/n)^3 + 2(-1 + 4k/n)]4/n$$

$$= -8 + \frac{32}{n^2} \sum_{k=1}^{n} k + \frac{4}{n^4} \sum_{k=1}^{n} (-n + 4k)^3$$

$$= -8 + \frac{32}{n^2} S_n{}^1 + \frac{4}{n^4} \sum_{k=1}^{n} [-n^3 + 12n^2 k - 48nk^2 + 64k^3]$$

$$= -8 + \frac{32}{n^2} S_n{}^1 + \frac{4}{n^4} \left[-n^4 + 12n^2 \sum_{k=1}^{n} k - 48n \sum_{k=1}^{n} n^2 + 64 \sum_{k=1}^{n} k^3 \right]$$

$$= -8 + \frac{32}{n^2} S_n{}^1 - 4 + \frac{48}{n^2} S_n{}^1 - \frac{192}{n^3} S_n{}^2 + \frac{256}{n^4} S_n{}^3$$

$$= -12 + \frac{80}{n^2} \frac{n(n+1)}{2} - \frac{192}{n^3} \frac{n(n+1)(2n+1)}{6} + \frac{256}{n^4} \left[\frac{n(n+1)}{2} \right]^2$$

$$= -12 + 40(1 + 1/n) - 32(1 + 1/n)(2 + 1/n) + 64(1 + 1/n)^2$$

Hence $\{R_n\} \to -12 + 40 - 64 + 64 = 28 = \int_{-1}^{3} (x^3 + 2x)\,dx$.

6-2 The Riemann sum R_n is given by

$$R_n = \sum_{k=0}^{n-1} (ar^k)^p (ar^{k+1} - ar^k) = a^{p+1}(r - 1) \sum_{k=0}^{n-1} r^{(p+1)k}$$

In this equation we recognize the sum of a geometric progression with ratio $r^{p+1} \neq 1$. Thus

$$R_n = a^{p+1}(r - 1) \frac{r^{n(p+1)} - 1}{r^{p+1} - 1} = a^{p+1} \left[\left(\frac{b}{a} \right)^{p+1} - 1 \right] \frac{r - 1}{r^{p+1} - 1}$$

$$= (b^{p+1} - a^{p+1}) \frac{1}{(r^{p+1} - 1)/(r - 1)} = \frac{b^{p+1} - a^{p+1}}{\displaystyle\sum_{k=0}^{p} r^k}$$

$$= \frac{(b^{p+1} - a^{p+1})}{\displaystyle\sum_{k=0}^{p} [(b/a)^{1/n}]^k}$$

Now $\{(b/a)^{1/n}\} \to 1$ (see Problem 2-5b), and so

$$\{R_n\} \to \frac{b^{p+1} - a^{p+1}}{p+1} = \int_a^b x^p \, dx$$

6-3   Using the trigonometric identity $2 \sin A \sin B = \cos (A - B) - \cos (A + B)$, we have

$$R_h = \frac{h}{2 \sin (h/2)} \sum_{k=1}^n 2 \sin (h/2) \sin (a + kh)$$

$$= \frac{h}{2 \sin (h/2)} \sum_{k=1}^n \{\cos [a + (k - 1/2)h] - \cos [a + (k + 1)h]\}$$

$$= \frac{h}{2 \sin (h/2)} [\cos (a + h/2) - \cos (a + nh + h/2)]$$

$$= \frac{h}{2 \sin (h/2)} [\cos (a + h/2) - \cos (b + h/2)]$$

Now as $n \to \infty$, $h \to 0$, and

$$\left\{ \frac{h/2}{\sin (h/2)} [\cos (a + h/2) - \cos (b + h/2)] \right\} \to \cos a - \cos b = -(\cos b - \cos a)$$

$$= \int_a^b \sin x \, dx$$

6-4   (a)   We have

$$\int_0^{2\pi} \cos mx \sin nx \, dx = (1/2) \int_0^{2\pi} [\sin (m - n)x + \sin (m + n)x] \, dx$$

$$= \begin{cases} \left. -\frac{1}{2(m - n)} \cos (m - n)x - \frac{1}{2(m + n)} \cos (m + n)\, x \right|_0^{2\pi} = 0 \quad \text{if} \quad m \neq n \\[2mm] \left. \frac{\sin 2nx}{4n} \right|_0^{2\pi} = 0 \quad \text{if} \quad m = n \end{cases}$$

If $m \neq n$, $\displaystyle\int_0^{2\pi} \cos mx \cos nx \, dx = (1/2) \int_0^{2\pi} [\cos (m - n)x + \cos (m + n) \, x] \, dx$

$$= \left. \frac{1}{2(m - n)} \sin (m - n)x + \frac{1}{2(m + n)} \sin (m + n)x \right|_0^{2\pi} = 0, \text{ while}$$

if $m = n$, $\displaystyle\int_0^{2\pi} \cos^2 nx \, dx = \int_0^{2\pi} [1 + \cos 2nx]/2 \, dx$

$$= \left. x/2 + \sin 2nx/4n \right|_0^{2\pi} = \pi. \quad \text{Similarly, if } m \neq n,$$

$$\int_0^{2\pi} \sin mx \sin nx \, dx = (1/2) \int_0^{2\pi} [\cos (m - n)x - \cos (m + n)x] \, dx$$

$$= \left. \frac{1}{2(m - n)} \sin (m - n)x - \frac{1}{2(m + n)} \sin (m + n)x \right|_0^{2\pi} = 0, \text{ while}$$

if $m = n$, $\displaystyle\int_0^{2\pi} \sin^2 nx \, dx = \int_0^{2\pi} [1 - \cos 2nx]/2 \, dx$

$$= x/2 - \sin 2nx/4n \Big|_0^{2\pi} = \pi$$

(b)   With the aid of the results of part (a) together with the identities $\displaystyle\int_0^{2\pi} \sin nx \, dx = 0 = \int_0^{2\pi} \cos nx \, dx$, we have

$$\int_0^{2\pi} f(x) \, dx = \frac{a_0}{2} \int_0^{2\pi} dx + \sum_{m=1}^{n} \int_0^{2\pi} (a_m \cos mx + b_m \sin mx) \, dx = a_0 \pi$$

$$\int_0^{2\pi} f(x) \cos kx \, dx = \frac{a_0}{2} \int_0^{2\pi} \cos kx \, dx + \sum_{m=1}^{n} \int_0^{2\pi} (a_m \cos mx \cos kx$$

$$+ \, b_m \sin mx \cos kx) \, dx = \begin{cases} \pi a_k & \text{if } k \le n \\ 0 & \text{if } k > n \end{cases}$$

and

$$\int_0^{2\pi} f(x) \sin kx \, dx = \frac{a_0}{2} \int_0^{2\pi} \sin kx \, dx$$

$$+ \sum_{m=1}^{n} \int_0^{2\pi} (a_m \cos mx \sin kx + b_m \sin mx \sin kx) \, dx = \begin{cases} \pi b_k & \text{if } k \le n \\ 0 & \text{if } k > n \end{cases}$$

Remark   This problem is of interest in the theory of Fourier series.

6-5   (a)   We have $\displaystyle\int_{-a}^{a} f(x^2) \, dx = \int_0^{a} f(x^2) \, dx + \int_{-a}^{0} f(x^2) \, dx$.   In the last of these integrals let $x = -t$.   Then $dx = -dt$, and $\displaystyle\int_{-a}^{0} f(x^2) \, dx = -\int_{a}^{0} f[(-t)^2] \, dt = \int_0^{a} f(t^2) \, dt$.   Now in an integral it is immaterial what we call the variable. Thus $\displaystyle\int_0^{a} f(t^2) \, dt = \int_0^{a} f(x^2) \, dx$, and so $\displaystyle\int_{-a}^{a} f(x^2) \, dx = 2 \int_0^{a} f(x^2) \, dx$.

(b)   Here $\displaystyle\int_{-a}^{a} xf(x^2) \, dx = \int_0^{a} xf(x^2) \, dx + \int_{-a}^{0} xf(x^2) \, dx$.   Again, in the last integral make the substitution $x = -t$.   Then $\displaystyle\int_{-a}^{0} xf(x^2) \, dx = \int_{a}^{0} tf[(-t)^2] \, dt = -\int_0^{a} tf(t^2) \, dt = -\int_0^{a} xf(x^2) \, dx$.   Thus $\displaystyle\int_{-a}^{a} xf(x^2) \, dx = 0$.

6-6   Let $x = a + b - y$.   Then $dx = -dy$, and so

$$\int_a^{b} F(x) \, dx = -\int_b^{a} F(a+b-y) \, dy = \int_a^{b} F(a+b-y) \, dy = \int_a^{b} F(a+b-x) \, dx$$

6-7   (a)   Integrate $\displaystyle\int_0^{x} f_1(t) \, dt$ by parts as follows: let $u = f_1(t)$, and $dv = dt$, or $du = f(t) \, dt$, and $v = t$.

Then

$$\int_0^{x} f_1(t) \, dt = tf_1(t) \Big|_0^{x} - \int_0^{x} tf(t) \, dt = xf_1(x) - \int_0^{x} tf(t) \, dt$$

$$= x \int_0^{x} f(t) \, dt - \int_0^{x} tf(t) \, dt = \int_0^{x} (x - t)f(t) \, dt$$

It is easy to see that a similar integration by parts will result in the formula $f_n(x) = \int_0^x (x-t)f_{n-2}(t)\,dt$.   In this equation let $u = f_{n-2}(t)$ and $dv = (x-t)\,dt$. Then $du = f_{n-3}(t)\,dt$, $v = -(x-t)^2/2$, and

$$\int_0^x (x-t)f_{n-2}(t)\,dt = -(x-t)^2 f_{n-2}(t)/2 \Big|_0^x + \int_0^x (x-t)^2 f_{n-3}(t)/2\,dt$$
$$= \int_0^x (x-t)^2 f_{n-3}(t)/2\,dt$$

Another integration by parts would give $f_n(x) = \int_0^x (x-t)^3 f_{n-4}(t)/3!\,dt$, and it is clear that we can continue in this way and obtain

$$f_n(x) = \int_0^x \frac{(x-t)^{n-1}}{(n-1)!} f(t)\,dt$$

(b)   If $f(x) = x^m$, then, in the notation of part (a), $\int_0^1 (1-x)^n x^m\,dx = n!f_{n+1}(1)$.

Now

$$f_1(t) = \int_0^t x^m\,dx = t^{m+1}/(m+1),\ f_2(t) = \int_0^t m!x^{m+1}/(m+1)!\,dx$$
$$= m!\,t^{m+2}/(m+2)!,\ \ldots,\ f_k(t) = m!\,t^{m+k}/(m+k)!$$

Thus $n!f_{n-1}(1) = m!\,n!/(m+n+1)!$.

Remark   The integrals appearing in this problem are instances of *convolutions* of functions.   We know from Chapter 2 how to define the convolution of two sequences.   It is the sequence whose general term is

$$\sum_{k=0}^n a_{n-k}b_k$$

The analog of this expression for functions f and g both having domain $[0, \infty)$ is the integral

$$\int_0^x f(x-t)g(t)\,dt$$

If it exists for all $x \geq 0$, we define it to be the convolution of f and g, and denote it by $f*g(x)$.

We shall see in Chapter 7 that convolutions of sequences arise in a natural way. Convolutions of functions also arise in a natural way, but to see precisely how would take us beyond the scope of this book.   Accordingly, we do no more than call attention to the possibility of being able to define this operation for functions as well as for sequences.

6-8   (a)   We can pick a sequence of partitions whose norms have limit zero and form sequences of Riemann sums $\{R_n\}$ and $\{T_n\}$ relative to them wherein f and g are evaluated at the same points.   But then

$$R_n = \sum_{k=1}^n f(z_k)(x_k - x_{k-1}) \leq \sum_{k=1}^n g(z_k)(x_k - x_{k-1}) = T_n$$

since $x_k - x_{k-1} > 0$. If we pass to the limit in $R_n \le T_n$, we get $\int_a^b f(x)\,dx \le \int_a^b g(x)\,dx$.

(b) We can form a sequence of Riemann sums $\{R_n\}$ such that $\{R_n\} \to \int_a^b f(x)\,dx$. Now the triangle inequality gives us

$$|R_n| = \left| \sum_{k=1}^n f(z_k)(x_k - x_{k-1}) \right| \le \sum_{k=1}^n |f(z_k)|(x_k - x_{k-1}) = R_n'$$

Inasmuch as the absolute value function is continuous we have $\{|R_n|\} \to \left| \int_a^b f(x)\,dx \right|$. Further, $\{R_n'\} \to \int_a^b |f(x)|\,dx$. The inequality $\left| \int_a^b f(x)\,dx \right| \le \int_a^b |f(x)|\,dx$ thus follows by passing to the limit in the inequality $|R_n| \le R_n'$.

Remark The inequality in part (b) can be regarded as an extension of the triangle inequality to integrals. Many of the inequalities of Chapter 1 can also be so extended by an argument similar to that of part (b), but we shall not pursue this. The reader may care to do so himself for the Cauchy–Schwartz inequality and Minkowski's inequality.

6-9 If $\alpha = 0$, then $\int_{-1}^1 dx/(1 - 2\alpha x + \alpha^2)^{1/2} = 2$.   Otherwise,

$$\int_{-1}^1 dx/(1 + \alpha^2 - 2\alpha x)^{1/2} = (-1/\alpha)(1 + \alpha^2 - 2\alpha x)^{1/2} \Big|_{-1}^1$$

$$= (1/\alpha)(1 + \alpha^2 + 2\alpha)^{1/2} - (1/\alpha)(1 + \alpha^2 - 2\alpha)$$

$$= (1/\alpha)[(1 + \alpha)^2]^{1/2} - (1/\alpha)[(1 - \alpha)^2]^{1/2}$$

$$= (1/\alpha)|1 + \alpha| - (1/\alpha)|1 - \alpha|$$

$$= \begin{cases} (1/\alpha)(1 + \alpha) - (1/\alpha)(1 - \alpha) = 2 \text{ if } |\alpha| \le 1 \\ (1/\alpha)(1 + \alpha) - (1/\alpha)(\alpha - 1) = 2/\alpha \text{ if } \alpha > 1 \\ -(1/\alpha)(1 + \alpha) - (1/\alpha)(1 - \alpha) = -2/\alpha \text{ if } \alpha < -1 \end{cases}$$

6-10 Integration by parts gives

$$U_n = \int_0^{\pi/2} \sin^n x\,dx = \int_0^{\pi/2} \sin^{n-1} x \sin x\,dx$$

$$= -\sin^{n-1} x \cos x \Big|_0^{\pi/2} + (n-1)\int_0^{\pi/2} \sin^{n-2} x \cos^2 x\,dx$$

$$= (n-1)\int_0^{\pi/2} \sin^{n-2} x\,dx - (n-1)\int_0^{\pi/2} \sin^n x\,dx = (n-1)U_{n-2} - (n-1)U_n$$

Thus $nU_n = (n-1)U_{n-2}$.   Hence if n is an odd integer,

$$U_n = \frac{n-1}{n} U_{n-2} = \frac{n-1}{n}\frac{n-3}{n-2} U_{n-4} = \cdots = \frac{n-1}{n}\frac{n-3}{n-2} \cdots \frac{6}{7}\frac{4}{5}\frac{2}{3} \int_0^{\pi/2} \sin x\,dx$$

$$= \frac{n-1}{n}\frac{n-3}{n-2} \cdots \frac{6}{7}\frac{4}{5}\frac{2}{3}$$

while if n is an even integer,

$$U_n = \frac{n-1}{4}\frac{n-3}{n-2}\cdots\frac{5}{6}\frac{3}{4}\frac{1}{2}\int_0^{\pi/2}dx = \frac{n-1}{n}\frac{n-3}{n-2}\cdots\frac{5}{6}\frac{3}{4}\frac{1}{2}\frac{\pi}{2}$$

Remark This problem has some interesting consequences. We start with

$$\frac{\pi}{2} = \frac{U_{2m}}{U_{2m+1}}\frac{2\cdot2\,4\cdot4}{1\cdot3\,3\cdot5}\cdots\frac{2m\cdot2m}{(2m-1)(2m+1)}$$

for $m = 1, 2, \ldots$. Now for $0 < x < \pi/2$, $0 < \sin^{2m+1}x \le \sin^{2m}x \le \sin^{2m-1}x$, and so, integrating this inequality from 0 to $\pi/2$, $0 < U_{2m+1} \le U_{2m} \le U_{2m-1}$, or

$$1 \le \frac{U_{2m}}{U_{2m+1}} \le \frac{U_{2m-1}}{U_{2m+1}} = 1 + \frac{1}{2m}$$

Thus $\{U_{2m}/U_{2m+1}\} \to 1$, and we have *Wallis' product*

$$\left\{\frac{2}{1}\cdot\frac{2}{3}\cdot\frac{4}{3}\cdot\frac{4}{5}\cdots\frac{2m}{2m-1}\frac{2m}{2m+1}\right\} \to \frac{\pi}{2}$$

A variant of this, which will be useful later, can be obtained by noting that $\{2m/(2m+1)\} \to 1$. Hence

$$\left\{\frac{2^2}{3^2}\frac{4^2}{5^2}\cdots\frac{(2m-2)^2}{(2m-1)^2}2m\right\} \to \frac{\pi}{2}$$

or

$$\left\{\frac{2}{3}\frac{4}{5}\cdots\frac{2m-2}{2m-1}(2m)^{1/2}\right\} \to \left(\frac{\pi}{2}\right)^{1/2}$$

or

$$\left\{\frac{2^2\cdot4^2\cdots(2m-2)^2}{(2m-1)!}(2m)^{1/2}\right\} \to \left(\frac{\pi}{2}\right)^{1/2}$$

or

$$\left\{\frac{2^2\cdot4^2\cdots(2m)^2}{(2m)!}\frac{(2m)^{1/2}}{2m}\right\} \to \left(\frac{\pi}{2}\right)^{1/2}$$

From the last relation we finally get

$$\left\{\frac{(m!)^2\,2^{2m}}{(2m)!\,m^{1/2}}\right\} \to \pi^{1/2}$$

It will be observed that we have evaluated the limit in Problem 2-8b in passing.

6-11   If $0 < x < 1/2$, then $(1 - x^{2n})^{1/2} < 1$, and $(1 - x^{2n})^{1/2} > (1 - x^2)^{1/2}$. Thus $1 < 1/(1 - x^{2n})^{1/2} < 1/(1 - x^2)^{1/2}$, and so $1/2 = \int_0^{1/2}dx < \int_0^{1/2}dx/(1 - x^{2n})^{1/2} < \int_0^{1/2}dx/(1 - x^2)^{1/2} = \sin^{-1}(1/2) = \pi/6 < 0.524$.

6-12 If $0 < x < \phi < \pi/2$, then $(1 - \sin^2 \alpha \sin^2 \phi)^{1/2} < (1 - \sin^2 \alpha \sin^2 x)^{1/2} < 1$, or $1 < 1/(1 - \sin^2 \alpha \sin^2 x) < 1/(1 - \sin^2 \alpha \sin^2 \phi)^{1/2}$. Hence

$$\phi = \int_0^\phi dx < \int_0^\phi dx/(1 - \sin^2 \alpha \sin^2 \phi)^{1/2} < [1/(1 - \sin^2 \alpha \sin^2 \phi)^{1/2}] \int_0^\phi dx$$

$$= \phi/(1 - \sin^2 \alpha \sin^2 \phi)^{1/2}$$

6-13 As $x \to \infty$, $\int_0^x e^{t^2} \, dt / e^{x^2}$ is an ∞/∞ indeterminate form. By L'Hospital's rule then

$$\lim_{x \to \infty} \frac{\int_0^x e^{t^2} \, dt}{e^{x^2}} = \lim_{x \to \infty} \frac{e^{x^2}}{2xe^{x^2}} = \lim_{x \to \infty} \frac{1}{2x} = 0$$

6-14 If $|\alpha| \neq |\beta|$, $\int_0^T \sin \alpha x \sin \beta \, x dx = (1/2) \int_0^T [\cos (\alpha - \beta)x - \cos (\alpha + \beta)x] \, dx$

$$= \frac{1}{2(\alpha - \beta)} \sin (\alpha - \beta)x - \frac{1}{2(\alpha + \beta)} \sin (\alpha + \beta)x \Big|_0^T$$

$$= \frac{1}{2(\alpha - \beta)} \sin (\alpha - \beta)T - \frac{1}{2(\alpha + \beta)} \sin (\alpha + \beta)T$$

and so $\lim_{T \to \infty} (1/T) \int_0^T \sin \alpha x \sin \beta x \, dx = 0$ if $|\alpha| \neq |\beta|$. If $|\alpha| = |\beta| \neq 0$,

$$\int_0^T \sin \alpha x \sin \beta x \, dx = \begin{cases} \int_0^T \sin^2 |\alpha| x \, dx \text{ if } \alpha\beta > 0 \\ -\int_0^T \sin^2 |\alpha| x \, dx \text{ if } \alpha\beta < 0 \end{cases}$$

Now $\int_0^T \sin^2 |\alpha| x \, dx = (1/2) \int_0^T [1 - \cos 2 |\alpha| x] \, dx = T/2 - \sin 2 |\alpha| T/4|\alpha|$, and so

$$\lim_{T \to \infty} \frac{1}{T} \int_0^T \sin \alpha x \sin \beta x \, dx = \begin{cases} 1/2 \text{ if } |\alpha| = |\beta| \text{ and } \alpha\beta > 0 \\ -1/2 \text{ if } |\alpha| = |\beta| \text{ and } \alpha\beta < 0 \end{cases}$$

6-15 (a) $\lim_{h \to 0} \int_{-a}^a h/(h^2 + x^2) \, dx = \lim_{h \to 0} [\tan^{-1} (a/h) - \tan^{-1} (-a/h)]$

$$= \pi/2 - (-\pi/2) = \pi$$

(b) We have

$$\int_{-a}^a \frac{h}{h^2 + x^2} f(x) \, dx = \int_{-a}^a \frac{h}{h^2 + x^2} [f(x) - f(0)] \, dx + f(0) \int_{-a}^a \frac{h}{h^2 + x^2} \, dx$$

Thus by part (a),

$$\lim_{h \to 0} \int_{-a}^a \frac{h}{h^2 + x^2} f(x) \, dx = \pi f(0) + \lim_{h \to 0} \int_{-a}^a \frac{h}{h^2 + x^2} [f(x) - f(0)] \, dx$$

Now if we let $x = hy$, then

$$\int_{-a}^{a} \frac{h}{h^2 + x^2} [f(x) - f(0)] \, dx = \int_{-a/h}^{a/h} \frac{1}{1 + y^2} [f(hy) - f(0)] \, dy$$

Because f is continuous on $[-a, a]$, we can find $M > 0$ such that $|f(z) - f(0)| \leq M$ for all z in $[-a, a]$, and given $\varepsilon/2 > 0$ we can find δ such that $0 < \delta < a$ and if $|z| < \delta$, then $|f(z) - f(0)| < \varepsilon/2\pi$. Thus

$$\left| \int_{-a/h}^{a/h} \frac{1}{1 + y^2} [f(hy) - f(0)] \, dy \right| \leq \int_{-\delta/|h|}^{\delta/|h|} \frac{1}{1 + y^2} |f(hy) - f(0)| \, dy$$

$$+ \int_{\delta/|h|}^{a/|h|} \frac{1}{1 + y^2} |f(hy) - f(0)| \, dy + \int_{-a/|h|}^{-\delta/|h|} \frac{1}{1 + y^2} |f(hy) - f(0)| \, dy$$

$$\leq \frac{\varepsilon}{2\pi} \int_{-\delta/|h|}^{-\delta/|h|} \frac{1}{1 + y^2} \, dy + M \int_{\delta/|h|}^{a/|h|} \frac{1}{1 + y^2} \, dy + M \int_{-a/|h|}^{-\delta/|h|} \frac{1}{1 + y^2} \, dy$$

$$= \frac{\varepsilon}{2\pi} \left[\tan^{-1}\left(\frac{\delta}{|h|}\right) - \tan^{-1}\left(\frac{-\delta}{|h|}\right) \right] + 2M \left[\tan^{-1}\left(\frac{a}{|h|}\right) - \tan^{-1}\left(\frac{\delta}{|h|}\right) \right]$$

Finally, we can find $H > 0$ such that if $|h| < H$, then $\tan^{-1}(\delta/h) > \pi/2 - \varepsilon/4M$, or $0 < \tan^{-1}(a/|h|) - \tan^{-1}(\delta/|h|) < \pi/2 - \pi/2 + \varepsilon/4M = \varepsilon/4M$, and so if $|h| < H$,

$$\left| \int_{-a}^{a} \frac{h}{h^2 + x^2} [f(x) - f(0)] \, dx \right| < \frac{\varepsilon}{2\pi} \left[\tan^{-1}\left(\frac{\delta}{|h|}\right) - \tan^{-1}\left(\frac{-\delta}{|h|}\right) \right] + \frac{2M\varepsilon}{4M}$$

$$< \frac{\varepsilon}{2\pi} \pi + \frac{\varepsilon}{2} = \varepsilon$$

In other words,

$$\lim_{h \to 0} \int_{-a}^{a} \frac{h}{h^2 + x^2} [f(x) - f(0)] \, dx = 0$$

and hence

$$\lim_{h \to 0} \int_{-a}^{a} \frac{h}{h^2 + x^2} f(x) \, dx = \pi f(0)$$

Remark The technique of dividing the range of integration and using different methods on the different subranges resulting, as in part (b), is frequently useful and, indeed, necessary.

6-16 (a) Suppose that $\int_{0}^{\infty} f(x) \, dx$ does not converge. Since, from Problem 6-8b, we have $\left| \int_{0}^{t} f(x) \, dx \right| \leq \int_{0}^{t} |f(x)| \, dx \leq \int_{0}^{\infty} |f(x)| \, dx < \infty$, $\int_{0}^{t} f(x) \, dx$ must oscillate, and so we can find sequences $\{a_n\}$ and $\{b_n\}$ such that $\{a_n\} \to \infty$, $\{b_n\} \to \infty$, $a_n < b_n < a_{n+1}$ for $n = 1, 2, \ldots$, and $\left\{ \int_{0}^{a_n} f(x) \, dx \right\} \to a$,

$\left\{ \int_0^{b_n} f(x)\,dx \right\} \to b$, with $a \neq b$. Let $\delta > 0$ be such that $|a - b| - \delta = \alpha > 0$.

Since $\left\{ \left| \int_0^{a_n} f(x)\,dx - \int_0^{b_n} f(x)\,dx \right| \right\} \to |a - b|$, we can find N_1 such that for all $n > N_1$ we have $\alpha < \left| \int_0^{a_n} f(x)\,dx - \int_0^{b_n} f(x)\,dx \right|$. On the other hand, inasmuch as $\left\{ \int_0^{a_n} |f(x)|\,dx \right\} \to \int_0^{\infty} |f(x)|\,dx$, we can find N_2 such that for all $n > N_2$ we have $\int_{a_n}^{\infty} |f(x)|\,dx < \alpha$, and so $\int_{a_n}^{b_n} |f(x)|\,dx < \alpha$ as well for all $n > N_2$. But then if $n >$ the larger of N_1 and N_2, we have

$$\alpha < \left| \int_{a_n}^{b_n} f(x)\,dx \right| \le \int_{a_n}^{b_n} |f(x)|\,dx < \alpha$$

This contradiction means that $\int_0^{\infty} f(x)\,dx$ indeed converges.

(b) For all t we have, from Problem 6-8a, $\int_0^t f(x)\,dx \le \int_0^t g(x)\,dx$. But $\int_0^t g(x)\,dx \le \int_0^{\infty} g(x)\,dx$, and if $t_1 < t_2$, then $\int_0^{t_2} f(x)\,dx - \int_0^{t_1} f(x)\,dx = \int_{t_1}^{t_2} f(x)\,dx \ge 0$. Thus $\int_0^t f(x)\,dx$ is, as a function of t, bounded and increasing. Hence $\lim_{t \to \infty} \int_0^t f(x)\,dx$ exists, or $\int_0^{\infty} f(x)\,dx$ converges.

6-17 (a) Since secant is discontinuous at $\pi/2$,

$$\int_0^{\pi/2} \sec x\,dx = \lim_{t \to \pi/2 -} \int_0^t \sec x\,dx = \lim_{t \to \pi/2 -} \log |\sec x + \tan x| \Big|_0^t$$

$$= \lim_{t \to \pi/2 -} \log |\sec t + \tan t| = \log | \lim_{t \to \pi/2 -} (\sec t + \tan t)|$$

and this limit does not exist. The integral is divergent.

(b) Inasmuch as log 0 does not exist,

$$\int_0^1 x \log x\,dx = \lim_{t \to 0+} \int_t^1 x \log x\,dx = \lim_{t \to 0+} \left[(x^2/2) \log x \Big|_t^1 - (1/2) \int_t^1 x\,dx \right]$$

$$= \lim_{t \to 0+} [(x^2/2) \log x - x^2/4] \Big|_t^1 = -1/4 - \lim_{t \to 0+} (t^2/2) \log t$$

Now $\lim_{t \to 0+} (t^2/2) \log t = 0$ (see Problem 4-46a), and so $\int_0^1 x \log x\,dx = -1/4$.

(c) Here the integrand is discontinuous at a, and so

$$\int_0^a dx/(a^2 - x^2)^{1/2} = \lim_{t \to a-} \int_0^t dx/(a^2 - x^2)^{1/2} = \lim_{t \to a-} \sin^{-1}(x/a) \Big|_0^t$$

$$= \lim_{t \to a-} \sin^{-1}(t/a) = \pi/2$$

(d)   Here we have $\int_0^1 dx/(e^x - e^{-x}) = \lim_{t \to 0+} \int_t^1 dx/(e^x - e^{-x})$. Now if we let $x = \log u$, then

$$\int_t^1 dx/(e^x - e^{-x}) = \int_t^1 e^x/(e^{2x} - 1)\, dx = \int_{e^t}^e [u/(u^2 - 1)u]\, du$$

$$= \tfrac{1}{2} \log |(u - 1)/(u + 1)| \Big|_{e^t}^e$$

$$= \tfrac{1}{2} \log [(e - 1)/(e + 1)] - \tfrac{1}{2} \log [(e^t - 1)/(e^t + 1)]$$

Hence

$$\int_0^1 dx/(e^x - e^{-x}) = \tfrac{1}{2} \log [(e - 1)/(e + 1)] - \tfrac{1}{2} \lim_{t \to 0+} \log [(e^t - 1)/(e^t + 1)]$$

and since the limit does not exist, the integral is divergent.

6-18   (a)   We have

$$\int_0^\infty e^{-x} \sin x\, dx = \lim_{t \to \infty} \int_0^t e^{-x} \sin x\, dx = \lim_{t \to \infty} [(-1/2)][e^{-x} \cos x + e^{-x} \sin x] \Big|_0^t$$

$$= 1/2 - (1/2) \lim_{t \to \infty} [e^{-t} \cos t + e^{-t} \sin t] = 1/2$$

(b)   Here

$$\int_0^\infty [x/(1 + x^2)]\, dx = \lim_{t \to \infty} \int_0^t [x/(1 + x^2)]\, dx = \lim_{t \to \infty} (1/2) \log (1 + x^2) \Big|_0^t$$

$$= \lim_{t \to \infty} (1/2) \log (1 + t^2)$$

and this limit does not exist.   The integral does not exist either.

(c) $\int_0^\infty dx/(a^2 + b^2 x^2) = \lim_{t \to \infty} (1/b^2) \int_0^t dx\, /[(a/b)^2 + x^2]$

$$= \lim_{t \to \infty} (1/b^2)(b/a) \tan^{-1} [(b/a)x] \Big|_0^t = (1/ab) \lim_{t \to \infty} \tan^{-1} [(b/a)t] = \pi/2ab$$

(d) $\int_e^\infty dx/x \log x = \lim_{t \to \infty} \int_e^t dx/x \log x = \lim_{t \to \infty} \log (\log x) \Big|_e^t = \lim_{t \to \infty} \log (\log t)$
Since $\lim_{t \to \infty} \log (\log t)$ does not exist, the integral is divergent.

6-19   (a)   Since the integrand is discontinuous in the interior of $[0, 3a]$, we have

$$\int_0^{3a} [2x/(x^2 - a^2)^{2/3}]\, dx$$

$$= \lim_{t \to a-} \int_0^t [2x/(x^2 - a^2)^{2/3}]\, dx + \lim_{u \to a+} \int_u^{3a} [2x/(x^2 - a^2)^{2/3}]\, dx$$

$$= \lim_{t \to a-} 3(x^2 - a^2)^{1/2} \Big|_0^t + \lim_{u \to a+} 3(x^2 - a^2)^{1/3} \Big|_u^{3a}$$

$$= \lim_{t \to a-} 3(t^2 - a^2)^{1/3} - 3(-a^2)^{1/3} + 3(9a^2 - a^2)^{1/3} - \lim_{u \to a+} 3(u^2 - a^2)^{1/3}$$

$$= 3a^{2/3} + 6a^{2/3} = 9a^{2/3}$$

(b)   The integrand being discontinuous at a, we have

$$\int_0^{2a} dx/(a-x)^2 = \lim_{t \to a-} \int_0^t dx/(a-x)^2 + \lim_{u \to a+} \int_u^{2a} dx/(a-x)^2$$

$$= \lim_{t \to a-} 1/(a-x) \Big|_0^t + \lim_{u \to a+} 1/(a-x) \Big|_u^{2a} = \lim_{t \to a-} 1/(a-t) - 1/a$$

$$+ 1/a - \lim_{u \to a+} 1/(a-u)$$

Since neither limit exists, the integral is divergent.

6-20   (a)   Inasmuch as the integrand is discontinuous at 0, we have

$$\int_0^\infty dx/(x+1)x^{1/2} = \lim_{t \to 0+} \int_t^1 dx/(x+1)x^{1/2} + \lim_{u \to \infty} \int_1^u dx/(x+1)x^{1/2}$$

Let $v^2 = x$, or $dx = 2v\, dv$.   Then

$$\int dx/(x+1)x^{1/2} = \int [2v/(v^2+1)v]\, dv = 2 \tan^{-1} v = 2 \tan^{-1} (x^{1/2})$$

Thus

$$\lim_{t \to 0+} \int_t^1 dx/(1+x)x^{1/2} = \lim_{t \to 0+} 2 \tan^{-1} (x^{1/2}) \Big|_t^1 = \pi/2 - 2 \lim_{t \to 0+} \tan^{-1} (t^{1/2})$$

$$= \pi/2, \text{ and } \lim_{u \to \infty} \int_1^u dx/(1+x)x^{1/2} = \lim_{u \to \infty} 2 \tan^{-1} (x^{1/2}) \Big|_1^u$$

$$= 2 \lim_{u \to \infty} \tan^{-1} (u^{1/2}) - \pi/2 = \pi/2. \quad \text{Therefore,}$$

$$\int_0^\infty dx/(1+x)x^{1/2} = \pi$$

(b)   Here

$$\int_a^\infty [x/(x^2-a^2)^{1/2}]\, dx = \lim_{t \to a+} \int_t^b [x/(x^2-a^2)^{1/2}]\, dx + \lim_{u \to \infty} \int_b^u [x/(x^2-a^2)^{1/2}]\, dx$$

$$= \lim_{t \to a+} (x^2-a^2)^{1/2} \Big|_t^b + \lim_{u \to \infty} (x^2-a^2)^{1/2} \Big|_b^u = (b^2-a^2)^{1/2}$$

$$- \lim_{t \to a+} (t^2-a^2)^{1/2} + \lim_{u \to \infty} (u^2-a^2)^{1/2} - (b^2-a^2)^{1/2}$$

and inasmuch as $\lim_{u \to \infty} (u^2-a^2)^{1/2}$ does not exist, the integral diverges.

6-21   (a) $\int_{-\infty}^\infty \sin x\, dx = \lim_{t \to -\infty} \int_t^0 \sin x\, dx + \lim_{u \to \infty} \int_0^u \sin x\, dx = \lim_{t \to -\infty} (-\cos x) \Big|_t^0$

$$+ \lim_{u \to \infty} (-\cos x) \Big|_0^u = -1 + \lim_{t \to -\infty} \cos t - \lim_{u \to \infty} \cos u + 1$$

Since neither limit exists, the integral is divergent.

(b) If we let $x = \log u$, then

$$\int dx/(e^x + e^{-x}) = \int [e^x/(e^{2x} + 1)]\, dx$$

$$= \int [u/(u^2 + 1)u]\, du = \tan^{-1} u = \tan^{-1} e^x$$

Thus

$$\int_{-\infty}^{\infty} dx/(e^x + e^{-x}) = \lim_{t \to -\infty} \int_{t}^{0} dx/(e^x + e^{-x}) + \lim_{v \to \infty} \int_{0}^{v} dx/(e^x + e^{-x})$$

$$= \lim_{t \to -\infty} \tan^{-1} e^x \Big|_{t}^{0} + \lim_{v \to \infty} \tan^{-1} e^x \Big|_{0}^{v} = \pi/4 - \lim_{t \to -\infty} \tan^{-1} e^t$$

$$+ \lim_{v \to \infty} \tan^{-1} e^v - \pi/4 = \pi/2$$

6-22 (a) The given reasoning is faulty because it ignores the discontinuity of the integrand at 0. The correct statement is

$$\int_{-1}^{1} dx/x = \lim_{t \to 0-} \int_{-1}^{t} dx/x + \lim_{u \to 0+} \int_{u}^{1} dx/x = \lim_{t \to 0-} \log |x| \Big|_{-1}^{t} + \lim_{u \to 0+} \log |x| \Big|_{u}^{1}$$

$$= \lim_{t \to 0-} \log |t| - \lim_{u \to 0+} \log |u|$$

and so, since neither of these limits exist, the integral diverges.

(b) Here the fallacy lies in the fact that t and $-t$ do not approach ∞ and $-\infty$ independently of one another. The statement should be

$$\int_{-\infty}^{\infty} [4x^3/(1 + x^4)]\, dx = \lim_{t \to \infty} \int_{0}^{t} [4x^3/(1 + x^4)]\, dx + \lim_{u \to -\infty} \int_{u}^{0} [4x^3/(1 + x^4)]\, dx$$

$$= \lim_{t \to \infty} \log (1 + x^4) \Big|_{0}^{t} + \lim_{u \to -\infty} \log (1 + x^4) \Big|_{u}^{0}$$

$$= \lim_{t \to \infty} \log (1 + t^4) - \lim_{u \to -\infty} \log (1 + u^4)$$

and neither limit exists. The integral diverges.

6-23 We are here interested in

$$\lim_{t \to \infty} \int_{0}^{t} \left(\frac{1}{(1 + ax^2)^{1/2}} - \frac{\alpha}{x + 1} \right) dx$$

Now

$$\int_{0}^{t} [1/(1 + ax^2)^{1/2} - \alpha/(x + 1)]\, dx$$

$$= [(1/a^{1/2}) \log |x + (1/a + x^2)^{1/2}| - \alpha \log |x + 1|] \Big|_{0}^{t}$$

$$= (1/a^{1/2}) \log [t + (1/a + t^2)^{1/2}] - \alpha \log (t + 1) - (1/a^{1/2}) \log (1/a^{1/2})$$

$$= \log\{[t + (1/a + t^2)^{1/2}]1/a^{1/2}/(t + 1)^{\alpha}\} + (1/a^{1/2}) \log a^{1/2}$$

Thus our improper integral converges if and only if

$$\lim_{t \to \infty} \log \left(\frac{[t + (1/a + t^2)^{1/2}]^{1/a^{1/2}}}{(t+1)^\alpha} \right)$$

exists, or if and only if

$$\lim_{t \to \infty} \frac{[t + (1/a + t^2)^{1/2}]^{1/a^{1/2}}}{(t+1)^\alpha}$$

exists and is not zero. If $\alpha > 1/a^{1/2}$, it is easy to see that this last limit is zero, while if $\alpha < 1/a^{1/2}$, it fails to exist. However, when $\alpha = 1/a^{1/2}$, we have

$$\lim_{t \to \infty} \left[\frac{t + (1/a + t^2)^{1/2}}{t+1} \right]^{1/a^{1/2}} = \lim_{t \to \infty} \left[\frac{1 + (1/at^2 + 1)^{1/2}}{1 + 1/t} \right] 1/a^{1/2} = 2^{1/a^{1/2}}$$

Therefore, the integral converges only when $\alpha = 1/a^{1/2}$, and its value is $\log 2^{1/a^{1/2}} + (1/a^{1/2}) \log a^{1/2} = (1/a^{1/2}) \log [2(a)^{1/2}]$.

6-24   (a) $\displaystyle \int_1^\infty dx/x^\alpha = \lim_{t \to \infty} \int_1^t dx/x^\alpha$

$$= \begin{cases} \lim_{t \to \infty} \dfrac{x^{1-\alpha}}{1-\alpha} \bigg|_1^t & \text{if } \alpha \neq 1 \\[3mm] \lim_{t \to \infty} \log x \bigg|_1^t & \text{if } \alpha = 1 \end{cases}$$

Now $\lim_{t \to \infty} \log x \big|_1^t = \lim_{t \to \infty} \log t$ does not exist. The integral thus diverges if $\alpha = 1$. Otherwise, $\lim_{t \to \infty} x^{1-\alpha}/(1-\alpha) \big|_1^t = 1/(\alpha - 1) + \lim_{t \to \infty} t^{1-\alpha}/(1-\alpha)$, and this limit exists only if $1 - \alpha < 0$, or $\alpha > 1$. Hence $\int_1^\infty dx/x^\alpha$ converges only if $\alpha > 1$.

(b) $\displaystyle \int_0^1 dx/x^\alpha = \lim_{t \to 0+} \int_t^1 dx/x^\alpha$

$$= \begin{cases} \lim_{t \to 0+} \dfrac{x^{1-\alpha}}{1-\alpha} \bigg|_t^1 & \text{if } \alpha \neq 1 \\[3mm] \lim_{t \to 0+} \log x \bigg|_t^1 & \text{if } \alpha = 1 \end{cases}$$

Since $\lim_{t \to 0+} \log x \big|_t^1 = - \lim_{t \to 0+} \log t$ does not exist, the integral diverges if $\alpha = 1$. Otherwise, $\lim_{t \to 0+} x^{1-\alpha}/(1-\alpha) \big|_t^1 = 1/(1-\alpha) - \lim_{t \to 0+} t^{1-\alpha}/(1-\alpha)$, and the limit exists only if $1 - \alpha > 0$, or $\alpha < 1$. Therefore, $\int_0^1 dx/x^\alpha$ converges only if $\alpha < 1$.

6-25   By repeated application of L'Hospital's rule, we have, for any integer n

$\lim\limits_{x \to \infty} x^n/e^x = \lim\limits_{x \to \infty} x^{n-1}/e^x = \cdots = \lim\limits_{x \to \infty} n!/e^x = 0.$ Hence for any integer n, we can find $X_n > 1$ such that if $x > X_n$, then $x^{n+2}e^{-x}/x^2 < 1/x^2$.   Since (see the previous problem) $\int_1^\infty dx/x^2$ converges, we have, from Problem 6-16b, that $\int_{X_n}^\infty x^n e^{-x}\, dx$ converges. Inasmuch as the integrand is continuous on $[0, X_n]$, we can conclude that $\int_0^\infty x^n e^{-x}\, dx$ converges for all positive integers n.

Remark   This result shows that $\int_0^\infty x^\alpha e^{-x}\, dx$ converges for any $\alpha > 0$ because there exists an integer $n > \alpha$, and, for $x > 1$, $x^\alpha e^{-x} < x^n e^{-x}$.   Further if $0 < \alpha < 1$, and $x > 1$, then $x^{\alpha - 1}e^{-x} < e^{-x}$, and it is easy to see from the previous problem that $\int_0^1 x^{\alpha - 1}e^{-x}\, dx$ converges.   Hence $\int_0^\infty x^{\alpha - 1}e^{-x}\, dx$ converges for all $\alpha > 0$.   This integral defines the important *gamma function* Γ by putting

$$\Gamma(\alpha) = \int_0^\infty x^{\alpha - 1}e^{-x}\, dx$$

for $\alpha > 0$.   Among the many properties of this interesting function, we point out the following.   An integration by parts gives us

$$\Gamma(\alpha + 1) = \int_0^\infty x^\alpha e^{-x}\, dx = -x^\alpha e^{-x}\Big|_0^\infty + \alpha \int_0^\infty x^{\alpha - 1}e^{-x}\, dx = \alpha \int_0^\infty x^{\alpha - 1}e^{-x}\, dx = \alpha\Gamma(\alpha)$$

We thus have the famous functional equation (see also the remark following the solution of Problem 3-19) of the gamma function

$$\Gamma(\alpha + 1) = \alpha\Gamma(\alpha)$$

If α is a positive integer n, this functional equation implies that $\Gamma(n + 1) = n!$ For an extensive treatment of this function, see the monograph, *The Gamma Function* by Emil Artin.

6-26   Notice first that $\int_0^1 [x^{\alpha - 1}/(1 + x)]\, dx$ diverges if $\alpha \le 0$ because $1/(1 + x) \ge 1/2$ if $0 \le x \le 1$, or $x^{\alpha - 1}/(1 + x) \ge x^{\alpha - 1}/2$, and $\int_0^1 x^{\alpha - 1}\, dx$ diverges if $\alpha \le 0$ by Problem 6-24b.   Further, $\int_1^\infty [x^{\alpha - 1}/(1 + x)]\, dx$ diverges if $\alpha \ge 1$ since then $x^{\alpha - 1}/(1 + x) \ge 1/(1 + x)$ for $x \ge 1$, and $\int_1^\infty dx/(x + 1)$ clearly diverges.   We can thus concentrate on $0 < \alpha < 1$.   For $0 \le x \le 1$, $x^{\alpha - 1}/(1 + x) \le x^{\alpha - 1}$, and $\int_0^1 x^{\alpha - 1}\, dx$ converges if $0 < \alpha < 1$ by Problem 6-24b again.   Next, for $x \ge 1$ we have $x^{\alpha - 1}/(1 + x) = [x/(1 + x)]x^{\alpha - 2} < x^{\alpha - 2}$, and since $\int_1^\infty x^{\alpha - 2}\, dx$ converges if $0 < \alpha < 1$ by Problem 6-24a, it follows that $\int_1^\infty [x^{\alpha - 1}/(1 + x)]\, dx$ converges if $0 < \alpha < 1$.   Thus our improper integral converges if and only if $0 < \alpha < 1$.

6-27   Observe that the question is meaningful because for any $\alpha > 0$ the integral is convergent.   This is true because of the inequality $1/(1 + \alpha x^{1+\delta}) < 1/\alpha x^{1+\delta}$, and

Problem 6-24a. Now given $\varepsilon/2 > 0$, we have

$$\int_0^{\varepsilon/2} dx/(1 + \alpha x^{1/\delta}) < \int_0^{\varepsilon/2} dx = \varepsilon/2$$

Next,

$$\int_{\varepsilon/2}^\infty dx/(1 + \alpha x^{1+\delta}) < (1/\alpha) \int_{\varepsilon/2}^\infty dx/x^{1+\delta} = (1/\alpha\delta)(2/\varepsilon)^\delta$$

and so if $\alpha > (1/\delta)(2/\varepsilon)^{\delta+1} = \alpha_0$, then $\int_{\varepsilon/2}^\infty dx/(1 + \alpha x^{1+\delta}) < \varepsilon/2$. Hence if $\alpha > \alpha_0$, then $\int_0^\infty dx/(1 + \alpha x^{1+\delta}) < \varepsilon$, or $\lim_{\alpha \to \infty} \int_0^\infty dx/(1 + \alpha x^{1+\delta}) = 0$.

6-28 The convergence of this integral depends on that of $\int_{\pi/2}^\infty [\sin x/x^{1/2}]\, dx$ since it follows from the inequality $0 \le \sin x/x^{1/2} \le 1/x^{1/2}$ if $0 \le x \le \pi/2$ and Problem 6-24b that $\int_0^{\pi/2} [\sin x/x^{1/2}]\, dx$ converges. Now if we integrate by parts,

$$\int_{\pi/2}^t \frac{\sin x}{x^{1/2}}\, dx = -\frac{\cos x}{x^{1/2}}\Big|_{\pi/2}^t - \frac{1}{2}\int_{\pi/2}^t \frac{\cos x}{x^{3/2}}\, dx = -\cos t/t^{1/2} - (1/2)\int_{\pi/2}^t \cos x/x^{3/2}\, dx$$

Inasmuch as $\lim_{t \to \infty} \cos t/t^{1/2} = 0$, we see that $\int_{\pi/2}^\infty [\sin x/x^{1/2}]\, dx$ converges if $\int_{\pi/2}^\infty [\cos x/x^{3/2}]\, dx$ does. But $|\cos x| \le 1$, and so it follows from Problem 6-24b that $\int_{\pi/2}^\infty [\cos x/x^{3/2}]\, dx$ converges.

Remark It can be shown in a similar way that $\int_0^\infty [\sin x/x^p]dx$ converges for $0 < p < 2$. If $p = 0$, the integral clearly diverges. If $p > 2$, then, as above, $\int_{\pi/2}^\infty [\sin x/x^p]\, dx$ converges, but $\int_0^{\pi/2} [\sin x/x^p]\, dx$ converges only if $p < 2$. The last assertion is true because inasmuch as $\lim_{x \to 0} \sin x/x = 1$, the integral $\int_0^{\pi/2} [\sin x/x^p]\, dx$ behaves as far as convergence is concerned in the same way as $\int_0^{\pi/2} dx/x^{p-1}$, which by Problem 6-24b, converges if $p - 1 < 1$, and diverges if $p - 1 \ge 1$. It can also be shown that $\int_0^\infty [\cos x/x^p]\, dx$ converges for $0 < p < 1$.

Special cases of interest are the *Dirichlet integral* $\int_0^\infty [\sin x/x]\, dx$, important in the theory of Fourier series, and the *Fresnel integrals* $F_1 = (1/2) \int_0^\infty [\sin u/u^{1/2}]\, du$, and $F_2 = (1/2) \int_0^\infty [\cos u/u^{1/2}]\, du$, which occur in optics. The usual forms of F_1 and F_2, obtained by the substitution $u = x^2$, are $F_1 = \int_0^\infty \sin(x^2)\, dx$, and $F_2 = \int_0^\infty \cos(x^2)\, dx$. The Fresnel integrals thus show that an improper infinite integral may converge even if the integrand does not have zero limit at infinity. In fact, the integrand does not even have to be bounded in a convergent infinite integral. For an example, let $t^2 = x$ in the second form of F_2. Then $F_2 = \int_0^\infty 2t \cos(t^4)\, dt$, and

$2t \cos (t^4)$ is not bounded as can be seen by letting $t = (n\pi)^{1/4}$ for $n = 0, 1, 2, \ldots$.

6-29 The two curves intersect at $(0, 0)$ and $((ab^2)^{1/3}, (a^2b)^{1/3})$. For $0 \le x \le (ab^2)^{1/2}$, $(ax)^{1/2} \ge x^2/b$, and so the area between the parabolas is

$$\int_0^{(ab^2)^{1/3}} [(ax)^{1/2} - x^2/b] \, dx = [2(a)^{1/2}/3]x^{3/2} - x^3/3b \Big|_0^{(ab^2)^{1/3}}$$

$$= [2(a)^{1/2}/b]a^{1/2}b - ab^2/3b = ab/3$$

6-30 The equation of the line from $(0, 0)$ to (x_0, y_0) is $y = (y_0/x_0)x$. The area in question is thus

$$\int_0^a (y_0/x_0)x \, dx + \int_a^{x_0} [(y_0/x_0)x - (x^2 - a^2)^{1/2}] \, dx$$

$$= \int_0^{x_0} (y_0/x_0)x \, dx - \int_a^{x_0} (x^2 - a^2)^{1/2} \, dx$$

$$= (y_0/x_0)x^2/2 \Big|_0^{x_0} - (x/2)(x^2 - a^2)^{1/2} + (a^2/2) \log |x + (x^2 - a^2)^{1/2}| \Big|_a^{x_0}$$

$$= x_0 y_0/2 - (x_0/2)(x_0^2 - a^2)^{1/2} + (a^2/2) \log (x_0 + (x_0^2 - a^2)^{1/2})$$

$$- (a^2/2) \log a = (a^2/2) \log (x_0 + y_0) - (a^2/2) \log a$$

$$= (a^2/2) \log [(x_0 + y_0)/a]$$

6-31 The loop is between $x = 0$ and $x = 2$, and the curve is symmetric through the x axis. Thus the area of the loop is $2 \int_0^2 x^{1/2} |x - 2| \, dx = 2 \int_0^2 (2x^{1/2} - x^{3/2}) \, dx = 2[(4/3)x^{3/2} - (2/5)x^{5/2} \Big|_0^2 = 32(2)^{1/2}/15$.

6-32 The loop is between $x = 0$ and $x = 4$, and the curve is symmetric through the x axis. The area of the loop is thus $2 \int_0^4 [x^2(4 - x)^{1/2}/2] \, dx$. Let $x = 4 - y$. Then $\int_0^4 [x^2(4 - x)^{1/2}/2] \, dx = -\int_4^0 (4 - y)^2 y^{1/2} \, dy = \int_0^4 [16y^{1/2} - 8y^{3/2} + y^{5/2}] dy = (32/3)y^{3/2} - (16/5)y^{5/2} + (2/7)y^{7/2} \Big|_0^4 = 2048/105$.

6-33 The area is

$$\int_0^{2\pi} a(1 - \cos \theta)a(1 - \cos \theta) \, d\theta = a^2 \int_0^{2\pi} (1 - \cos \theta)^2 \, d\theta$$

$$= a^2 \int_0^{2\pi} [1 - 2 \cos \theta + \cos^2 \theta] d\theta = 2\pi a^2 + \pi a^2$$

(see Problem 6-4a). $= 3\pi a^2$

6-34 We want to evaluate $\int_0^{2\pi} a(1 - \cos \theta)a \, d\theta = a^2 \int_0^{2\pi} (1 - \cos \theta) \, d\theta = 2\pi a^2$.

6-35 Parametric equations of the ellipse are $x = a \sin \theta$, $y = b \cos \theta$. Since this

ellipse is symmetric through the origin, its area is $4 \int_0^{\pi/2} b \cos\theta a \cos\theta \, d\theta =$
$4ab \int_0^{\pi/2} \cos^2\theta \, d\theta = 2ab \int_0^{\pi/2} (1 + \cos 2\theta) \, d\theta = \pi ab + ab \sin 2\theta \Big|_0^{\pi/2} = \pi ab.$

6-36   Parametric equations of this curve are $x = a \sin^3\theta$, $y = a \cos^3\theta$, and so the area is

$$4 \int_0^{\pi/2} a \cos^3\theta \, 3a \sin^2\theta \cos\theta \, d\theta$$

$$= 12a^2 \int_0^{\pi/2} \cos^4\theta \sin^2\theta \, d\theta = (3a^2/2) \int_0^{\pi/2} (1 + \cos 2\theta)^2 (1 - \cos 2\theta) \, d\theta$$

$$= (3a^2/2) \int_0^{\pi/2} (1 + \cos 2\theta - \cos^2 2\theta - \cos^3 2\theta) \, d\theta$$

$$= 3\pi a^2/4 - (3a^2/2) \int_0^{\pi/2} \cos^2 2\theta \, d\theta + (3a^2/2) \int_0^{\pi/2} \cos 2\theta \, (1 - \cos^2 2\theta) \, d\theta$$

$$= 3\pi a^2/4 - (3a^2/4) \int_0^{\pi/2} (1 + \cos 4\theta) \, d\theta + (3a^2/2) \int_0^{\pi/2} \sin^2 2\theta \cos 2\theta \, d\theta$$

$$= 3\pi a^2/4 - 3\pi a^2/8 - (3a^2/16) \sin 4\theta \Big|_0^{\pi/2} + (a^2/4) \sin^3 2\theta \Big|_0^{\pi/2} = 3\pi a^2/8$$

6-37   The curve is a 3-leaved rose, and its area is

$$(6/2) \int_0^{\pi/6} a^2 \cos^2 3\theta \, d\theta = (3a^2/2) \int_0^{\pi/6} (1 + \cos 6\theta) \, d\theta = \pi a^2/4$$

6-38   The area is $2(1/2) \int_0^{\pi} a^2 (1 + \cos\theta)^2 \, d\theta = a^2 \int_0^{\pi} [1 + 2\cos\theta + \cos^2\theta] \, d\theta =$
$\pi a^2 + 2a^2 \sin\theta \Big|_0^{\pi} + (a^2/2) \int_0^{\pi} (1 + \cos 2\theta) \, d\theta = 3\pi a^2/2.$

6-39   The curves are leminscates, one with axis $\theta = 0$, the other with axis $\theta = \pi/4$, and they intersect on the line $\theta = \pi/8$. Hence the area in question is

$$2(1/2) \int_0^{\pi/8} \sin 2\theta \, d\theta + 2(1/2) \int_{\pi/8}^{\pi/4} \cos 2\theta \, d\theta = 1 - 1/2^{1/2}$$

6-40   (a)   The area is $(1/2) \int_0^{2\pi} a^2\theta^2 \, d\theta = 4\pi^3 a^2/3.$

(b)   The additional area is $(a^2/2) \int_{2\pi}^{4\pi} \theta^2 \, d\theta - 4\pi^3 a^2/3 = 8\pi^3 a^2.$

(c)   The additional area is $(a^2/2) \int_{4\pi}^{6\pi} \theta^2 \, d\theta - 28\pi^3 a^2/3 = 16\pi^3 a^2.$

6-41   Inasmuch as the ellipse is symmetric through the x axis, the volume is

$$2\pi \int_0^{a} b^2 (1 - x^2/a^2) \, dx = 4\pi ab^2/3$$

6-42 Due to symmetry through the y axis, the volume is

$$2\pi \int_0^a (a^{2/3} - x^{2/3})^3 \, dx = 2\pi \int_0^a [a^2 - 3a^{4/3}x^{2/3} + 3a^{2/3}x^{4/3} - x^2] \, dx = 32\pi a^2/105$$

6-43 We want to evaluate

$$\pi \int_0^{2\pi} a^2(1 - \cos \theta)^2 a(1 - \cos \theta) \, d\theta = \pi a^3 \int_0^{2\pi} (1 - \cos \theta)^3 \, d\theta$$

$$= \pi a^3 \int_0^{2\pi} [1 - 3\cos \theta + 3\cos^2 \theta - \cos^3 \theta] \, d\theta$$

$$= 2\pi^2 a^3 - 3\pi a^2 \int_0^{2\pi} \cos \theta \, d\theta + (3\pi a^3/2) \int_0^{2\pi} (1 + \cos 2\theta) d\theta$$

$$- \pi a^3 \int_0^{2\pi} \cos \theta(1 - \sin^2 \theta) \, d\theta = 5\pi^2 a^3$$

6-44 The parabolas intersect when $x = 1$, $y = \pm 2$. Thus the volume about the x axis is $\pi \int_0^1 4x \, dx + \pi \int_1^5 (5 - x) \, dx = 10\pi$. Also, the parabolas are symmetric through the x axis. Therefore, the volume about the y axis is

$$2\pi \int_0^2 [(5 - y^2)^2 - y^4/16] \, dy = 2\pi \int_0^2 [25 - 10y^2 + 15y^4/16] \, dy = 176\pi/3$$

6-45 The volume here is given by

$$2\pi \int_0^{1/2^{1/2}} xe^{-x^2} \, dx = 2\pi(-e^{-x^2}/2 \Big|_0^{1/2^{1/2}} = \pi[1 - 1/e^{1/2}]$$

6-46 The solid can be generated by rotating the part of the $x^2 + y^2 = R^2$ for which $r \le x \le R$ about the y axis. The volume is thus

$$4\pi \int_r^R x(R^2 - x^2)^{1/2} \, dx = (-4\pi/3)(R^2 - x^2)^{3/2} \Big|_r^R = (4\pi/3)(R^2 - r^2)^{3/2}$$

6-47 If the equation of the circle is $(x - b)^2 + y^2 = a^2$, the volume V in question is $V = 4\pi \int_{b-a}^{b+a} x[a^2 - (x - b)^2]^{1/2} \, dx$. Make the change of variable $x - b = y$. Then

$$V = 4\pi \int_{-a}^a (y + b)(a^2 - y^2)^{1/2} \, dy$$

$$= 4\pi \int_{-a}^a y(a^2 - y^2)^{1/2} \, dy + 4\pi b \int_{-a}^a (a^2 - y^2)^{1/2} \, dy$$

$$= (-4\pi/3)(a^2 - y^2)^{3/2} \Big|_{-a}^a + 4\pi b[y(a^2 - y^2)^{1/2}/2 + a^2 \sin^{-1} (y/a)/2 \Big|_{-a}^a$$

$$= 2\pi^2 a^2 b$$

6-48 Here we want to evaluate

$$2\pi \int_0^{2\pi} a(\theta - \sin \theta) \, a(1 - \cos \theta) a(1 - \cos \theta) \, d\theta$$

$$= 2\pi a^3 \int_0^{2\pi} (\theta - \sin \theta)(1 - 2 \cos \theta + \cos^2 \theta) \, d\theta$$

$$= 2\pi a^3 \int_0^{2\pi} [\theta - 2\theta \cos \theta + \theta\cos^2 \theta - \sin \theta + 2 \sin \theta \cos \theta - \sin \theta \cos^2 \theta] \, d\theta$$

$$= 4\pi^3 a^3 + 2\pi a^3 \cos \theta \Big|_0^{2\pi} + (2\pi a^3/3) \cos^3 \theta \Big|_0^{2\pi}$$

$$+ 2\pi a^3 \int_0^{2\pi} \sin 2\theta \, d\theta - 4\pi a^3 \int_0^{2\pi} \theta \cos \theta \, d\theta + 2\pi a^3 \int_0^{2\pi} \theta \cos^2 \theta \, d\theta$$

Now the second and third expressions are zero,

$$\int_0^{2\pi} \sin 2\theta \, d\theta = (-1/2) \cos 2\theta \Big|_0^{2\pi} = 0$$

$$\int_0^{2\pi} \theta \cos \theta \, d\theta = \theta \sin \theta \Big|_0^{2\pi} - \int_0^{2\pi} \sin \theta \, d\theta = 0$$

and $\int_0^{2\pi} \theta \cos^2 \theta \, d\theta = (1/2) \int_0^{2\pi} (\theta + \theta \cos 2\theta) \, d\theta$

$$= \pi^2 + \theta \sin 2\theta/4 \Big|_0^{2\pi} - (1/4) \int_0^{2\pi} \sin 2\theta \, d\theta = \pi^2$$

Therefore, the volume is $4\pi^3 a^3 + 2\pi^3 a^3 = 6\pi^3 a^3$.

6-49 Let the equation of the intersection of one cylinder with the x-y plane be $x^2 + y^2 = r^2$. The crosssection of our solid by a plane perpendicular to the x-y plane and the x axis is a square of side $2(r^2 - x^2)^{1/2}$ and area $4(r^2 - x^2)$. The volume sought is thus $4 \int_{-r}^{r} (r^2 - x^2) \, dx = 16r^3/3$.

6-50 The base of this solid can be taken to be the semicircle whose boundaries are the x axis and the curve $y = (r^2 - x^2)^{1/2}$. A cross section by a plane perpendicular to the xy plane and the x axis is a right triangle of base $(r^2 - x^2)^{1/2}$ and height $(r^2 - x^2)^{1/2} \tan \theta$. Thus the cross-sectional area is $\tan \theta \, (r^2 - x^2)/2$, and the volume in question is $(\tan \theta/2) \int_{-r}^{r} (r^2 - x^2) \, dx = 2r^3 \tan \theta/3$.

6-51 The length is

$$\int_0^{\pi/3} [1 + (-\sin x/\cos x)^2]^{1/2} \, dx = \int_0^{\pi/3} [1 + \tan^2 x]^{1/2} \, dx$$

$$= \int_0^{\pi/3} \sec x \, dx = \log |\sec x + \tan x| \Big|_0^{\pi/3} = \log (2 + 3^{1/2})$$

6-52   Here the length is

$$\int_0^p [1 + (k/p)^2]^{1/2}\, dx = (1/p)\int_0^p [p^2 + x^2]^{1/2}\, dx$$

$$= (1/p)[(x/2)[p^2 + x^2]^{1/2} + p^2 \log |x + [p^2 + x^2]^{1/2}|/2\,\Big|_0^p$$

$$= p/2^{1/2} + p \log (1 + 2^{1/2})/2$$

6-53   By implicit differentiation, we have $dy/dx = -(b/a)^{2/3}(y/x)^{1/3}$, and so

$$\left[1 + \left(\frac{dy}{dx}\right)^2\right]^{1/2} = \left[1 + \left(\frac{b}{a}\right)^{4/3}\frac{y^{2/3}}{x^{2/3}}\right]^{1/2} = \frac{a^{4/3}x^{2/3} + b^2(y/b)^{2/3}}{a^{2/3}x^{1/3}}$$

$$= \frac{(a^2(x/a)^{2/3} + b^2[1 - (x/a)^{2/3}])^{1/2}}{a^{2/3}x^{1/3}}$$

$$= \frac{[b^2 + (a^2 - b^2)(x/a)^{2/3}]^{1/2}}{a^{2/3}x^{1/2}}$$

The length L we want is then

$$L = \int_0^a ([b^2 + (a^2 - b^2)(x/a)^{2/3}]^{1/2}/a^{2/3}x^{1/3})\, dx$$

Let $(x/a)^{2/3} = u$.   Then $2/3\, a^{2/3}x^{1/3} = du$, and so

$$L = (3/2)\int_0^1 [b^2 + (a^2 - b^2)u]^{1/2}\, du = [b^2 + (a^2 - b^2)u]^{3/2}/(a^2 - b^2)\,\Big|_0^1$$

$$= [b^2 + (a^2 - b^2)]^{3/2}/(a^2 - b^2) - (b^2)^{3/2}/(a^2 - b^2) = (a^3 - b^3)/(a^2 - b^2)$$

$$= (a^2 + ab + b^2)/(a + b)$$

Notice that when $a = b$ our curve is the hypocycloid $x^{2/3} + y^{2/3} = a^{2/3}$, and the above calculation does not apply.   But in this case, $L = (3/2)\int_0^1 (a^2)^{1/2}\, du = 3a/2$.

6-54   We have, by implicit differentiation, $dy/dx = -2e^x/(e^{2x} - 1)$, and so

$$1 + (dy/dx)^2 = 1 + 4e^{2x}/(e^{2x} - 1)^2$$

$$= [e^{4x} - 2e^{2x} + 1 + 4e^{2x}]/[e^{2x} - 1]^2$$

$$= [(e^{2x} + 1)/(e^{2x} - 1)]^2$$

The length L is thus $L = \int_a^b (e^{2x} + 1)/(e^{2x} - 1)\, dx$.   Put $x = \log u$, $dx = du/u$.   Then

$$L = \int_{e^a}^{e^b} [(u^2 + 1)/u(u^2 - 1)]\, du$$

$$= \int_{e^a}^{e^b} [1/(u + 1) + 1/(u - 1) - 1/u]\, du$$

$$= \log |u + 1| + \log |u - 1| - \log u\,\Big|_{e^a}^{e^b}$$

$$= \log [(e^{2b} - 1)/(e^{2a} - 1)] + a - b$$

6-55 Here we want to find $\int_0^\alpha [a^2(\theta \cos \theta)^2 + a^2(\theta \sin \theta)^2]^{1/2}\, d\theta = \int_0^\alpha a\theta\, d\theta = a\alpha^2/2.$

6-56 The length is

$$\int_0^{2\pi} [a^2(1 - \cos \theta)^2 + a^2 \sin^2 \theta]^{1/2}\, d\theta = a \int_0^{2\pi} [1 - 2 \cos \theta + \cos^2 \theta + \sin^2 \theta]^{1/2}\, d\theta$$

$$= a \int_0^{2\pi} [2(1 - \cos \theta)]^{1/2}\, d\theta$$

$$= a \int_0^{2\pi} [4 \sin^2 (\theta/2)]^{1/2}\, d\theta$$

$$= 2a \int_0^{2\pi} \sin (\theta/2)\, d\theta = -4a[\cos (\theta/2)\Big|_0^{2\pi} = 8a$$

6-57 Since the curve is symmetric about $\theta = 0$, the length is

$$2 \int_0^\pi [a(1 + \cos \theta)^2 + (-a \sin \theta)^2]^{1/2}\, d\theta = 2a \int_0^\pi [1 + 2 \cos \theta + \cos^2 \theta + \sin^2 \theta]^{1/2}\, d\theta$$

$$= 2a \int_0^\pi [2(1 + \cos \theta)]^{1/2}\, d\theta$$

$$= 2a \int_0^\pi [4 \cos^2 (\theta/2)]^{1/2}\, d\theta$$

$$= 4a \int_0^\pi \cos (\theta/2)\, d\theta = 8a$$

6-58 We want to find

$$\int_0^{2\pi} [a^2\theta^2 + a^2]^{1/2}\, d\theta = a \int_0^{2\pi} [1 + \theta^2]^{1/2}\, d\theta$$

$$= a[(\theta/2)[1 + \theta^2]^{1/2} + (1/2) \log |\theta + [1 + \theta^2]^{1/2}|\Big|_0^{2\pi}$$

$$= \pi a[1 + 4\pi^2]^{1/2} + (a/2) \log (2\pi + [1 + 4\pi^2]^{1/2}).$$

6-59 We have $d\theta/d\rho = 1/ae^{a\theta}$, and so the length is

$$\int_0^{\rho_0} \{(e^{a\theta})^2[1/(ae^{a\theta})^2] + 1\}^{1/2}\, d\rho = ([a^2 + 1]^{1/2}/a) \int_0^{\rho_0} d\rho = \rho_0[a^2 + 1]^{1/2}/a$$

Remark For ρ to be zero, we must have $\theta = -\infty$. Hence the spiral must wind about the origin infinitely many times.   Nevertheless, it has finite length.

6-60 The area is

$$2\pi \int_0^{4p} (2px)^{1/2}\{1 + [(p/2x)^{1/2}]^2\}^{1/2}\, dx = 2(2)^{1/2}p^{1/2}\pi \int_0^{4p} [x + p/2]^{1/2}\, dx$$

$$= 2(2p)^{1/2}\pi(2/3)(x + p/2)^{3/2}\Big|_0^{4p} = 52\pi p^2/3$$

6-61   About the x axis the area S is given by

$$S = 2\pi \int_{-a}^{a} y[1 + (-b^2x/a^2y)^2]^{1/2} \, dx = (2\pi b/a) \int_{-a}^{a} (a^2y^2/b^2 + b^2x^2/a^2]^{1/2} \, dx$$

$$= (2\pi b/a) \int_{-a}^{a} [a^2 - a^2x^2/a^2 + b^2x^2/a^2]^{1/2} \, dx$$

$$= (2\pi b/a) \int_{-a}^{a} [a^2 - (a^2 - b^2)x^2/a^2]^{1/2} \, dx$$

Now let $u = [a^2 - b^2]^{1/2}x/a$, $du = ([a^2 - b^2]^{1/2}/a) \, dx$.   Then

$$S = \frac{2\pi b}{[a^2 - b^2]^{1/2}} \int_{-[a^2-b^2]^{1/2}}^{[a^2-b^2]^{1/2}} [a^2 - u^2]^{1/2} \, du$$

$$= \frac{2\pi b}{[a^2 - b^2]^{1/2}} \left[\frac{u}{2} [a^2 - u^2]^{1/2} + \frac{a^2}{2} \sin^{-1} \frac{u}{a} \right]_{-[a^2-b^2]^{1/2}}^{[a^2-b^2]^{1/2}}$$

$$= 2\pi b^2 + [2\pi a^2 b/[a^2 - b^2]^{1/2}] \sin^{-1} ([a^2 - b^2]^{1/2}/a)$$

$[= 2\pi b^2 + (2\pi ab/e) \sin^{-1} e$, where $e = [a^2 - b^2]^{1/2}/a$ is the so-called "eccentricity" of the ellipse].   About the y axis, the area S is given by

$$S = 2\pi \int_{-b}^{b} x[1 + (-a^2y/b^2x)^2]^{1/2} \, dy$$

$$= (2\pi a/b) \int_{-b}^{b} [b^2x^2/a^2 + a^2y^2/b^2]^{1/2} \, dy = (2\pi a/b) \int_{-b}^{b} [b^2 + (a^2 - b^2)y^2/b^2]^{1/2} \, dy$$

Let $[a^2 - b^2]^{1/2}y/b = u$, $([a^2 - b^2]/b) \, dy = du$.   Then

$$S = \frac{2\pi a}{[a^2 - b^2]^{1/2}} \int_{-[a^2-b^2]^{1/2}}^{[a^2-b^2]^{1/2}} [b^2 + u^2]^{1/2} \, du$$

$$= \frac{2\pi a}{[a^2 - b^2]^{1/2}} \left[\frac{u}{2} [b^2 + u^2]^{1/2} + \frac{b^2}{2} \log |u + [b^2 + u^2]^{1/2}| \right]_{-[a^2-b^2]^{1/2}}^{[a^2-b^2]^{1/2}}$$

$$= 2\pi a^2 + \frac{\pi ab^2}{[a^2 - b^2]^{1/2}} [\log (a + [a^2 - b^2]^{1/2}) - \log (a - [a^2 - b^2]^{1/2})]$$

$$= 2\pi a^2 + (\pi b^2/e) \log [(1 + e)/(1 - e)]$$

6-62   The loop is between $x = 0$ and $x = 3a$.   Implicit differentiation gives

$$\frac{dy}{dx} = \frac{(3a - x)^2 - 2x(3a - x)}{6(ax)^{1/2}(3a - x)} = \frac{1}{2(a)^{1/2}} \frac{a - x}{x^{1/2}}$$

Thus the surface area sought is

$$2\pi \int_0^{3a} y[1 + (a - x)^2/4ax]^{1/2}\, dx = 2\pi \int_0^{3a} [x^{1/2}(3a - x)[x^2 + 2ax + a^2]^{1/2}/6a(x)^{1/2}]\, dx$$

$$= (\pi/3a) \int_0^{3a} (3a - x)(a + x)\, dx$$

$$= (\pi/3a) \int_0^{3a} [3a^2 + 2a - x^2]\, dx = 3\pi a^2$$

6-63 The area here is

$$2\pi \int_0^{2\pi} a(1 - \cos\theta)[a^2(1 - \cos\theta)^2 + a^2 \sin^2\theta]^{1/2}\, d\theta$$

$$= 2\pi a^2 \int_0^{2\pi} (1 - \cos\theta)[1 - 2\cos\theta + \cos^2\theta + \sin^2\theta]^{1/2}\, d\theta$$

$$= 2\pi a^2 \int_0^{2\pi} (1 - \cos\theta)[2(1 - \cos\theta)]^{1/2}\, d\theta = 8\pi a^2 \int_0^{2\pi} \sin^3(\theta/2)\, d\theta$$

$$= 8\pi a^2 \int_0^{2\pi} \sin(\theta/2)[1 - \cos^2(\theta/2)]\, d\theta$$

$$= 8\pi a^2 [-2\cos(\theta/2) + (2/3)\cos^3(\theta/2)]\Big|_0^{2\pi} = 64\pi a^2/3$$

6-64 The area is

$$\int_0^1 (1/x^{3/4})\, dx = \lim_{t \to 0+} \int_t^1 (1/x^{3/4})\, dx = \lim_{t \to 0+} 4x^{1/4}\Big|_t^1 = 4$$

On the other hand, the volume is $\pi \int_0^1 (1/x^{3/2})\, dx$ which is divergent by Problem 6-24b. Thus a region of finite area may generate a solid of infinite volume.

6-65 The volume is

$$\pi \int_1^\infty (1/x^2)\, dx = \lim_{t \to \infty} \pi \int_1^t (1/x^2)\, dx = \pi \lim_{t \to \infty} (-1/x)\Big|_1^t = \pi$$

The surface area is $2\pi \int_1^\infty (1/x)[1 + (-1/x)^2]^{1/2}\, dx$. But $[1 + 1/x^4]^{1/2}/x \geq 1/x$, and, by Problem 6-24a, $\int_1^\infty dx/x$ diverges. Hence the surface area of this solid is infinite even though the volume is finite. Further, as opposed to Problem 6-64, the area of the generating region is also infinite.

6-66 The volume is $2\pi \int_0^\infty xe^{-x^2}\, dx = \pi \lim_{t \to \infty} \int_0^t 2xe^{-x^2}\, dx = \pi \lim_{t \to \infty} [-e^{-x^2}\Big|_0^t = \pi$.

6-67 The area is

$$2\pi \int_0^\infty e^{-x}[1 + e^{-2x}]^{1/2}\, dx = 2\pi \lim_{t \to \infty} \int_0^t e^{-x}[1 + e^{-2x}]^{1/2}\, dx$$

Now let $e^x = u$.   Then

$$\int_0^t e^{-x}[1 + e^{-2x}]^{1/2}\, dx = \int_1^{e^t} ([u^2 + 1]^{1/2}/u^3)\, du$$

$$= \left[-\frac{[u^2 + 1]^{1/2}}{2u^2} - \frac{1}{2} \log \left(\frac{1 + [u^2 + 1]^{1/2}}{u} \right) \right] \Bigg|_1^{e^t} = \frac{1}{2^{1/2}} - \frac{\log (1 + 2^{1/2})}{2} - \frac{[e^{2t} + 1]^{1/2}}{2e^{2t}}$$

$$+ \frac{1}{2} \log \left(\frac{1 + [e^{2t} + 1]^{1/2}}{e^t} \right)$$

Now $\lim_{t \to \infty} [e^{2t} + 1]^{1/2}/e^{2t} = 0$, and $\lim_{t \to \infty} \log ([1 + (e^{2t} + 1)^{1/2}]/e^t = \log 1 = 0$.   Hence
$\int_0^\infty e^{-x}[1 + e^{-2x}]^{1/2}\, dx = 1/2^{1/2} + \log (1 + 2^{1/2})/2$,   and   so   the   surface   area   is
$\pi[2^{1/2} + \log (1 + 2^{1/2})]$.

6-68   Since $dy/dx = -y^{1/3}/x^{1/3}$, the area is

$$2(2\pi) \int_0^a y[1 + y^{2/3}/x^{2/3}]^{1/2}\, dx = 4\pi \int_0^a (a^{2/3} - x^{2/3})^{3/2}\, a^{1/3}x^{-1/3}\, dx =$$

$$4\pi a^{1/3} \lim_{t \to 0+} \int_t^a (a^{2/3} - x^{2/3})^{3/2}x^{-1/3}\, dx = 4\pi a^{1/3} \lim_{t \to 0+} [(-3/5)(a^{2/3} - x^{2/3})^{5/2}\Big|_t^a$$

$$= 12\pi a^2/5$$

6-69   Paramateric equations of this curve are $x = 3at/(1 + t^3)$ and $y = 3at^2/(1 + t^3)$.
The loop is described as t varies between ∞ and 0.   Since

$$dx = [3a(1 - 2t^3)/(1 + t^3)^2]\, dt$$

the area is $9a^2 \int_\infty^0 [t^2(1 - 2t^3)/(1 + t^3)^3]\, dt$

$$= -9a^2 \lim_{h \to \infty} \int_0^h [t^2(1 - 2t^3)/(1 + t^3)^3]\, dt$$

$$= -9a^2 \lim_{h \to \infty} \int_0^h \left(\frac{t^2}{(1 + t^3)^3} + \frac{1}{(1 + t^3)^2} \left[-2t^2 + \frac{2t^2}{1 + t^3} \right] \right)\, dt$$

$$= -9a^2 \lim_{h \to \infty} \left[-\frac{1}{2} \frac{1}{(1 + t^3)^2} + \frac{2}{3} \frac{1}{1 + t^3} \right|_0^h = \frac{3}{2} a^2$$

6-70(a) $$\frac{1}{n} \sum_{k=0}^n \cos \left(\frac{kx}{n} \right)$$

is a Riemann sum of the function $\cos xt$ on $[0, 1]$.   Inasmuch as the function
is continuous, we must have

$$\left\{ \frac{1}{n} \sum_{k=0}^n \cos \left(\frac{kx}{n} \right) \right\} \to \int_0^1 \cos xt\, dt = \frac{\sin xt}{x} \bigg|_0^1 = \frac{\sin x}{x}$$

(b) As in part (a),

$$\left\{\frac{1}{n}\sum_{k=0}^{n-1}\left(\frac{k}{n}\right)^\alpha\right\} \to \int_0^1 x^\alpha \, dx = \frac{x^{\alpha+1}}{\alpha+1}\bigg|_0^1 = \frac{1}{\alpha+1}$$

(c) Since

$$\sum_{k=0}^{n-1}\frac{n}{n^2+k^2} = \frac{1}{n}\sum_{k=0}^{n-1}\frac{1}{1+(k/n)^2}$$

we have, for the same reason as in parts (a) and (b),

$$\left\{\frac{1}{n}\sum_{k=0}^{n-1}\frac{1}{1+(k/n)^2}\right\} \to \int_0^1 \frac{1}{1+x^2}\, dx = \tan^{-1} x \bigg|_0^1 = \frac{\pi}{4}$$

(d) We have $\log (n!/n^n)^{1/n} = (1/n)[\log n! - n \log n]$

$$= \frac{1}{n}\left[\sum_{k=0}^{n}\log k - n\log n\right] = \frac{1}{n}\sum_{k=1}^{n}[\log k - \log n] = \frac{1}{n}\sum_{k=1}^{n}\log\left(\frac{k}{n}\right)$$

Now

$$\frac{1}{n}\sum_{k=1}^{n}\log\left(\frac{k}{n}\right)$$

may be regarded as a Riemann sum approximating the improper integral

$$\int_0^1 \log x \, dx = \lim_{t\to 0+}\int_t^1 \log x \, dx = \lim_{t\to 0+}\left[x\log x - x\right]\bigg|_t^1$$

$$= -1 + \lim_{t\to 0+}[t - t\log t] = -1$$

Hence

$$\{\log (n!/n^n)^{1/n}\} \to -1 \quad \text{or} \quad \{(n!/n^n)^{1/n}\} \to 1/e$$

(Alternately, see Problem 2-14.)

Remark We may put part (d) in the form $n! = b_n n^n e^{-n}$, where $\{b_n^{1/n}\} \to 1$. If it were true that $\{b_n\}$ had a limit as well as $\{b_n^{1/n}\}$, we would have an asymptotic formula for $n!$. Let us see if we can determine whether or not $\{b_n\}$ has a limit. Now $b_{n+1}/b_n = e(n/(n+1))^n$, and so

$$\log (b_{n+1}/b_n) = 1 + n \log [n/(n+1)]$$
$$= 1 + n \log [1 - 1/(n+1)] > 1 + n[-1/(n+1)]/[1 - 1/(n+1)]$$

(see Problem 4-56) $= 1 - n[1/(n+1)]/n/(n+1) = 0$. Thus $b_{n+1}/b_n > 1$, and so $\{b_n\}$ is increasing. Hence $\{b_n\}$ has a limit if it is bounded. To determine if it is bounded, let us examine $\log b_n = \log n! + n - n \log n$

$$= \sum_{k=2}^{n-1}\log k + \frac{1}{2}\log \quad + \frac{1}{2}\log n + n - n\log n$$

But if $z > 0$, then $\log z = \log [1 + (z-1)] \le z - 1$. (Problem 4-56 again.) Thus if $x > 0$ and $k > 0$, then $\log (x/k) \leq x/k - 1$, or $\log x \leq x/k - 1 + \log k$. [The last inequality is obvious geometrically because $y = x/k - 1 + \log k$ is the equation of the tangent line to the graph of $y = \log k$ at the point $(k, \log k)$.]   Hence

$$\int_{k-1/2}^{k+1/2} \log x \, dx \leq \int_{k-1/2}^{k+1/2} [x/k - 1 + \log k] \, dx = \log k$$

Because $\log$ is an increasing function, we also have $(1/2) \log n \geq \int_{n-1/2}^{n} \log x \, dx$, and so

$$\log b_n > \sum_{k=2}^{n-1} \int_{k-1/2}^{k+1/2} \log x \, dx + \int_{n-1/2}^{n} \log x \, dx + \frac{\log n}{2} + n - n \log n$$

$$= \int_{3/2}^{n} \log x \, dx + \frac{\log n}{2} + n - n \log n$$

$$= x \log x - x \Big|_{3/2}^{n} + \frac{\log n}{2} + n - n \log n = \frac{\log n}{2} + \frac{3}{2}\left[1 - \log\left(\frac{3}{2}\right)\right]$$

Thus $\{\log b_n\}$ is not bounded, and so neither is $\{b_n\}$.   Hence $\{b_n\}$ has no limit.

Instead of abandoning this subject, a close examination of what we have done above suggests consideration of the sequence $c_n = n! \, e^n / n^{n+1/2}$.   Then

$$c_{n+1}/c_n = e[n/(n+1)]^{n+1/2}$$

and

$$\log (c_{n+1}/c_n) = 1 + (n + 1/2) \log [n/(n+1)] = 1 - (n + 1/2) \log (1 + 1/n)$$

Now we have

$$0 \leq [x - (1 + 1/2n)]^2 = x^2 - 2(1 + 1/2n)x + (1 + 1/2n)^2 \text{ or, if } x > 0,$$

$$\frac{2(1 + 1/2n) - x}{(1 + 1/2n)^2} \leq \frac{1}{x}$$

[This last inequality is also obvious geometrically, because

$$y = [2(1 + 1/2n) - x]/(1 + 1/2n)^2$$

is the equation of the tangent line to the graph of $y = 1/x$ at the point

$(1 + 1/2n, \, 1/[1 + 1/2n])$.]

Thus

$$\log (1 + 1/n) = \int_{1}^{1+1/n} dx/x \geq \int_{1}^{1+1/n} \left[\frac{2}{1 + 1/2n} - \frac{x}{(1 + 1/2n)^2}\right] dx = \frac{1}{n + 1/2}$$

and so $1 - (n + 1/2) \log (1 + 1/n) \leq 1 - (n + 1/2)/(n + 1/2) = 0$. Hence $c_{n+1}/c_n \leq 1$, or $\{c_n\}$ is a decreasing sequence and so, inasmuch as $c_n \geq 0$, it has limit $C \geq 0$.   This fact is of use to us only if $C > 0$.   If we can show that $c_n > S > 0$, it will follow that C is indeed positive.   To this end, note that $c_n = b_n / n^{1/2}$, and so $\log c_n = \log b_n - \log n/2$. But we have seen that $\log b_n - \log n/2 \geq (3/2)[1 - \log 3/2]$. Now

$(3/2)[1 - \log 3/2] > 0$ because $3/2 < e$, or $\log 3/2 < \log e = 1$. Thus $\log c_n \geq 0$, or $c_n \geq 1$, and we have shown that $\{c_n\} \to C \geq 1$.

To find C, recall, from the remark following the solution of Problem 6-10, that $\{(n!)^2 2^{2n}/(2n)! \, n^{1/2}\} \to \pi^{1/2}$. But

$$\{(n!)^2 2^{2n}/(2n)! \, n^{1/2}\} = \{c_n^2 n^{2n+1} e^{-2n} 2^{2n}/c_{2n}(2n)^{2n+1/2} e^{-2n} n^{1/2}\}$$
$$= \{c_n^2/c_{2n}(2)^{1/2}\} \to C^2/C(2)^{1/2}$$

Thus $C = (2\pi)^{1/2}$, and we have established *Stirling's formula*, useful both theoretically and computationally,

$$\left\{ \frac{n!}{n^{n+1/2}e^{-n}} \right\} \to (2\pi)^{1/2}$$

or, as it is usually written

$$n! \sim (2\pi)^{1/2} n^{n+1/2} e^{-n}$$

It should be kept in mind that that symbol $\sim$ introduced here does not necessarily mean that $n! - (2\pi)^{1/2} n^{n+1/2} e^{-n}$ is small in some sense, or even bounded. (As a matter of fact, $n! - (2\pi)^{1/2} n^{n+1/2} e^{-n}$ is unbounded.) All that is meant by the symbol $\sim$ is that the ratio of the two sides tends to one. This is a standard notation for expressing asymptotic relations like Stirling's formula. Thus $n^2 \sim n^2 + n$, and $n^2 \sim n^2 + 1/n$. It is also used for functions as well as for sequences. Thus $e^x \sim e^x + x$ as $x \to \infty$, or $\sin x \sim x$ as $x \to 0$.

CHAPTER 7

7-1 In all parts denote the series by $\sum a_n$.
  (a) We have $a_1 = 1/3$, and

$$a_n = n/(2n + 1) - (n-1)/(2n - 1) = 1/(2n - 1)(2n + 1)$$

for $n > 1$. Since $\{n/(2n + 1)\} \to 1/2$, this series converges and has sum $1/2$.
  (b) Here $a_1 = 0$, and $a_n = \log(n + 1) - \log n = \log(1 + 1/n)$ for $n > 1$. Inasmuch as $\{\log(n + 1)\} \to \infty$, the series diverges.
  (c) We have $a_1 = -1/2$, and $a_n = (-1)^n/2^n - (-1)^{n-1}/2^{n-1} = (-1)^n 3/2^n$ for $n > 1$. This series converges and has sum zero because $\{(-1)^n/2^n\} \to 0$.

7-2 For each of these series, the sequence of terms does not have limit zero, and so each series diverges. In part (a) the terms oscillate between 1 and -1. In (b) we have $\{(1/2)^{1/n}\} \to 1$ from Problem 2-5b. In (c),

$$\{1/n \log(1 + 1/n)\} = \{1/\log[(1 + 1/n)^n]\} \to 1/\log e = 1$$

Finally, in (d) we have

$$\{n^{n-1/n}/(n - 1/n)^n\} = \{(1/n^{1/n})[1/(1 - 1/n^2)^n]\}$$
$$= \{1/n^{1/n}\}\{1/(1 + 1/n)^n\}\{1/(1 - 1/n)^n\} \to (1/1)\left(\frac{1}{e}\right)(1/e^{-1}) = 1$$

since $\{n^{1/n}\} \to 1$ from Problem 2-5e.

7-3 (a) We have $n^2 + a^2 > n^2$, or $1/(n^2 + a^2) < 1/n^2$, and so the given series converges.

(b) It is clear that $\{n/(n^2+1)/1/n\} \to 1$, and so we can find N such that for $n > N$, $n/(n^2+1) > 1/2n$. Our series thus diverges.

(c) Since $2 + \cos n \le 3$, we get an immediate comparison with the convergent series of part (a). This series converges.

(d) Since $\{\sin(\pi/n)/\pi/n\} \to 1$, it is clear we have $\sin(\pi/n) > \pi/2n$ for all sufficiently large n, and so this series diverges.

(e) Here $\sin^2[\pi(n+1/n)] = \sin^2(\pi/n)$, and $\{[\sin(\pi/n)/(\pi/n)]^2\} \to 1$. Thus we have $\sin^2[\pi(n+1/n)] < (3/2)(\pi^2/n^2)$ for all sufficiently large n, and so the series converges.

(f) We have $[1+n^2]^{1/2} - n = 1/([1+n^2]^{1/2} + n)$, and $\{1/([1+n^2]^{1/2}+n)/1/n\} \to 1/2$. From this limit it follows, as in part (b), that the series diverges.

(g) From Problem 2-5e, we have $\{n^{1/n}\} \to 1$. Thus $\{1/n^{1+1/n}/1/n\} \to 1$, and, as we have seen, this fact means that the series diverges.

(h) We know from Problem 4-49a that $\{\log n/n^\delta\} \to 0$ for all $\delta > 0$. Hence $\log n < n^{1/\alpha}$ for all sufficiently large n, and so $1/(\log n)^\alpha > 1/n$ for all sufficiently large n. The series diverges.

(i) As in part (h), we have $1/\log n > 1/n$ for all sufficiently large n, or $1/(\log n)^{1/n} > 1/n^{1/n}$. But $\sum 1/n^{1/n}$ is divergent because $\{1/n^{1/n}\} \to 1$. Our series diverges.

(j) We have $(\log n)^{\log n} = e^{\log n[\log(\log n)]} = n^{\log(\log n)}$, and $\log(\log n) > 2$ for all sufficiently large n, or $1/(\log n)^{\log n} < 1/n^2$ for all sufficiently large n, and our series is seen to be convergent.

7-4 Since both the functions $f(x) = 1/x(\log x)^\alpha$, and $g(x) = 1/x\log x[\log(\log x)]^\alpha$ are the products of positive, decreasing functions, it follows that both are decreasing. We may thus apply the integral test.

(a) $\displaystyle\int_2^t \frac{1}{x(\log x)^\alpha}\,dx = \begin{cases} (\log x)^{1-\alpha}/(1-\alpha)\Big|_2^t, & \text{if } \alpha \ne 1 \\[2mm] \log(\log x)\Big|_2^t, & \text{if } \alpha = 1 \end{cases}$

It is now easy to see that $\int_2^\infty [1/x(\log x)^\alpha]\,dx$ converges if $\alpha > 1$, diverges if $\alpha \le 1$. The same is true of our series.

(b) $\displaystyle\int_3^t \frac{1}{x\log x[\log(\log x)]^\alpha}\,dx = \begin{cases} [\log(\log x)]^{1-\alpha}/(1-\alpha)\Big|_3^t, & \text{if } \alpha \ne 1 \\[2mm] \log(\log(\log x))\Big|_3^t, & \text{if } \alpha = 1 \end{cases}$

As in part (a), it follows that the integral converges if $\alpha > 1$, diverges if $\alpha \le 1$. The same is then true of the series.

7-5 (a) This series converges because $\{[(n^{1/n}-1)^n]^{1/n}\} = \{n^{1/n} - 1\} \to 0 < 1$.

(b) Here $\{(n/2^n)^{1/n}\} = \{n^{1/n}/2\} \to 1/2$, and so this series converges.

(c) We have $\{(2^n n!/n^n)^{1/n}\} = \{2(n!/n^n)^{1/n}\} \to 2/e$ by Problem 6-70d. Since $e > 2$, $2/e < 1$, and so our series converges.

(d)   As in part (c), $\{(3^n n!/n^n)^{1/n}\} \to 3/e > 1$.   The series diverges.

(e)   We have $\{(\alpha^n n^2)^{1/n}\} = \{\alpha(n^{1/n})^2\} \to \alpha \cdot 1^2 = \alpha$.   This series thus converges if $\alpha < 1$, diverges if $\alpha > 1$.   If $\alpha = 1$, the series obviously diverges.

7-6   In parts (a)–(d), denote the nth term by a_n.

(a)   Here $\{a_{n+1}/a_n\} = \{(n+1)/(2n+3)\} \to 1/2 < 1$, and so the series converges.

(b)   We have $\{a_{n+1}/a_n\} = \{(n+1)/\alpha\} \to \infty$.   Hence the series diverges for all $\alpha > 0$.

(c)   Here $\{a_{n+1}/a_n\} = \{(n+1)^2/(2n+2)(2n+1)\} \to 1/4 < 1$.   The series converges.

(d)   In this case

$$\{a_{n+1}/a_n\} = \{[(n+1)^{n+1}/(n+1)n^n][n/(n+1)]^2\}$$
$$= \{(1+1/n)^n\}\{[n/(n+1)]^2\} \to e \cdot 1^2 = e > 1$$

and so this series diverges for all $\alpha > 0$.

(e)   We know from the remark following the solution of Problem 2-10 that $\{1/a_{n+1}/1/a_n\} = \{a_n/a_{n+1}\} \to 2/[1+5^{1/2}]$.   But $1 + 5^{1/2} > 3$, or $2/[1+5^{1/2}] < 2/3 < 1$, and so this series converges.

7-7   (a)   Since $\{\sin^2(\pi/n)/(\pi/n)^2\} \to 1$, we have $\sin^2(\pi/n) < 2\pi^2/n^2$ for all sufficiently large n, and so the series converges.

(b)   Here $1/bn - 1/an = [(a-b)/ab]1/n$, and $(a-b)/ab > 0$.   Our series diverges.

(c)   For this series $\{[2n/2^n\alpha^n]^{1/n}\} = \{2^{1/n}n^{1/n}/2\alpha\} \to 1/2\alpha$.   It thus follows that the series converges if $\alpha > 1/2$, diverges if $\alpha < 1/2$.   When $\alpha = 1/2$, the series obviously diverges.

(d)   We have $\{n^{n+1/n}/(n+1/n)^n\} = \{n^{1/n}\}\{1/(n+1/n^2)^n\}$.   Now $\{n^{1/n}\} \to 1$, $\{(1+1/n^2)^{n^2}\} \to e$, and $(1+1/n^2)^{n^2} > (1+1/n^2)^n$.   We can find N such that if $n > N$, then $(1+1/n^2)^{n^2} < e+1$.   Hence if $n > N$, then $1/(1+1/n^2)^n > 1/(1+1/n^2)^{n^2} > 1/(e+1)$.   Therefore, $\{n^{n+1/n}/(n+1/n)^{1/n}\}$ cannot have limit zero, and so the series diverges.

(e)   Since $(n+1)/n^2 = 1/n + 1/n^2 > 1/n$, this series diverges.

(f)   Here $(n+1)/n^3 = 1/n^2 + 1/n^3 < 2/n^2$, and so this series converges.

(g)   Because $n+1 \le 2n$, the convergence of this series follows from the convergence of the series of part (c) when $\alpha = 1$.

(h)   If $p \ge q$, the series diverges because its sequence of terms cannot have limit zero.   If $q > p$, then $\{n^p/(n^q+a)/1/n^{q-p}\} = \{n^q/(n^q+a)\} \to 1$, and so we have $1/2n^{q-p} < n^p/(n^q a) < 3+/2n^{q-p}$.   If $q-p > 1$, the right-hand inequality implies that the series converges, if $0 < q-p \le 1$, the left-hand inequality implies that the series diverges.

(i)   To begin, L'Hospital's rule gives us $\lim\limits_{x\to\infty} x/e^{x^2} = \lim\limits_{x\to\infty} 1/2xe^{x^2} = 0$.   In particular, we have $\{n/e^{n^2}\} \to 0$.   Hence $1/e^{n^2} < 1/2n$ for all sufficiently large n, or $1/n - 1/e^{n^2} > 1/2n$ for all sufficiently large n.   Our series thus diverges.

(j)   We have $\{1/(a+bn)^\alpha/1/(bn)^\alpha\} = \{1/(a/bn+1)^\alpha\} \to 1$.   Hence $1/2b^\alpha n^\alpha < 1/(a+bn)^\alpha < 3/2b^\alpha n^\alpha$ for all sufficiently large n.   Consequently, this series diverges if $0 < \alpha \le 1$, converges if $\alpha > 1$.

(k)   If a_n denotes the general term of this series, then

$$\{a_{n+1}/a_n\} = \{2/(n+1)\}\{[1 + (n+1)^2]/(1+n^2)\} \to 0 \cdot 1 = 0 < 1$$

The series thus converges.

(l)   Since $(1/2)^{n/2} = (1/2^{1/2})^n$, and $1/2^{1/2} < 1$, this series is a convergent geometric series.

(m)   If the general term of this series is denoted by a_n, then $\{a_{n+1}/a_n\} = \{1/(n+1)\}\{(1 + 1/n)^\alpha\} \to 0 \cdot 1^\alpha = 0 < 1$, and so the series converges for all α.

(n)   Because $\{(n+1)/(n+2)\} \to 1$, we have for all sufficiently large n that $(n+1)/2^n(n+2) < (3/2)(1/2^n)$, and so our series converges.

(o)   Since $\log (n+1) - \log n = \log (1 + 1/n)$, and $\log (1 + 1/n) < 1/n$ by Problem 4-56, we have $[\log (n+1) - \log n]/(\log n)^2 < 1/n(\log n)^2$.   But

$$\sum_{n=2}^{\infty} \frac{1}{n(\log n)^2}$$

converges by Problem 7-4, and so our series converges.

(p)   Observe first that $(n^\alpha + 1)^{1/2} - (n^\alpha)^{1/2} = 1/[(n^\alpha + 1)^{1/2} + (n^\alpha)^{1/2}]$.   Now

$$\{1/[(n^\alpha + 1)^{1/2} + (n^\alpha)^{1/2}]/1/(n^\alpha)^{1/2}\} = \{1/[(1 + 1/n)^\alpha + 1]\} \to 1/2$$

Hence $1/4n^{\alpha/2} < (n^\alpha + 1)^{1/2} - (n^\alpha)^{1/2} < 1/n^{\alpha/2}$ for all sufficiently large n.   These inequalities imply that the series converges if $\alpha/2 > 1$, or $\alpha > 2$, and diverges if $\alpha/2 \le 1$ or $\alpha \le 2$.

(q)   If $\alpha \le 1$, then, because $\log n/n^\alpha > 1/n^\alpha$ for $n \ge 3$, the series diverges. Suppose that $\alpha > 1$.   We may then write $\alpha = 1 + 2\delta$, where $\delta > 0$.   Now $\{\log n/n^\delta\} \to 0$ (see Problem 4-44a).   Hence $\log n/n^\alpha < 1/n^{1+\delta}$ for all sufficiently large n.   Therefore, the series converges if $\alpha > 1$.

7-8   (a)   Since $\{(2n-1)/(9n+1)\} \to 2/9$, the series diverges.

(b)   Here $\{\log [n/(n+1)]\} \to \log 1 = 0$.   Since $n/(n+1) < 1$, $\log [n/(n+1)] < 0$, and so $|\log [n/(n+1)]| = -\log [n/(n+1)] = \log [(n+1)/n]$.   Now

$$\log [(n+1)/n] - \log [(n+2)/(n+1)] = \log [(n+1)^2/n(n+2)]$$
$$= \log [(n^2 + 2n + 1)/(n^2 + 2n)] > \log 1 = 0.$$

Thus the terms decrease to zero in absolute value, and so the series converges. As for absolute convergence, we have from Problem 4-56 that $\log [(n+1)/n] = \log (1 + 1/n) > 1/n/(1 + 1/n) = 1/(n+1)$.   Thus, this series is not absolutely convergent.

(c)   As in Problem 7-3h, $\{\log n/n\} \to 0$.   Now

$$\log n/n - \log (n+1)/(n+1) = [(n+1) \log n - n \log (n+1)]/n(n+1)$$
$$= \log [n^{n+1}/(n+1)^n]/n(n+1)$$
$$= \log [n/(1 + 1/n)^n]/n(n+1)$$

But

$$\{1/(1 + 1/n)^n\} \to 1/e, \text{ and so } \{n/(1 + 1/n)^n\} \to \infty$$

Thus $\log [n/(1 + 1/n)^n] > \log 1 > 0$ for all sufficiently large n, and so the absolute values of the terms of this series decrease to zero.   The series thus converges. Since $\log n/n > 1/n$ for $n \geq 3$, the series is only conditionally convergent.

(d) Obviously $\{1/n^{\alpha}\} \to 0$, and $1/n^{\alpha} > 1/(n+1)^{\alpha}$. Thus the series converges for all $\alpha > 0$. As we know, it converges absolutely if $\alpha > 1$, conditionally if $\alpha \leq 1$.

7-9 Since Σa_n converges, we must have $\{a_n\} \to 0$, or $\{1/a_n\} \to \infty$. Thus $\{1/a_n\}$ does not have limit zero, and so $\Sigma 1/a_n$ diverges.

7-10 From Problem 1-16c, $(a_n a_{n+1})^{1/2} \leq (a_n + a_{n+1})/2$. Since $\Sigma a_n/2$ and $\Sigma a_{n+1}/2$ both converge, so does $\Sigma(a_n + a_{n+1})/2$. Thus $\Sigma(a_n a_{n+1})^{1/2}$ converges.

7-11 We have, by Problem 1-16c again, $(a_n/n^{1+\delta})^{1/2} \leq [a_n + 1/n^{1+\delta}]/2$, and since both $\Sigma a_n/2$ and $\Sigma 1/2n^{1+\delta}$ both converge, so does their sum, and hence so does $\Sigma(a_n/n^{1+\delta})^{1/2}$.

7-12 Since $\{a_n\}$ decreases and $a_n > 0$, it must be the case that $\{a_n\} \to L \geq 0$. Now $\{b_{n+1}/b_n\} = \{a_{n+1}\} \to L$. If $L < 1$, Σb_n converges, while if $L > 1$, it diverges. Hence we only have to settle the behavior of Σb_n when $L = 1$. In this case, since $\{a_n\}$ is strictly decreasing, we must have $a_{n+1} > 1$, or $b_{n+1}/b_n > 1$ for all n, and so Σb_n diverges when $L = 1$.

7-13 Put

$$T_n = \sum_{k=n+1}^{\infty} a_k$$

for $n = 1, 2, \ldots$. Since Σa_k converges, we must have $\{T_n\} \to 0$, and so $\{a_{n+1} + a_{n+2} + \cdots + a_{2n}\} \to 0$ as well. Now $a_{n+1} + a_{n+2} + \cdots + a_{2n} \geq n a_{2n}$, and so $\{2n a_{2n}\} \to 0$. Inasmuch as $(2n+1)a_{2n+1} \leq [(2n+1)/2n][2n a_{2n}]$, we also have $[(2n+1)a_{2n+1}] \to 0$. Consequently, $\{n a_n\} \to 0$.

Remark This problem gives us a criterion by means of which series whose terms are decreasing and positive may be tested for divergence. For instance, consider $\Sigma(\alpha^{1/n} - 1)$, where $\alpha > 1$. It is easy to see that $\{\alpha^{1/n} - 1\}$ is decreasing and positive. Now $n(\alpha^{1/n} - 1) = [\alpha^{1/n} - \alpha^0]/[1/n - 0]$, and we see that the limit of $\{n(\alpha^{1/n} - 1)\}$ is the derivative of $f(x) = \alpha^x$ evaluated at $x = 0$. That is, $\{n(\alpha^{1/n} - 1)\} \to \log \alpha$, and since $\log \alpha > 0$, $\Sigma(\alpha^{1/n} - 1)$ diverges. At the same time, this condition is not sufficient for such series as the divergent series $\Sigma 1/n \log n$ of Problem 7-4a shows.

7-14 We have $|a_n b_n| \leq [a_n^2 + b_n^2]/2$ by the inequality of Problem 1-16c. Now $\Sigma [a_n^2 + b_n^2]/2$ is the sum of two convergent series, and so $\Sigma |a_n b_n|$ is convergent. Hence $\Sigma a_n b_n$ is convergent as well.

7-15 (a) Since Σa_n converges, we must have $\{a_n\} \to 0$, and so $1 + a_n > 1/2$ for all sufficiently large n. Thus $|a_n/(1 + a_n)| < 2 |a_n|$ for such n, and so $\Sigma a_n/(1 + a_n)$ converges absolutely.

(b) Because $\{a_n\} \to 0$, $\{|a_n|\}$ must be bounded. Let $M > |a_n|$ for all n. Then $a_n^2 = |a_n^2| = |a_n|^2 < M |a_n|$, and so Σa_n^2 is absolutely convergent.

(c) This follows from part (b) together with part (a).

7-16   Since Σa_n has sum S, the sequence of partial sums $\{S_n\}$ has limit S. But then, by Problem 2-13b, the sequence of arithmetic means of $\{S_n\}$ also has limit S; that is

$$\left\{\frac{1}{m}\sum_{n=1}^{m}S_n\right\}\to S$$

Remark   Consider the sequence of partial sums $\{S_n\}$ of the divergent series $\Sigma(-1)^{n-1}$. It is easy to see that $S_{2n-1}=1$, and $S_{2n}=0$ for all n. Consequently,

$$\sum_{n=1}^{m}S_n = \begin{cases} m/2 & \text{if } m \text{ is even} \\ (m+1)/2 & \text{if } m \text{ is odd} \end{cases}$$

and so

$$\left\{\frac{1}{m}\sum_{n=1}^{m}S_n\right\}\to\frac{1}{2}$$

Thus a divergent series can be such that its sequence of partial sums shares the property of Problem 7-16 with the sequence of partial sums of any convergent series; it is not "too" divergent.   A series, convergent, or divergent, whose sequence of partial sums $\{S_n\}$ has the property that

$$\left\{\frac{1}{m}\sum_{n=1}^{m}S_n\right\}\to S$$

is said to be *Cesaro summable* to S.   As this problem shows, a convergent series whose sum is S is also Cesaro summable to S.   But we have seen that a divergent series can be Cesaro summable.   Cesaro summability is important in the theory of Fourier series.   For a further discussion of Cesaro summability, as well as of other types of summability, see Chapter 8 of the book, *Theory and Application of Infinite Series* by K. Knopp.

7-17   We have

$$(S_n - I_n) - (S_{n+1} - I_{n+1}) = \int_n^{n+1} f(x)\,dx - a_{n+1} = \int_n^{n+1}[f(x)-f(n+1)]\,dx \geq 0$$

On the other hand,

$$a_1 = S_1 - I_1 \geq S_n - I_n$$

$$= a_n + \sum_{k=1}^{n-1}\left[a_k - \int_k^{k+1} f(x)\,dx\right] > \sum_{k=1}^{n-1}\int_k^{k+1}[f(k)-f(x)]\,dx \geq 0$$

Thus $\{S_n - I_n\}\to L$, where $a_1 \geq L \geq 0$.   If f is strictly decreasing, it is easy to see that $L > 0$.

Remark   One instance of this problem that is of particular interest is given by putting $f(x) = 1/x$ for x in $[1, \infty)$.   Then

$$\left\{\sum_{k=1}^{n}\frac{1}{k} - \log n\right\}$$

has a limit between 0 and 1. This limit is called *Euler's constant*, and it is denoted by γ. Thus

$$\left\{ \sum_{k=1}^{n} \frac{1}{k} - \log n \right\} \to \gamma$$

or otherwise stated, there is a sequence $\{a_n\} \to 0$ such that

$$\sum_{k=1}^{n} \frac{1}{k} = \log n + \gamma + a_n$$

This last identity can be used to illustrate an important point. Consider the series

$$1 + \frac{1}{3} - \frac{1}{2} + \frac{1}{5} + \frac{1}{7} - \frac{1}{4} + \frac{1}{9} + \frac{1}{11} - \frac{1}{6} + \cdots$$

which is obtained by alternately taking two positive terms and one negative term from the sequence $\{(-1)^{n-1}/n\}$. Let S_n denote the nth partial sum of this series. Then

$$S_{3n} = \sum_{k=1}^{2n} \frac{1}{2k-1} - \sum_{k=1}^{n} \frac{1}{2k} = \left[\sum_{k=1}^{4n} \frac{1}{k} - \sum_{k=1}^{2n} \frac{1}{2k} \right] - \frac{1}{2} \sum_{k=1}^{n} \frac{1}{k}$$

$$= \sum_{k=1}^{4n} \frac{1}{k} - \frac{1}{2} \sum_{k=1}^{2n} \frac{1}{k} - \frac{1}{2} \sum_{k=1}^{n} \frac{1}{k}$$

$$= \log 4n + \gamma + a_{4n} - \frac{1}{2}(\log 2n + \gamma + a_{2n}) - \frac{1}{2}(\log n + \gamma + a_n)$$

$$= \log 4 - \frac{1}{2} \log 2 + a_{4n} - \frac{1}{2}(a_{2n} + a_n)$$

Hence $\{S_{3n}\} \to (3/2) \log 2$, and inasmuch as the terms of this series have limit zero, this fact implies that the sum of the series is also $(3/2) \log 2$.

To better appreciate the point illustrated, we introduce some terminology. Let $\{a_n\}$ be any sequence, and let $\{b_n\}$ be a sequence such that (i) b_n is a positive integer for each n, (ii) $b_n = b_m$ only if $n = m$, and (iii) if k is any positive integer, then there is an integer n such that $k = b_n$. In other words, when the sequence $\{b_n\}$ is regarded as a function whose domain is the set of natural numbers then its range is the set of natural numbers, and it is a one-to-one function. Since the range of $\{b_n\}$ coincides with the domain of $\{a_n\}$, we can form the composition of these two functions; it will be the sequence $\{a_{b_n}\}$. Such a sequence will be called a *permutation* of $\{a_n\}$. Let us put $r_n = a_{b_n}$. Note that each a_n is some term of $\{r_n\}$, but that in general it does not appear as the nth term of $\{r_n\}$. The series Σr_n will be called a *rearrangement* of Σa_n.

It can be seen that the series $1 + 1/3 - 1/2 + 1/5 + 1/7 - 1/4 + \cdots$ is a rearrangement of the alternating harmonic series $\Sigma(-1)^{n-1}/n$. Now we know that $\Sigma(-1)^{n-1}/n = \log 2$, but we have seen that the rearrangement of it given above has sum $(3/2) \log 2$. Thus, it is possible to rearrange a convergent series and obtain a different sum. However, such can be the case only for a conditionally convergent series because it can be proved that *if Σa_n is absolutely convergent with* sum S, and if Σr_n is any rearrangement of Σa_n, then Σr_n is absolutely convergent, and $\Sigma r_n = S$. That is, any rearrangement of an absolutely convergent series has the same sum. This fact is one of the more important theoretical consequences of absolute convergence.

As far as conditionally convergent series are concerned, it can be proved that if Σa_n is conditionally convergent, and $-\infty < L < \infty$, then there exists a rearrangement Σr_n such that $\Sigma r_n = L$. Further, if $-\infty \leq \alpha < \beta \leq \infty$, then there exists a rearrangement Σr_n such that subsequences $\{n_k\}$ and $\{n_m\}$ can be found with the property that $\{S_{n_k}\} \to \alpha$ and $\{S_{n_m}\} \to \beta$, where S_n is the nth partial sum of Σr_n. Otherwise stated, there exist rearrangements of conditionally convergent series that exhibit any conceivable behavior as far as convergence or divergence is concerned.

Proofs of these two theorems can be found in the book of K. Knopp cited in the remark following the solution of the previous problem.

7-18   If $L < r < 1$, we can find an integer m such that if $n \geq m$, then $a_{n+1}/a_n < r$. In particular, we have $a_{m+1} < a_m r$, $a_{m+2} < a_{m+1} < a_m r^2$, $a_{m+3} < a_{m+2} r < a_m r^3$, and, in general, $a_{m+k} < a_{m+1} r^k$. Consequently,

$$\sum a_n - \sum_{n=1}^{m} a_n = \sum_{k=1}^{\infty} a_{m+k} < a_m \sum_{k=1}^{\infty} r^k = a_m \frac{r}{1-r}$$

7-19   (a)   From Problem 1–2a,

$$\left\{ \sum_{n=1}^{m} \frac{1}{n(n+1)} \right\} = \left\{ 1 - \frac{1}{m+1} \right\} \to 1$$

(b)   From Problem 1–2b,

$$\left\{ \sum_{n=1}^{m} (-1)^{n+1} \frac{2n+1}{n(n+1)} \right\} = \left\{ 1 + \frac{(-1)^{m+1}}{m+1} \right\} \to 1$$

(c)   Here

$$\sum [2^n + 3^n]/6^n = \sum (1/3)^n + \sum (1/2)^n = 1/(1-1/3) - 1 + 1/(1-1/2) - 1 = 3/2$$

(d)   Since $1/(n^2 - 1) = 1/(n+1)(n-1) = 1/2(n-1) - 1/2(n+1)$, we have $1/(n^2 - 1) + 1/[(n+1)^2 - 1] = (1/2)[1/(n-1) - 1/(n+1) + 1/n - 1/(n+2)]$ Thus

$$\left\{ \sum_{n=2}^{2m+1} \frac{1}{n^2 - 1} \right\} = \left\{ \frac{1}{2}\left(1 + \frac{1}{2} \right) - \frac{1}{2}\left(\frac{1}{2m+2} + \frac{1}{2m+3} \right) \right\} \to \frac{3}{4}$$

(e)   Here $1/(\alpha + n)(\alpha + n + 1) = 1/(\alpha + n) - 1/(\alpha + n + 1)$, and so

$$\left\{ \sum_{n=0}^{m} \frac{1}{(\alpha + n)(\alpha + n + 1)} \right\} = \left\{ \frac{1}{\alpha} - \frac{1}{\alpha + m + 1} \right\} \to \frac{1}{\alpha}$$

7-20   In all of these, let a_n be the coefficient of x^n.
(a) $\{|a_{n+1}|/|a_n|\} |x| = \{n^\alpha/(n+1)^\alpha\} |x| \to |x|$. Thus the radius of convergence is one. When $x = 1$, we know that the series converges for all α from Problem 7-8d. When $x = -1$, we know that $\Sigma 1/n^\alpha$ converges for $\alpha > 1$, diverges for $\alpha \leq 1$.

(b) Here $\{|a_{n+1}|/|a_n|\}|x| = \{1/(n+1)\}|x| \to 0$. Therefore, this series converges for all x.

(c) We have $\{|a_{n+1}|/|a_n|\}|x| = \{n+1\}|x| \to \infty$ for all $x \neq 0$. This series converges only for $x = 0$.

(d) $\{|a_{n+1}|/|a_n|\}|x| = \{[(n+1)/n]^2\}|x| \to |x|$. The radius of convergence is thus one. When $x = \pm 1$, the resulting series both diverge because their sequences of terms do not have limit zero.

(e) Here $\{|a_{n+1}|/|a_n|\}|x-a| = \{(n+1)/(n+2)\}|x-a|/2 \to |x-a|/2$. Thus the series converges for $|x-a| < 2$, or $a - 2 < x < a + 2$. When $x = a + 2$, the series becomes $\Sigma(-1)^n/(n+1)$, which obviously converges. When $x = a - 2$, the series becomes $\Sigma 1/(n+1)$, which diverges. This series thus converges for $a - 2 < x \leq a + 2$.

(f) We have $\{|a_{n+1}|/|a_n|\}|x| = \{[\log n/\log (n+1)]^2\}|x-a| \to |x-a|$, and so the series converges for $|x-a| < 1$, or $a - 1 < x < a + 1$. When $x = a + 1$, we know, from Problem 7-3h, that the series diverges. When $x = a - 1$, the series is an alternating series whose terms have limit zero and decrease in absolute value. This series converges for $a - 1 \leq x < a + 1$.

(g) From Problem 6-70d, $\{a_n^{1/n}\}|x| \to e\,|x|$. The radius of convergence is thus $1/e$. When $x = \pm 1/e$, we have to consider $\Sigma n^n/e^n n!$ and $\Sigma(-1)^n n^n/e^n n!$. To determine the convergence or divergence of these series, we use Stirling's formula (see the remark following the solution of Problem 6-70), and the fact that $\{(1 + 1/n)^n\}$ is increasing and has limit e (see Problem 2-8d). Let $b_n = n^n/e^n n!$. From Stirling's formula, it follows that $\{b_n\} \to 0$. Now $b_{n+1}/b_n = (1/e)(1 + 1/n)^n < 1$. Thus $\{b_n\}$ decreases to zero, and so $\Sigma(-1)^n b_n$ converges. On the other hand, from Stirling's formula, $b_n > 1/2(2\pi n)^{1/2}$ for all sufficiently large n, and so Σb_n diverges. The series converges for $-1/e \leq x < 1/e$.

7-21 Suppose that $f(x) = \Sigma a_n x^n$ for x in $(-a, a)$, $a > 0$. If f is even, then $f(x) = f(-x)$, or $0 = f(x) - f(-x) = \Sigma a_n[x^n - (-x)^n]$ for all x in $(-a, a)$. But if n is odd, then $x^n - (-x)^n = 2x^n \neq 0$. Consequently, a_n must be zero if n is odd, and so the Maclaurin's series of f will contain only even powers. If f is odd, then $f(x) = -f(-x)$, or $0 = f(x) + f(-x) = \Sigma a_n[x^n + (-x)^n]$ for all x in $(-a, a)$. Now $x^n + (-x)^n = 2x^n$ if n is even, and this implies that $a_n = 0$ if n is even.

7-22 (a) Since $1/(a-x) = (1/a)[1/(1-x/a)]$, we have

$$1/(a-x) = (1/a)\Sigma(x/a)^n = \Sigma x^n/a^{n+1}$$

for $|x| < a$.

(b) Here

$$bx^m/(a-x^k) = (bx^m/a)[1/(1-x^k/a)] = (bx^m/a)\Sigma(x^k/a)^n = \Sigma bx^{nk+m}/a^{n+1}$$

and the interval of convergence is $|x^k| < a$ or $|x| < a^{1/k}$.

(c) We have $x/(1 + x - 2x^2) = (1/3)[1/(1-x)] - (1/3)[1/(1+2x)]$. Now

$1/(1-x) = \Sigma x^n$ for $|x| < 1$, and $1/(1+2x) = \Sigma(-2)^n x^n$ for $|2x| < 1$. Thus

$$x/(1+x-2x^2) = (1/3)\Sigma x^n - (1/3)\Sigma(-2)^n x^n$$
$$= (1/3)\Sigma[1 - (-2)^n]x^n$$

for $|x| < 1/2$.
(d) We have

$$[26x^2 - 18x + 3]/(1-2x)(1-3x)(1-4x)$$
$$= 1/(1-2x) + 1/(1-3x) + 1/(1-4x) = \Sigma(2x)^n + \Sigma(3x)^n + \Sigma(4x)^n$$
$$= \Sigma(2^n + 3^n + 4^n)x^n$$

for $|x| < 1/4$.

7-23 (a) Since $(a+x)^p = a^p(1+x/a)^p$, we have

$$(a+x)^p = a^p \Sigma \binom{p}{n}(x/a)^n = \Sigma \binom{p}{n}x^n/a^{n-p} \text{ for } |x| < a$$

(b) For this function we have

$$bx^m(a+x^k)^p = bx^m a^p(1+x/a)^p = a^p bx^m \Sigma \binom{p}{n}(x^k/a)^n = \Sigma(b/a^{n-p})\binom{p}{n}x^{nk+m}$$

for $|x^k| < a$, or $|x| < a^{1/k}$.
(c) Since $(1-x)^p(1+x)^p = (1-x^2)^p$, we have

$$(1-x)^p(1+x)^p = \Sigma \binom{p}{n}(-x^2)^n = \Sigma(-1)^n \binom{p}{n}x^{2n}$$

for $|x^2| < 1$, or $|x| < 1$.
(d) Here

$$[(a^2+b^2)x^2 - 2(a+b)x + 2]/(1-ax)^2(1-bx)^2$$
$$= 1/(1-ax)^2 + 1/(1-bx)^2 = (1-ax)^{-2} + (1-bx)^{-2}$$

$$= \Sigma \binom{-2}{n}(-ax)^n + \Sigma \binom{-2}{n}(-bx)^n = \Sigma(-1)^n(a^n) + b^n \binom{-2}{n}x^n$$

for $|ax| < 1$ and $|bx| < 1$, or $x <$ the smaller of $1/a, 1/b$. Now

$$\binom{-2}{n} = \frac{-2(-2-1)\cdots(-2-n+1)}{n!} = (-1)^n \frac{(n+1)!}{n!} = (-1)^n(n+1)$$

Therefore, $f(x) = \Sigma(n+1)(a^n + b^n)x^n$ for $|x| <$ the smaller of $1/a, 1/b$.

7-24 (a) Observe that $D[1/m(1-x^m)] = x^{m-1}/(1-x^m)^2$. Now $1/m(1-x^m) = (1/m)\Sigma(-x^m)^n = (1/m)\Sigma(-1)^n x^{mn}$ for $|x^m| < 1$, or $|x| < 1$. Consequently, $f(x) = \Sigma(-1)^n nx^{mn-1}$ for $|x| < 1$.

(b) We have $\sin^{-1} x = \int_0^x dy/(1-y^2)^{1/2}$, and

$$1/(1-y^2)^{1/2} = \sum \binom{-1/2}{n}(-y^2)^n \text{ for } |y^2| < 1 \text{ or } |y| < 1. \quad \text{Now}$$

$$\binom{-1/2}{n} = \frac{(-1/2)(-1/2-1)\cdots(-1/2-n+1)}{n!}$$

$$= (-1)^n \frac{1}{2^n} \frac{1 \cdot 3 \cdot 5 \cdots (2n-1)}{n!} = (-1)^n \frac{1 \cdot 3 \cdot 5 \cdots (2n-1)}{2 \cdot 4 \cdot 6 \cdots (2n)}$$

Thus

$$\frac{1}{(1-y^2)^{1/2}} = 1 + \sum_{n=1}^{\infty} \frac{1 \cdot 3 \cdot 5 \cdots (2n-1)}{2 \cdot 4 \cdot 6 \cdots (2n)} y^{2n}$$

and it follows that

$$\sin^{-1} x = y \Big|_0^x + \sum_{n=1}^{\infty} \frac{1 \cdot 3 \cdot 5 \cdots (2n-1)}{2 \cdot 4 \cdot 6 \cdots (2n)} \frac{y^{2n+1}}{2n+1} \Big|_0^x$$

$$= x + \sum_{n=1}^{\infty} \frac{1 \cdot 3 \cdot 5 \cdots (2n-1)}{2 \cdot 4 \cdot 6 \cdots (2n)} \frac{x^{2n+1}}{2n+1}$$

for $|x| < 1$.

(c) Since

$$D \log [x + (1+x^2)^{1/2}] = 1/(1+x^2)^{1/2} = \sum \binom{-1/2}{n} x^{2n}$$

for $|x^2| < 1$, or $|x| < 1$, we have

$$\log [x + (1+x^2)^{1/2}] =$$

$$\log [x + (1+x^2)^{1/2}] - \log 1 = \int_0^x \sum \binom{-1/2}{n} y^{2n} \, dy = \sum \binom{-1/2}{n} \frac{x^{2n+1}}{2n+1}$$

for $|x| < 1$.

(d) Here

$$D \log ([1 + (1+x)^{1/2}]/2) = (1/2)(1/(1+x)^{1/2})(1/[1+(1+x)^{1/2}])$$

$$= (1/2)(1/(1+x)^{1/2})([1-(1+x)^{1/2}]/-x) = (-1/2x)[(1+x)^{-1/2}-1]$$

$$= \frac{-1}{2x} \sum_{n=1}^{\infty} \binom{-1/2}{n} x^n = \sum_{n=1}^{\infty} -\binom{-1/2}{n} \frac{x^{n-1}}{2}$$

for $|x| < 1$. Therefore,

$$f(x) = f(x) - f(0) = \int_0^x \sum_{n=1}^{\infty} -\binom{-1/2}{n} \frac{y^{n-1}}{2} \, dy = \sum_{n=1}^{\infty} -\binom{-1/2}{n} \frac{x^n}{2n}$$

for $|x| < 1$.

7-25 Since $1/(1-p) = \Sigma p^n$ for $|p| < 1$, we have $D[1/(1-p)] = 1/(1-p)^2 = \Sigma np^{n-1}$, or $p/(1-p)^2 = \Sigma np^n$. From this last equation we have $p/q^2 = \Sigma np^n$, and $D[p/(1-p)^2] = [(1-p)^2 + 2p(1-p)]/(1-p)^4 = \Sigma n^2 p^{n-1}$, or $(1+p)/(1-p)^3 = \Sigma n^2 p^{n-1}$. From the last identity we get $p(1+p)/(1-p)^3 = \Sigma n^2 p^n$, or $(p^2+p)/q^3 = \Sigma n^2 p^n$.

7-26 Since $e^x = \Sigma x^n/n!$, we have $De^x = e^x = \Sigma nx^{n-1}/n!$ for all x. Thus $e = \Sigma n/n!$, and $xe^x = \Sigma nx^n/n!$. Now $Dxe^x = e^x + xe^x = \Sigma n^2 x^{n-1}/n!$, and so $2e = \Sigma n^2/n!$ and $e^x(x + x^2) = \Sigma n^2 x^n/n!$ Hence $De^x(x + x^2) = e^x(x + x^2) + e^x(1 + 2x) = \Sigma n^3 x^{n-1}/n!$, and so $5e = \Sigma n^3/n!$ In general it seems as though we must have $\Sigma n^k x^{n-1}/n! = e^x p_k(x)$, where p_k is a polynomial of degree k with integer coefficients. We will prove this by induction. As we have seen, the statement is true for $k = 1$ [with $p_1(x) = x$]. Assume then that it is true for some integer k. Then $xe^x p_k(x) = \Sigma n^k x^n/n!$, and so $D[xe^x p_k(x)] = e^x[xp_k(x) + p_k(x) + xp_k'(x)] = \Sigma n^{k+1} x^{n-1}/n!$ Now

$$xp_k(x) + p_k(x) + xp_k'(x)$$

is clearly a $(k+1)$st degree polynomial with integer coefficients; if we denote it by $p_{k+1}(x)$, then $e^x p_{k+1}(x) = \Sigma n^{k+1} x^{n-1}/n!$ The induction is complete. Once we have the formula $e^x p_k(x) = \Sigma n^k x^{n-1}/n!$, it follows, because $p_k(1)$ is an integer, that $\Sigma n^k/n!$ is an integral multiple of e.

7-27 (a) From Exercise 7-14, we have $\tan^{-1} x^2 = \Sigma(-1)^{n-1}(x^2)^{2n+1}/(2n+1) = \Sigma(-1)^{n-1} x^{4n+2}/(2n+1)$ for $|x^2| < 1$, or $|x| < 1$.

     (b) Here $e^{-x^2/2} = \Sigma(-x^2/2)^n/n! = \Sigma(-1)^n x^{2n} n!$ for all x.

     (c) $\cos x^{1/2} = \Sigma(-1)^n(x^{1/2})^{2n}/(2n)! = \Sigma(-1)^n x^n/(2n)!$ for all $x \geq 0$.

     (d) Since $\log [1/(1-x)] = -\log [1 + (-x)]$, it follows from Eq. (7-12) that

$$\log \frac{1}{1-x} = -\sum_{n=1}^{\infty} \frac{(-1)^{n-1}(-x)^n}{n} = \sum_{n=1}^{\infty} \frac{(-1)^{2n}x^n}{n} = \sum_{n=1}^{\infty} \frac{x^n}{n}$$

for $|x| < 1$.

7-28 (a) From the equation $\log [(1+x)/(1-x)] = \log (1+x) - \log (1-x)$, it follows from Problem 7-27d that

$$\log \frac{1+x}{1-x} = \sum_{n=1}^{\infty} \frac{(-1)^{n-1}x^n}{n} + \sum_{n=1}^{\infty} \frac{x^n}{n} = 2 \sum_{n=0}^{\infty} \frac{x^{2n+1}}{2n+1}$$

for $|x| < 1$.

     (b) We have

$$f(x) = (x+a)^p(x+b)^{-p} = (a/b)^p(x/a + 1)^p(x/b + 1)^{-p}$$

$$= (a/b)^p \left[\sum \binom{p}{n}(x/a)^n\right]\left[\sum \binom{-p}{n}(x/b)^n\right]$$

$$= \left(\frac{a}{b}\right)^p \sum \left[\sum_{k=0}^{n} \frac{1}{a^k}\binom{p}{k}\frac{1}{b^{n-k}}\binom{-p}{n-k}\right]x^n$$

for $|x| < a$ and $|x| < b$, or $|x| <$ the smaller of a and b.

(c) Since

$$x/(1-x)(1-x^2) = (-1/4)1/(1-x) + (1/2)1/(1-x)^2 + (-1/4)1/(1+x)$$

we have

$$f(x) = \sum -x^n/4 + \sum \binom{-2}{n}(-x)^n/2 + \sum -(-x)^n/4$$

$$= \sum (1/2)[n+1 - (1+(-1)^n)/2]x^n$$

for $|x| < 1$. [See the solution of Problem 7-23d for the evaluation of $\binom{-2}{n}$.]

(d) Here, for $|x| < 1$,

$$\frac{\log(1+x)}{1+x} = \left(\sum_{n=1}^{\infty} \frac{(-1)^{n-1}x^n}{n}\right)\left(\sum_{n=0}^{\infty} (-1)^n x^n\right) = \sum_{n=1}^{\infty} a_n x^n$$

where $\{a_n\}$ is the convolution of the sequences $0, 1, -1/2, 1/3, \ldots$ and $1, -1, 1, -1, \ldots$. The first n terms of these sequences are (counting the initial term as the zeroth term) $0, 1, -1/2, \ldots, (-1)^{n-1}/n$ and $1, -1, \ldots, (-1)^n$. Thus $a_n = 0 \cdot (-1)^n + (-1)^{n-1}1 + (-1)^{n-2}(-1/2) + \cdots + 1 \cdot (-1)^n/n =$

$$\sum_{k=1}^{n} (-1)^{n-k} \frac{(-1)^{k-1}}{k} = (-1)^{n-1} \sum_{k=1}^{n} \frac{1}{k}$$

Hence

$$\frac{\log(1+x)}{1+x} = \sum_{n=1}^{\infty} (-1)^{n-1} \left[\sum_{n=1}^{n} \frac{1}{k}\right] x^n$$

(e) We know that $\tan^{-1} x = \Sigma(-1)^n x^{2n+1}/(2n+1)$ for $|x| < 1$, and that

$$\log(1+x^2) = \sum_{n=1}^{\infty} \frac{(-1)^{n-1}}{n} x^{2n}$$

for $|x^2| < 1$, or $|x| < 1$. Hence for $|x| < 1$, $\tan^{-1} x \log(1+x^2) = \sum a_n x^n$, where $\{a_n\}$ is the convolution of the sequences $0, 1, 0, -1/3, 0, 1/5, \ldots$ and $0, 0, 1, 0, -1/2, 0, 1/3, \ldots$.   Let b_n denote the nth term of the former sequence, c_n denote the nth term of the latter sequence. Then

$$a_n = \sum_{k=0}^{n} b_k c_{n-k}$$

If k is even, then $b_k = 0$, while if $n - k$ is odd, then $c_{n-k} = 0$. Now if n is even and k is odd, then $n - k$ is odd, and so $b_k c_{n-k} = 0$ for all $k \leq n$ when n is even; hence $a_n = 0$ when n is even. Let n be odd. If k is odd, then $a_k b_{n-k} \neq 0$ because $n - k$ is even. Therefore,

$$a_{2m+1} = \sum_{k=0}^{m} b_{2k-1} c_{2(m-k)} = \sum_{k=0}^{m-1} \frac{(-1)^k}{2k+1} \frac{(-1)^{m-k-1}}{m-k}$$

$$= (-1)^{m-1} \sum_{k=0}^{m-1} \frac{1}{(2k+1)(m-k)}$$

But

$$\frac{1}{(2k+1)(m-k)} = \frac{2}{2m+1}\frac{1}{2k+1} + \frac{1}{2m+1}\frac{1}{m-k}$$

Consequently,

$$a_{2m+1} = \frac{2(-1)^{m-1}}{2m+1}\sum_{k=0}^{m-1}\left[\frac{1}{2k+1} + \frac{1}{2(m-k)}\right] = \frac{2(-1)^{m-1}}{2m+1}\sum_{k=1}^{2m}\frac{1}{k}$$

and so

$$\tan^{-1}x\,\log(1+x^2) = \sum_{n=1}^{\infty}\frac{2(-1)^{n-1}}{2n+1}\left(\sum_{k=1}^{2n}\frac{1}{k}\right)x^{2n+1}$$

for $|x| < 1$.

7-29 (a) Notice that $D[\log(1+x)]^2 = 2\log(1+x)/(1+x)$. But from Problem 7-28d,

$$\frac{2\log(1+x)}{1+x} = 2\sum_{n=1}^{\infty}(-1)^{n-1}\left(\sum_{k=1}^{n}\frac{1}{k}\right)x^n$$

for $|x| < 1$. Therefore,

$$[\log(1+x)]^2 = [\log(1+x)]^2 - [\log 1]^2 = \int_0^x \sum_{n=1}^{\infty}(-1)^{n-1}2\left(\sum_{k=1}^{n}\frac{1}{k}\right)y^n\,dy$$

$$= \sum_{n=1}^{\infty}(-1)^{n-1}\left(\sum_{k=1}^{n}\frac{1}{k}\right)\frac{2}{n+1}x^{n+1}$$

for $|x| < 1$.

(b) We have, after some computation, $Df(x) = [(x+a)/(x+b)]^{1/2}$. As we saw in Problem 7-28b,

$$\left(\frac{x+a}{x+b}\right)^{1/2} = \left(\frac{a}{b}\right)^{1/2}\sum\left[\sum_{k=0}^{n}\frac{1}{a^k}\binom{1/2}{k}\frac{1}{b^{n-k}}\binom{-1/2}{n-k}\right]x^n = \left(\frac{a}{b}\right)^{1/2}\sum a_n x^n$$

for $|x| <$ the smaller of a and b. Consequently,

$$f(x) - f(0) = \int_0^x (a/b)^{1/2}\sum a_n y^n\,dy = (a/b)^{1/2}\sum a_n x^{n+1}/(n+1).$$

In other words,

$$(x+a)^{1/2}(x+b)^{1/2} + (a-b)\log[(x+a)^{1/2} + (x+b)^{1/2}] = (ab)^{1/2} +$$

$$(a-b)\log[a^{1/2} + b^{1/2}] + \left(\frac{a}{b}\right)^{1/2}\sum\left[\sum_{k=0}^{n}\frac{1}{a^k}\binom{1/2}{k}\frac{1}{b^{n-k}}\binom{-1/2}{n-k}\right]\frac{x^{n+1}}{n+1}$$

for $|x| <$ the smaller of a and b.

(c) Here it turns out that $Df(x) = [(a-x)/(b+x)]^{1/2}$. Now just as in part (b),

$$\left(\frac{a-x}{b+x}\right)^{1/2} = \left(\frac{a}{b}\right)^{1/2} \sum \left[\sum_{k=0}^{n} \frac{(-1)^k}{a^k} \binom{1/2}{k} \frac{1}{b^{n-k}} \binom{-1/2}{n-k}\right] x^n$$

for $|x| <$ the smaller of a and b. It follows that

$$(a-x)^{1/2}(b+x)^{1/2} + (a+b) \sin^{-1}[(x+b)/(a+b)]^{1/2}$$

$$= (ab)^{1/2} + (a+b) \sin^{-1}[b/(a+b)]^{1/2}$$

$$+ \left(\frac{a}{b}\right)^{1/2} \sum \left[\sum_{k=0}^{n} \frac{(-1)^k}{a^k} \binom{1/2}{k} \frac{1}{b^{n-k}} \binom{-1/2}{n-k}\right] \frac{x^{n+1}}{n+1}$$

for $|x| <$ the smaller of a and b.

7-30 (a) $e^x e^y = (\sum x^n/n!)(\sum y^n/n!)$

$$= \sum \left[\sum_{k=0}^{n} \frac{x^k}{k!} \frac{y^{n-k}}{(n-k)!}\right] = \sum \frac{1}{n!} \left[\sum_{k=0}^{n} \frac{n!}{k!(n-k)!} x^k y^{n-k}\right]$$

$$= \sum \frac{1}{n!} \left[\sum_{k=0}^{n} \binom{n}{k} x^k y^{n-k}\right] = \sum \frac{(x+y)^n}{n!} = e^{x+y}$$

(b) $2 \sin x \cos x = 2(\Sigma(-1)^n x^{2n+1}/(2n+1)!)(\Sigma(-1)^n x^{2n}/(2n)!) = \sum_{n=1}^{\infty} 2a_n x^n$

where $\{a_n\}$ is the convolution of the sequences $0, 1, 0, -1/3!, 0, \ldots,$ $(-1)^n/(2n+1)!, 0, \ldots$ and $1, 0, -1/2!, 0, \ldots, (-1)^n/(2n)!, 0, \ldots$. It is easy to see that $a_{2n} = 0$. On the other hand,

$$a_{2n+1} = \frac{(-1)^n}{(2n+1)!} + \frac{(-1)^{n-1}}{(2n-1)!} \frac{-1}{2!} + \cdots + \frac{(-1)^{n-k}}{[2(n-k)+1]!} \frac{(-1)^k}{(2k)!} + \cdots$$

$$+ \frac{(-1)^n}{(2n)!} = \frac{(-1)^n}{(2n+1)!} \sum_{k=0}^{n} \binom{2n+1}{k}$$

Let

$$\sum_{k=0}^{n} \binom{2n+1}{k} = \alpha$$

To evaluate α, observe that if p and q are positive integers with $p > q$, then

$$\binom{p}{q} = \frac{p!}{q!(p-q)!} = \frac{p!}{(p-q)![p-(p-q)]!} = \binom{p}{p-q}$$

Therefore,

$$\binom{2n+1}{k} = \binom{2n+1}{2n+1-k}$$

and so

$$\sum_{k=0}^{n}\binom{2n+1}{k} = \sum_{k=0}^{n}\binom{2n+1}{2n+1-k} = \sum_{m=n+1}^{2n+1}\binom{2n+1}{m}$$

Thus

$$2\alpha = \sum_{k=0}^{2n+1}\binom{2n+1}{k}$$

But

$$\sum_{k=0}^{2n+1}\binom{2n+1}{k} = 2^{2n+1}$$

(see the remark following the solution of Problem 1-3), and so $\alpha = 2^{2n}$. Hence $a_{2n+1} = (-1)^n 2^{2n}/(2n+1)!$, or

$$2\sin x \cos x = \sum_{n=0}^{\infty}\frac{(-1)^n}{(2n+1)!}2^{2n+1}x^{2n+1} = \sum_{n=0}^{\infty}\frac{(-1)^n}{(2n+1)!}(2x)^{2n+1} = \sin 2x$$

7-31   (a)   Since $\tan^{-1} x = \Sigma(-1)^n x^{2n+1}/(2n+1)$ for $|x| < 1$, we have $\tan^{-1} x/x = \Sigma(-1)^n x^{2n}/(2n+1)$, also for $|x| < 1$. It follows that $\int_0^t [\tan^{-1} x/x]\,dx = \Sigma(-1)^n t^{2n+1}/(2n+1)^2$ for $0 \le t < 1$. But $\Sigma(-1)^n/(2n+1)^2$ is obviously convergent and so, by Abel's theorem, $\int_0^1 [\tan^{-1} x/x]\,dx = \Sigma(-1)^n/(2n+1)^2$.

(b)   Here

$$\frac{\log(1+x)}{x} = \sum_{n=1}^{\infty}\frac{(-1)^{n-1}}{n}x^{n-1}$$

for $|x| < 1$.   Consequently,

$$\int_0^t\frac{\log(1+x)}{x}\,dx = \sum_{n=1}^{\infty}\frac{(-1)^{n-1}}{n^2}t^n$$

for $0 \le t < 1$.   Since

$$\sum_{n=1}^{\infty}\frac{(-1)^{n-1}}{n^2}$$

is convergent, we have

$$\int_0^1\frac{\log(1+x)}{x}\,dx = \sum_{n=1}^{\infty}\frac{(-1)^{n-1}}{n^2}$$

(c)   We have $x^{m-1}/(1+x^k) = x^{m-1}\Sigma(-1)^n x^{nk} = \Sigma(-1)^n x^{nk+m-1}$ for $|x| < 1$. Therefore, for $0 \le t < 1$, $\int_0^t [x^{m-1}/(1+x^k)]\,dx = \Sigma(-1)^n t^{nk+m}/(nk+m)$. As in parts (a) and (b), Abel's theorem implies that this equation continues to hold for $t = 1$.   That is,

$$\int_0^1\frac{x^{m-1}}{1+x^k}\,dx = \sum\frac{(-1)^n}{nk+m}$$

Remark The integral in part (c) can, at least in principle, be evaluated by the method of partial fractions. In other words, it is possible to find the sum of the series. Needless to say, such evaluations are easier for some values of m and k than for others. We list a few instances where it is easy to evaluate the integral, and the resulting series summations. When $m = 1$ and $k = 3$, we have

$$1 - \frac{1}{4} + \frac{1}{7} - \frac{1}{10} + \cdots = \frac{1}{3}\left(\frac{\pi}{3^{1/2}} + \log 2\right)$$

When $n = 2$ and $k = 3$, we have

$$\frac{1}{2} - \frac{1}{5} + \frac{1}{8} - \frac{1}{11} + \cdots = \frac{1}{3}\left(\frac{\pi}{3^{1/2}} - \log 2\right)$$

When $n = 1$ and $k = 4$, we have

$$1 - \frac{1}{5} + \frac{1}{9} - \frac{1}{13} + \cdots = \frac{1}{4(2)^{1/2}}\left[\pi + 2\log\left(2^{1/2} + 1\right)\right]$$

7-32 We have $(1 + x)^p(1 + x)^q = (1 + x)^{p+q}$. If $|x| < 1$, then

$$(1 + x)^{p+q} = \sum \binom{p+q}{n}x^n$$

and

$$(1 + x)^p(1 + x)^q = \left(\sum \binom{p}{n}x^n\right)\left(\sum \binom{q}{n}x^n\right)$$

$$= \sum\left[\sum_{k=0}^{n}\binom{p}{k}\binom{q}{n-k}\right]x^n$$

Since the two power series must be equal for $|x| < 1$, it follows that

$$\sum_{k=0}^{n}\binom{p}{k}\binom{q}{n-k} = \binom{p+q}{n}$$

We can write this identity as

$$\sum_{k=0}^{n}\binom{p}{k}\binom{-q}{n-k} = \binom{p-q}{n}$$

Now

$$\binom{-r}{m} = \frac{(-r)(-r-1)\cdots(-r-m+1)}{m!}$$

$$= (-1)^m\frac{(m+r-1)(m+r-1-1)\cdots(m+r-1-m+1)}{m!}$$

$$= (-1)^m\binom{m+r-1}{m}$$

Therefore,

$$\sum_{k=0}^{n} \binom{p}{k}\binom{-q}{n-k} = \sum_{k=0}^{n} \binom{p}{k}(-1)^{n-k}\binom{n-k+q-1}{n-k}$$

$$= (-1)^n \sum_{k=0}^{n} (-1)^{-k}\binom{p}{k}\binom{n-k+q-1}{n-k}$$

Since $(-1)^n = (-1)^{-n}$ and $(-1)^k = (-1)^{-k}$, we have

$$(-1)^n\binom{p-q}{n} = \sum_{k=0}^{n} (-1)^k\binom{p}{k}\binom{n-k+q-1}{n-k}$$

Remark   There are several interesting special cases of these identities, particularly when p or q are positive integers.   For instance, when $p = q = n$ in the first identity, inasmuch as

$$\binom{n}{k} = \binom{n}{n-k}$$

we have

$$\sum_{k=0}^{n} \binom{n}{k}^2 = \binom{2n}{n}$$

and when $q = 1$ in the second identity, we have

$$\sum_{k=0}^{n} (-1)^k\binom{p}{k} = (-1)^n\binom{p-1}{n}$$

7-33   (a)   We saw in Problem 7-24d that

$$\log\left(\frac{1 + (1+x)^{1/2}}{2}\right) = \sum_{n=1}^{\infty} -\binom{-1/2}{n}\frac{x^n}{2n} = \sum_{n=1}^{\infty} (-1)^{n-1}\frac{1 \cdot 3 \cdots (2n-1)}{2 \cdot 4 \cdots (2n)}\frac{x^n}{2n}$$

for $|x| < 1$.   Inasmuch as

$$\log\left([1 + (1+1)^{1/2}]/2\right) = \log\left[(1 + 2^{1/2})/2\right]$$

it is sufficient to prove that

$$\sum_{n=1}^{\infty} -\binom{-1/2}{n}\frac{1}{2n}$$

converges.   Thus the problem is reduced to applying the alternating series test.   Put

$$c_n = \frac{1 \cdot 3 \cdots (2n-1)}{2 \cdot 4 \cdots (2n)}\frac{1}{2n}$$

Then $c_n/c_{n+1} = (2n+2)^2/2n(2n+1) > 1$,   or $\{c_n\}$ is   decreasing.   Since $c_n < 1/2n$, we see further that $\{c_n\} \to 0$, and so

$$\sum_{n=1}^{\infty} (-1)^{n-1}c_n$$

converges.

(b)   Since

$$\sin^{-1} x = x + \sum_{n=1}^{\infty} \frac{1 \cdot 3 \cdots (2n-1)}{2 \cdot 4 \cdots (2n)} \frac{x^{2n+1}}{2n+1}$$

and $\sin^{-1} 1 = \pi/2$, we only have to prove that the series

$$1 + \sum_{n=1}^{\infty} \frac{1 \cdot 3 \cdots (2n-1)}{2 \cdot 4 \cdots (2n)} \frac{1}{2n+1}$$

converges in order to conclude that its sum is $\pi/2$.   To this end, put

$$S_m(x) = x + \sum_{n=1}^{m} \frac{1 \cdot 3 \cdots (2n-1)}{2 \cdot 4 \cdots (2n)} \frac{x^{2n+1}}{2n+1}$$

Now $\{S_m(x)\}$ is increasing for each x in $(0, 1]$ because

$$[1 \cdot 3 \cdots (2n-1)x^{2n+1}]/[2 \cdot 4 \cdots (2n)(2n+1)] > 0.$$

Thus, since $\{S_m(x)\} \to \sin^{-1} x$ for $0 < x < 1$, we have $S_m(x) < \sin^{-1} x < \sin^{-1} 1 = \pi/2$, or $S_m(x) < \pi/2$ for all x in $(0, 1)$ and each m.   It follows that $\lim_{x \to 1-} S_m(x) \le \pi/2$, but $\lim_{x \to 1-} S_m(x) = S_m(1)$ because S_m is a polynomial of degree $2m + 1$.   That is, $S_m(1) \le \pi/2$ for all m.   In other words, $\{S_m(1)\}$ is bounded as well as increasing, and so $\{S_m(1)\}$ has a limit, or the series we are interested in converges.

(c)   We know from Problem 7-29a that

$$\frac{1}{2} [\log (1 + x)]^2 = \sum_{n=1}^{\infty} (-1)^{n-1} \left[1 + \frac{1}{2} + \cdots + \frac{1}{n}\right] \frac{x^{n+1}}{n+1}$$

for $|x| < 1$.   Consequently, the identity will follow if we can prove that

$$\sum_{n=1}^{\infty} (-1)^{n-1} \left[1 + \frac{1}{2} + \cdots + \frac{1}{n}\right] \frac{1}{n+1}$$

converges.   We will do this by applying the alternating series test.   Put

$$a_n = \frac{1}{n+1} \sum_{k=1}^{n} \frac{1}{k}$$

Then

$$a_n - a_{n+1} = \frac{1}{n+1} \sum_{k=1}^{n} \frac{1}{k} - \frac{1}{n+2} \sum_{k=1}^{n+1} \frac{1}{k}$$

$$= \left(\frac{1}{n+1} - \frac{1}{n+2}\right) \sum_{k=1}^{n} \frac{1}{k} - \frac{1}{(n+1)(n+2)}$$

$$= \frac{1}{(n+1)(n+2)} \left[\sum_{k=1}^{n} \frac{1}{k} - 1\right] > 0$$

Hence $\{a_n\}$ decreases. Now, as we know from the remark following the solution of Problem 7-17, we can find N such that if $n > N$, then

$$\gamma - 1 < \sum_{k=1}^{n} \frac{1}{k} - \log n < \gamma + 1$$

where γ is Euler's constant. This means that for $n > N$ we have

$$\frac{\log n}{n+1} + \frac{\gamma - 1}{n+1} < a_n < \frac{\log n}{n+1} + \frac{\gamma + 1}{n+1}$$

Inasmuch as $\{\log n/(n+1)\} \to 0$, it follows that $\{a_n\} \to 0$. Thus

$$\sum_{n=1}^{\infty} (-1)^{n-1} a_n$$

is indeed convergent.

(d) We know from Problem 7-28e that

$$f(x) = \frac{1}{2} \tan^{-1} x \log (1 + x^2) = \sum_{n=1}^{\infty} (-1)^{n-1} \left(\sum_{k=1}^{2n} \frac{1}{k} \right) \frac{x^{2n+1}}{2n+1}$$

for $|x| < 1$. Inasmuch as $f(1) = (1/2) \tan^{-1} 1 \log 2 = \pi \log 2/8$, we only have to prove that

$$\sum_{n=1}^{\infty} (-1)^{n-1} \frac{1}{2n+1} \left(\sum_{k=1}^{2n} \frac{1}{k} \right)$$

converges. In the notation of the solution of part (c) this series is

$$\sum_{n=1}^{\infty} (-1)^{n-1} a_{2n}$$

As we know, $\{a_n\}$ decreases and has limit zero, and so $\{a_{2n}\}$ also decreases and has limit zero. Consequently,

$$\sum_{n=1}^{\infty} (-1)^{n-1} a_{2n}$$

converges by the alternating series test.

(e) In Exercise 7-16 we saw that

$$f(x) = \frac{1}{2} (\tan^{-1} x)^2 = \sum_{n=1}^{\infty} (-1)^{n-1} \left(\sum_{k=0}^{n-1} \frac{1}{2k+1} \right) \frac{x^{2n}}{2n}$$

for $|x| < 1$. Since $f(1) = (1/2)(\pi/4)^2 = \pi^2/32$, it is sufficient, once again, to prove that

$$\sum_{n=1}^{\infty} (-1)^{n-1} \left(\frac{1}{2n} \sum_{k=0}^{n-1} \frac{1}{2k+1} \right)$$

converges. As in parts (c) and (d), we do this by means of the alternating series test. Put

$$b_n = \frac{1}{2n} \sum_{k=0}^{n-1} \frac{1}{2k+1}$$

Then

$$b_n - b_{n+1} = \frac{1}{2n} \sum_{k=0}^{n-1} \frac{1}{2k+1} - \frac{1}{2n+2} \sum_{k=0}^{n} \frac{1}{2k+1}$$

$$= \left(\frac{1}{2n} - \frac{1}{2n+2} \right) \sum_{k=0}^{n-1} \frac{1}{2k+1} - \frac{1}{(2n+2)(2n+1)}$$

$$= \frac{2}{2n(2n+2)} \sum_{k=0}^{n-1} \frac{1}{2k+1} - \frac{1}{(2n+2)(2n+1)}$$

$$> \frac{1}{(2n+2)(2n+1)} \left[2 \sum_{k=0}^{n-1} \frac{1}{2k+1} - 1 \right] > 0$$

Hence $\{b_n\}$ decreases. We will thus be finished if we show that $\{b_n\} \to 0$. To this end, we again make use of the relation

$$\left\{ \sum_{k=1}^{m} \frac{1}{k} - \log m \right\} \to \gamma$$

We can find N such that if $n > N$, then

$$\gamma - 1 < \sum_{k=1}^{2n} \frac{1}{k} - \log 2n < \gamma + 1$$

Now

$$\sum_{k=1}^{2n} \frac{1}{k} = \sum_{k=0}^{n-1} \frac{1}{2k+1} + \sum_{k=1}^{n} \frac{1}{2k}$$

Therefore, if $n > N$, then

$$\log 2n + (\gamma - 1) - \frac{1}{2} \sum_{k=1}^{n} \frac{1}{k} < \sum_{k=0}^{n-1} \frac{1}{2k+1} < \log 2n + (\gamma + 1) + \frac{1}{2} \sum_{k=1}^{n} \frac{1}{k}$$

If we divide this inequality by $2n$ and employ the notation of parts (c) and (d), we have

$$\frac{\log 2n}{2n} + \frac{\gamma - 1}{2n} - \frac{1}{4} \frac{n+1}{n} a_n < b_n < \frac{\log 2n}{2n} + \frac{\gamma + 1}{2n} - \frac{1}{4} \frac{n+1}{n} a_n$$

Since $\{\log 2n/2n\} \to 0$ and $\{a_n\} \to 0$, it follows that $\{b_n\} \to 0$.

7-34   Consider first $\Sigma \binom{p}{n}$. If $p = -1$, then $\binom{-1}{n} = (-1)^n$, and so $\Sigma \binom{p}{n}$ diverges if $p = -1$. If p is a positive integer, or zero, then $\binom{p}{n} = 0$ for $n > p$, and there is nothing to prove (recall the remark following the solution of Problem 1-3). Assume then that $p \neq -1, 0, 1, 2, \ldots$ and consider

$$\left| \frac{\binom{p}{n+1}}{\binom{p}{n}} \right| = \left| \frac{p(p-1) \cdots (p-n)/(n+1)!}{p(p-1) \cdots (p-n+1)/n!} \right| = \left| \frac{p-n}{n+1} \right|$$

If $p+1 < 0$, then $p - n < -(n+1) < 0$, and $|p-n| > n+1$.   That is, if $p+1 < 0$, then $\left| \binom{p}{n+1} \right| > \left| \binom{p}{n} \right|$.   Hence if $p+1 < 0$, then $\left\{ \binom{p}{n} \right\}$ cannot have limit zero, and so $\Sigma \binom{p}{n}$ diverges if $p+1 < 0$.   If $p+1 > 0$, then $p - n > -(n+1)$, and $p - n < 0$ for all $n > p$.   Hence $|p-n| < n+1$, or

$$\left| \frac{\binom{p}{n+1}}{\binom{p}{n}} \right| < 1$$

Further, $\binom{p}{n+1} \Big/ \binom{p}{n}$ is negative for all $n > p$.   In other words $\Sigma \binom{p}{n}$ is ultimately an alternating series whose terms decrease in absolute value.   Consequently, by the alternating series test, it will converge if $\left\{ \binom{p}{n} \right\} \to 0$.   Now for $n > p$,

$$\log \left(\frac{\left| \binom{p}{n+1} \right|}{\left| \binom{p}{n} \right|} \right) = \log \frac{n-p}{n+1} = \log \left(1 - \frac{p+1}{n+1} \right) < -\frac{p+1}{n+1}$$

(the last inequality is from Problem 4-56).   Hence

$$\log \left| \binom{p}{n+1} \right| - \log \left| \binom{p}{n} \right| < -\frac{p+1}{n+1}$$

and so, for $k > p$,

$$\sum_{n=k}^{m} \left[\log \left| \binom{p}{n+1} \right| - \log \left| \binom{p}{n} \right| \right] < -(p+1) \sum_{n=k}^{m} \frac{1}{n+1}$$

But the sum on the left is telescoping, and so

$$\log \left| \binom{p}{m+1} \right| - \log \left| \binom{p}{k} \right| < -(p+1) \sum_{n=k}^{m} \frac{1}{n+1}$$

Inasmuch as the harmonic series diverges,

$$\left\{ \sum_{n=k}^{m} \frac{1}{n+1} \right\} \to \infty$$

for $k > p$ fixed.   In other words, $\left\{\log\left|\binom{p}{m+1}\right|\right\} \to -\infty$, or $\left\{\left|\binom{p}{n+1}\right|\right\} \to 0$.   To

sum up, $\Sigma\binom{p}{n}$ converges if $p + 1 > 0$, diverges if $p + 1 \leq 0$.   In view of the power

series expansion $(1 + x)^p = \Sigma\binom{p}{n}x^n$ for $|x| < 1$, we can conclude from Abel's

theorem that $2^p = \Sigma\binom{p}{n}$ if $p + 1 > 0$.

Turn now to $\Sigma(-1)^n\binom{p}{n}$.   By the second identity in the remark following the
solution of Problem 7-32 we have

$$\sum_{n=0}^{m} (-1)^n\binom{p}{n} = (-1)^n\binom{p-1}{n}$$

In other words, we can evaluate the partial sums of this series.   From our analysis
in the preceding paragraph, we know that $\left\{\binom{p-1}{n}\right\}$ can have no limit if $(p-1)$

$+ 1 \leq 0$, or $p \leq 0$, and that $\left\{\binom{p-1}{n}\right\} \to 0$ if $(p-1) + 1 > 0$, or $p > 0$.   Therefore,

$\Sigma(-1)^n\binom{p}{n} = 0$ if $p > 0$.

INDEX